Intelligent Systems Reference Library 44

Editors-in-Chief

Prof. Janusz Kacprzyk
Systems Research Institute
Polish Academy of Sciences
ul. Newelska 6
01-447 Warsaw
Poland
E-mail: kacprzyk@ibspan.waw.pl

Prof. Lakhmi C. Jain
School of Electrical and Information
Engineering
University of South Australia
Adelaide
South Australia SA 5095
Australia
E-mail: Lakhmi.jain@unisa.edu.au

T0142921

For further volumes:
http://www.springer.com/series/8578

Andreas Tolk (Ed.)

Ontology, Epistemology, and Teleology for Modeling and Simulation

Philosophical Foundations for Intelligent M&S Applications

 Springer

Editor
Dr. Andreas Tolk
Old Dominion University
Norfolk, VA
USA

ISSN 1868-4394
ISBN 978-3-642-44281-0
DOI 10.1007/978-3-642-31140-6
Springer Heidelberg New York Dordrecht London

e-ISSN 1868-4408
ISBN 978-3-642-31140-6 (eBook)

I like to dedicate this book to the memory of my friend and colleague

Dr. Zia-ur Rahman

January 19, 1962 - December 16, 2010

We both arrived at Old Dominion University at the same time and in comparable stages of our lives. I have spent several lunches and other breaks with Zia during which we discussed what it means to be true, what it means to be real, and what we as engineers can understand about this. These discussions touched not only ontological and epistemological foundations of our profession; they also touched faith and our shared values and our ethics. I learned so much and had so little time to give him something back. May this book contribute to his memory as a scholar, a colleague, and a friend!

May 2012   Andreas Tolk

Foreword

Modeling and simulation methods have permanently taken their place alongside the more traditional methods of theory and experiment in science and engineering, a fact that is now acknowledged by the National Science Foundation. This collection is a welcome addition to the growing body of literature that addresses methodological questions about these important techniques. A focus on methodological issues is required because many tenets developed for theoretical and experimental work are inapplicable to, or inappropriate for, simulations. For example, one of the issues that is addressed in this volume is the demand that simulation results be reproducible. What 'reproducibility' means is not at all straightforward in the case of simulations. Merely rerunning the simulation is largely pointless unless fraud is suspected. Sometimes, using a different language and operating system can be informative, not the least because when doing so, different coding techniques can be use to reach the same result. But in order to ensure that the results are appropriately robust, inventing a different and methodologically independent way of simulating the same kind of system would seem to be appropriate, just as using a different experimental technique to confirm the existence of a phenomenon or to estimate the value of a parameter in the experimental realm is desirable. This provides a kind of consilience that is grounded in different evidential sources but it requires a careful analysis of what counts as a different way of simulating the same system.

There are distinctive differences between the goals and methods that dominate the philosophy of science literature and the goals and methods that are appropriate to the kinds of projects, broadly construed, considered here. In addition to the usual scientific and engineering applications, there are commercial simulations that we increasingly encounter in the everyday world and that are as much a result of applied science as are any traditional scientific uses. Many simulation projects are inescapably multi-disciplinary in form and this requires the creation of a third methodological domain. The first domain is concerned with general scientific issues, such as testability and explanatory power; the second domain addresses "Methods of X" for some specific area X such as turbulent flow; and the third step must be the philosophy and methodology of multi-disciplinary activities of which modeling and simulation are central. The intelligent use of modeling and simulation science requires not just an appraisal of how well a chosen method works within a given model, but strategies for choosing the appropriate modeling techniques to attack a given problem. Although there are well-established optimization, Bayesian, numerical, and other methods available for use across different modeling subject matters, the appropriateness and scope of non-formal methods is still a matter of controversy. Should false but simple models or complex and accurate models be used? Bottom-up agent based models with emergent properties or top-down continuous models motivated by successes in another discipline? Is capturing qualitative features alone informative for this project or do we need to provide estimates of quantitative parameters? At what coarse-grained level should the system be modeled? While the answers to many such questions are undoubtedly subject-matter specific, answers that aim at some degree of generality can be found in many of the articles in this collection.

One of the key issues addressed in a number of these articles is the role played by methodological and representational ontologies. Computer science and allied disciplines use a different concept of ontology that does philosophy. In the former, the focus is on classification schemes whereas in the latter the goal is to identify what the representational schemes are representing. As the division between the virtual world and the material world becomes increasingly blurred, the separation between these two concepts of ontology decreases. Many models and the simulations that are based on them have a high degree of autonomy and one of the primary tasks is to find a vocabulary that will best serve the purposes of the modelers. However, as long as machine/human interfaces are required, the quest for a common vocabulary that is accessible both to humans and to computers must have a high priority for any intelligently designed model. There is an odd and problematical symmetry to choosing such representations. It is well known that any given theoretical representation is compatible with multiple philosophical ontologies. Yet the same object can be accessed by means of multiple different descriptions and if the simulation ontology is approached through an extensional classification, as it is in many standard programming languages, the space of computational ontologies becomes enormous, especially when it is no longer tightly constrained by a philosophical ontology. Yet this flexibility is one of the critical advantages of computer simulation models and the virtual worlds they represent. To take just one example, exploring the effects of an attack on warfighters by a fictional enemy using tactics which have not yet been employed in battle provides knowledge that it is practically impossible to gain in other ways. In this example, agent based modeling is the obvious choice, but the other three questions floated earlier all require answers to arrive at an effective modeling exercise.

The editor and contributors are to be congratulated on having assembled an informative and diverse set of approaches to these issues. Readers with a wide variety of interests will find much of value here, and the range of perspectives from the technical to the post-modern is unusual. Intellectual engagement with these ideas cannot help but advance intelligent modeling and simulation applications.

Paul Humphreys
University of Virginia

Ontology, Epistemology, and Teleology of Modeling and Simulation
Philosophical Foundations for Intelligent M&S Applications

Editor: Andreas Tolk

Chapter 1
Andreas Tolk
Truth, Trust, and Turing – Constraints for Modeling and Simulation

This chapter has been written as the introduction to the book "Ontology, Epistemology, and Teleology of Modeling and Simulation – Philosophical Foundations for Intelligent M&S Applications." It covers the main ideas important for modeling and simulation regarding its philosophical, computational, and conceptual aspects. What exists, how we come to know, and what we do with the knowledge are the guiding questions when the key terms are evaluated in the light of positivism, rationalism, and constructivism. Implications for a canon of research are described, and the constraints for modeling and simulation regarding truth, trust, and computability are derived. A short summary of the chapter contributions in the light of these ideas ends the chapter.

Chapter 2
Chris Partridge, Andy Mitchell, and Sergio de Cesare
Guidelines for Developing Ontological Architectures in Modeling and Simulation

This book is motivated by the belief that "a better understanding of ontology, epistemology, and teleology" is essential for enabling Modeling and Simulation (M&S) systems to reach the next level of 'intelligence'. This chapter focuses on one broad category of M&S systems where the connection is more concrete; ones where building an ontology – and, we shall suggest, an epistemology – as an integrated part of their design will enable them to reach the next level of 'intelligence'.

Within the M&S community, this use of ontology is at an early stage; so there is not yet a clear picture of what this will look like. In particular, there is little or no guidance on the kind of ontological architecture that is needed to bring the expected benefits.

This chapter aims to provide guidance by outlining some major concerns that shape the ontology and the options for resolving them. The hope is that paying attention to these concerns during design will lead to a better quality architecture, and so enable more 'intelligent' systems. It is also hoped that understanding these concerns will lead to a better understanding of the role of ontology in M&S.

Ontologies are formal specifications of concepts. They represent entities of a specific knowledge domain and the relationships that can hold between the entities. Ontologies are formal descriptions of the so called "body of knowledge" that composes a domain. Regardless of being implicitly or explicitly applied during the modeling, ontologies set the relation between formal signs used in computer simulations and "meaning" as a notion of human minds. Unfortunately, the essence of this relation is disputed, especially in modern epistemology, which deals with the "nature of knowledge" and the methods and limitations of gaining knowledge. Therefore, the chapter introduces first the debate which epistemological view is most appropriate for modeling and simulation. On the basis of this introduction ontologies are scrutinized with respect to their ability to capture knowledge. As a consequence of this analysis two main classes of ontologies for M&S are distinguished: Methodological and referential ontologies. Their values and limits are discussed in detail.

Models and simulations are immediately obfuscated by being what they are, abstract representations of reality. With reductionist parameters and defined algorithms, models and simulations obtain a definitiveness lacking in the reality they explain. Increased computational power has enabled the production of complex representations. This increased complexity makes understanding what is happening "behind the scenes" almost entirely unintelligible to consumers. At the same time, advancements in Animation enable practitioners to present the results at almost movie-like levels of production. This subtly transforms the ontological status of the results, making them appear as something one should view rather than something about which one should think. What happens when producers and consumers of models and simulations lose the self-certainty associated with their project? Such a situation calls into question the performative aspects of both groups' maneuvers. We situate the locus of this discussion around the notion of validity. Once considered essential, the quest for validity perhaps increasingly reveals a form of existential absurdity and, in a nihilistic twist of postmodern thought, the radical devaluation of one of the ideals of the philosophy of science.

A model may contribute to a phenomenon's explanation, despite having false assumptions, by offering a partial explanation of the phenomenon. The false assumptions may restrict the operation of laws that explain the phenomenon to exhibit their effect in the model. The laws retain their explanatory power after lifting the restrictions, although the model does not then describe their operation, so that they incompletely explain the phenomenon in a natural system. The paper shows that this view accommodates diverse models, makes precise the analogical explanations many theorists attribute to models, maintains the objectivity of explanation, and guides construction of models.

Chapter 6
Klaus G. Troitzsch
Theory Reconstruction of Several Versions of Modern Organization Theories

This chapter compares the technique of reconstructing theory under the 'non-statement view' with the design and implementation of simulation models. For this purpose it uses several different versions of the famous 'garbage can' model, redefines this theoretical attempt in terms of the 'non-statement view' and compares it to simulation models of different authors who replicated the original 'garbage can' model and built on it to extend it.

Chapter 7
Andreas Pyka and Simon Deichsel
Cutting Back Models and Simulations

Agent-based models (ABMs) range from purely theoretical exercises focussing on the patterns in the dynamics of interaction processes to modelling frameworks which are oriented closely at the replication of empirical cases. Advocates of the "Keep it descriptive, stupid!" (KIDS) approach openly recommend building models as empirically accurate as possible, they want to understand social processes from the bottom up.

This seems to be almost the direct opposite of Milton Friedman's famous and provocative methodological credo *"the more significant a theory, the more unrealistic the assumptions"*. Most methodologists and philosophers of science have harshly criticised Friedman's essay as inconsistent, wrong and misleading. By presenting arguments for a pragmatic reinterpretation of Friedman's essay, we will show why much of the philosophical criticism misses the point.

After that, we will use the developed arguments for contesting the claim that good simulations have to rely on descriptively accurate assumptions, which is, in a nutshell a plea for the "Keep it simple, stupid" (KISS) approach.

This plea is followed by a more general plea for dropping the philosophical idea of scientific realism. We give arguments challenging the idea that economic models should be "realistic" in the sense that they (more or less directly) represent mechanisms of the way the world works. We try to show that good economic modelling does not depend on seeing models as representing an external reality at all.

Chapter 8
Tuncer Ören and Levent Yilmaz
Philosophical Aspects of Modeling and Simulation

To examine philosophical foundations of Modeling and Simulation, we present and clarify relations between reality, representations of reality, and simulation. The role experimentation and experience are delineated along with purposes of simulation, knowledge generation via simulated experimentation, and ethics. In relation to experimentation, the need for computational reproducibility and replicability are emphasized to improve credibility of simulation studies.

Chapter 9
Andrew Collins and D'An Knowles Ball
Philosophical and Theoretic Underpinnings of Simulation Visualization Rhetoric and Their Practical Implications

Modeling and simulation has moved far beyond simple data representation into the world of visual communication over the past 15 years; ultimately, the acceptance of M&S within mainstream science and society will depend on the results that are produced visually. A simulation's function is of primary importance to its end result, but it cannot be denied that the discipline of M&S now prizes fancy graphics to communicate. Rhetorical methodological decisions have the greatest impact on the end user, and considerations that bring visual rhetoric to modeling and simulation should be examined as a necessity to application. This paper will expose the community to existing research on the rhetoric of visualization, demonstrate the importance of contemplating the philosophy of visualization, highlight and address current problems with simulation visualization, and bring visualization's inherent rhetoric to the forefront of consideration and utilization.

Chapter 10
Saikou Diallo, Jose Padilla, Ipek Bozkurt, and Andreas Tolk
Modeling and Simulation as a Theory Building Paradigm

This chapter makes the case that theory can be captured as a model, which can be implemented as a simulation. This allows composing and recomposing theory components to process new theory out of existing theory. While current modeling and simulation applications focus on simulation as a computational activity that algorithmically produces output data based on valid input data, therefore providing information, the proposed approach utilizes the information and combines the application thereof, which provides knowledge. Relevant work is evaluated, but existing approaches neither us the conceptualization as the central component nor are they applied to ill-defined problems, thus the proposed approach is innovative and closes existing gaps. To show the feasibility and validity, theory is represented as axiomatic structures that can be executed under bounded conditions. As such, the chapter presents a methodological approach for building theory out of existing theory using modeling and simulation.

Chapter 11
Levent Yilmaz and Tuncer Ören
Toward Replicability-Aware Modeling and Simulation: Changing the Conduct of M&S in the Information Age

The use of computational models in science end engineering is increasingly becoming pervasive. However, there is a growing credibility gap due to widespread, relaxed attitudes in communication of experiments, models, and validation of simulations used in computational research. Consequent disputes and article retractions due to unverified code and data suggest a pressing need for greater transparency. We introduce the e Portfolio concept, which is an ensemble documents that interweave the conceptual model, simulator design, experimental frames, and scientific workflow specifications. Strategies and potential mechanisms are delineated to enable authors, publishers, funding agencies, journals, and the broader scientific community to cooperate and establish a sustained model base, simulations, experiments, and documentation, so that scientists can build on each other's work and achievements.

Chapter 12
John Z. Elias
Immersed in Immersion: Simulation as Technology and Theory of Mind

Cognitive theories involving the notion of simulation have developed hand in hand with the advancement and pervasiveness of simulation technologies. This intimate interrelation suggests the promise of implementing simulation technology in cognitive research, as well as in the facilitation and manipulation of cognitive and affective mechanisms for learning and training. I describe the general interdependence of forms of technology and theories of mind, the former often furnishing metaphors for the latter, and offer a brief historical sketch leading up to the recent emergence of the centrality of simulation. I follow with a critical evaluation of the role of simulation in current cognitive theories, and relate these critiques to philosophical concerns about the ontological, epistemological, and methodological status of modeling and simulation as a research tool. I end with some illustrative examples from cognitive research and therapy, and point towards potential future applications.

Chapter 13
Roger Smith
On the Value of a Taxonomy in Modeling

Though modern science and business have created and adopted classification schemes, taxonomies, and operating rules that can be applied almost universally, the practice of building models and simulations remains unbounded by science. Like the arts, each practitioner has the freedom to create a model in any form that appears to offer a solution to a specific problem. A Periodic Table of modeling has not emerged. Practitioners do not rely on a framework of established, tested, and accepted modeling techniques to guide their work. Conversely, there are also no known poor methods for structuring a model which are not acceptable and which would bring censure from the professional community.

The unbounded nature of the current practice of modeling is supportive of an artistic approach to modeling that encourages creative freedom in imagining and building a unique new model. The environment is also convenient to modeling as a service in which a customer is allowed to direct the construction of a model in almost any direction that will address the problem, with few restrictions applied from known best practices. As expedient as these advantages are, they also allow inaccurate and inefficient approaches to be used without an objective or historic "model-of-modeling" as a reference. The current practice of modeling allows almost any approach while its measure of correctness is determined solely by the usefulness of the resulting product. This chapter is an attempt to begin the construction of a model-of-modeling which can serve as the Periodic Table for our profession.

Chapter 14
Kevin B. Korb, Nicholas Geard, and Alan Dorin
A Bayesian Approach to the Validation of Agent-Based Models

The rapid expansion of agent-based simulation modeling has left the theory of model validation behind its practice. Much of the literature emphasizes the use of empirical data for both calibrating and validating agent-based models. But a great deal of the practical effort in developing models goes into making sense of expert opinions about a modeling domain. Here we present a unifying view which incorporates both expert opinion and data in validating models, drawing upon Bayesian philosophy of science. We illustrate this in reference to a demographic model.

Chapter 15
Scott A. Douglass and Saurabh Mittal
A Framework for Modeling and Simulation of the Artificial

Artificial systems that generate contingency-based teleological behaviors in real-time are difficult to model. This chapter describes a modeling and simulation (M&S) framework designed specifically to reduce this difficulty. The described Knowledge-based Contingency-driven Generative Systems (KCGS) framework combines aspects of SES theory, DEVS-based general systems theory, net-centric heterogeneous simulation, knowledge engineering, cognitive modeling, and domain-specific language development using meta-modeling. The chapter outlines the theoretical and technical foundations of the KCGS framework as realized in the Cognitive Systems Specification Framework (CS2F), a subset of KCGS. Two executable models are described to illustrate how models of autonomous, goal-pursuing cognitive systems can be modeled and simulated in the framework. The technical content and agent descriptions in the chapter illustrate how the *M&S of the artificial* depends critically on *ontology, epistemology,* and *teleology* in the KCGS framework.

Chapter 16
Claudia Szabo and Yong Meng Teo
Semantic Validation of Emergent Properties in Component-based Simulation Models

Advances in composable modeling and simulation have facilitated the development and our understanding of more complex models. As a result, the representation, identification and validation of emergence is becoming of increasing importance because emergent properties can have a negative effect on the overall system behavior. Despite a plethora of definitions and methods, a practical approach to identify and validate emergent properties in newly composed simulation models remains a challenge. This chapter reviews current approaches and presents a new approach for identifying emergent properties in component-based systems. Using a simple example of a flock of birds model, we compare and contrast three main approaches: *grammar-based, variable-based and event-based.* Lastly, building on our previous work on formal semantic validation of model behavior, we present a new *objective-based approach* for semantic validation of emergent properties in composable simulation.

Chapter 17
Wenguang Wang, Weiping Wang, Qun Li, and Feng Yang
Ontological, Epistemological, and Teleological Perspectives on Service-Oriented Simulation Frameworks

This chapter investigates service-oriented simulation frameworks from the ontological, epistemological, and teleological perspectives. First, we give an overview of various specific frameworks that imply particular referential ontological, epistemological, and teleological perspectives for real world systems. Then we combine the partial considerations derived from the review into a unifying framework. It inspects the crossover between the disciplines of M&S, service-orientation, and software/systems engineering. From a methodological perspective, we show its ontological, epistemological, and teleological implications for abstract approaches. The unifying framework can, in turn, facilitate the classification, evaluation, selection, description, and prescription of the known or proposed frameworks. Thus, the referential and methodological perspectives build a systematical philosophical foundation of the service-oriented simulation paradigm.

Epilogue
Andreas Tolk
Modeling and Simulation as a Humble Approach

This epilogue makes the case to see modeling and simulation as a humble approach, i.e., bringing experts from relevant disciplines together to address significant questions, such as the search for ultimate truth. The approach is driven by interdisciplinary research that remains sensitive to disciplinary nuances while looking for theoretical linkages and connections. Intelligent M&S applications are identified to have significant potential to make a contribution.

Contents

List of Contributors

Ipek Bozkurt
University of Houston
United States

Sergio de Cesare
Brunel University
United Kingdom

Andrew Collins
Old Dominion University
United States

Simon Deichsel
University of Bremen
Germany

Saikou Y. Diallo
Old Dominion University
United States

Alan Dorin
Monash University
Australia

Scott A. Douglass
US Air Force Research
Lab
United States

John Z. Elias
University of Central
Florida
United States

Nicholas Geard
University of Melbourne
Australia

Brian L. Heath
Cardinal Health
United States

Marko Hofmann
University of the Federal
Armed Forces
Germany

Paul H. Humphreys
University of Virginia
United States

Ross A. Jackson
University of Phoenix
United States

D'An Knowles Ball
Old Dominion University
United States

Kevin B. Korb
Monash University
Australia

Qun Li
National University of
Defense Technology
China

Andrew Mitchell
BORO™ Solutions
United Kingdom

Saurabh Mittal
L-3 Communications
United States

Tuncer Ören
University of Ottawa
Canada

José J. Padilla
Old Dominion University
United States

Chris Partridge
BORO™ Solutions
United Kingdom

Andreas Pyka
University of Hohenheim
Germany

Roger Smith
Nicholson Center for
Surgical Advancement
United States

Claudia Szabo
University of Adelaide
Australia

Yong Meng Teo
National University of
Singapore
Singapore

Andreas Tolk
Old Dominion University
United States

Klaus G. Troitzsch
University of Koblenz-
Landau
Germany

Weiping Wang
National University of
Defense Technology
China

Wenguang Wang
National University of
Defense Technology
China

Paul Weirich
University of Missouri
United States

Feng Yang
National University of
Defense Technology
China

Levent Yilmaz
Auburn University
United States

Truth, Trust, and Turing – Implications for Modeling and Simulation

Andreas Tolk

Old Dominion University
Norfolk, VA, United States

Abstract. This chapter has been written as the introduction to the book "Ontology, Epistemology, and Teleology of Modeling and Simulation – Philosophical Foundations for Intelligent M&S Applications." It covers the main ideas important for modeling and simulation regarding its philosophical, computational, and conceptual aspects. What exists, how we come to know, and what we do with the knowledge are the guiding questions when the key terms are evaluated in the light of positivism, rationalism, and constructivism. Implications for a canon of research are described, and the constraints for modeling and simulation regarding truth, trust, and computability are derived. A short summary of the chapter contributions in the light of these ideas ends the chapter.

1   Introduction

This chapter shall introduce the reader to why and how this book has been written. Looking at other volumes featured in this Springer series making up the Intelligent Systems Reference Library (ISRL), the topic of using ontology, epistemology, and teleology as the philosophical foundations for intelligent modeling and simulation (M&S) applications is somewhat unique. Most other books feature intelligent methods to support better solutions, such as data mining, intelligent routines, and more. So what makes intelligent M&S applications special in comparison with other forms of computational intelligence?

In order to address this question, we first have to understand what M&S is about. For many professionals in the field, the answer is easy: *"M&S is a computational tool that helps to make better decisions, which can be technical or managerial in nature."* As the decisions and supporting solutions are typically very specific to the supported domain, M&S is accordingly seen as an enabler and not necessarily as an academic discipline of its own. This world view is supported by many conferences and journals that focus exclusively on such specific application domains. Examples include computer simulation approaches for climate models, soft matter sciences, biomolecular simulation, and others. This viewpoint results in defining M&S as a purely engineering discipline with the sole

A. Tolk (Ed.): Ontology, Epistemology, & Teleology for Model. & Simulation, ISRL 44, pp. 1–26.

purpose of helping other academic disciplines find and optimize workable solutions. The success story of such engineering solutions is particularly obvious in the military domain [1], and currently it seems to continue in healthcare applications as well [2]. As the focus is on solving the problem, M&S is perceived as a tool. Without the application domain, there is nothing of practical interest left in the M&S application.

While the application of M&S with a focus on making better decisions in application domains is very important, the viewpoint motivating the writing of this book is quite different. The various chapters currently available examine M&S applications as an object of scientific evaluation. The questions to be answered are, among others: *"What is M&S? What are the characteristics of M&S applications? What do we know, how do we gain new knowledge, and how do we act on this knowledge? Are there limits to what we can model, and are there limits to what we can simulate on computers? How do we know that we can trust the solution derived from an M&S application?"* In other words, instead of focusing on deriving applicable solutions by doing M&S engineering, we will focus on *M&S Science* to understand the general principles that build the foundations of M&S as a discipline. We don't want to ask "How do we find an optimal solution for a given specific problem using M&S?" but instead "What kind of problems are generally solvable and can be optimized using M&S?" Understanding such underlying general principles and being able to communicate them formally, in order to make them accessible for methods of computational intelligence, are necessary requirements to reach the next level of M&S applications as intelligent systems. Whether this is also sufficient is the topic of current research.

This alternative view on M&S should not lead to the interpretation that M&S Engineering is less valuable or desirable than M&S Science. We simply have to understand that these are mutually supportive contributions to the Body of Knowledge (BoK) of M&S [3]. Like Computer Science and Computer Engineering both uniquely contribute to a better understanding of what we can do with computers and how we can apply them to support better solutions, so we must look at M&S Science and M&S Engineering as two mutually supportive aspects both contributing to M&S as a discipline [4]. So far, however, the focus has been nearly exclusively on M&S Engineering; this singular focus means that we, as an academic community of common interest in M&S, need to catch up with the M&S Science components.

One of the foundations for all scientific aspects is that they are rooted in some philosophical worldview. As M&S supports a multitude of domains all based on different worldviews, addressing the question of the philosophical foundation is not easy and, as long as we stay in the engineering worldview, not necessarily something of interest. As long as M&S is just regarded as a tool whose only justification is the contribution to the supported discipline, the philosophical foundations of the supported discipline are often perceived to be sufficient. At the moment we evaluate M&S as our object of scientific interest, however, we need to better understand the specific worldview of M&S itself. This search is driven by the quest to better understand and contribute to a common definition of the ontology, epistemology, and teleology of M&S. In particular, this search focuses on the philosophical foundations for intelligent M&S applications as they are in the focus of interest of this series. What do we think is true? Why do we think it is true? How do we gain new knowledge? And how do we apply this new knowledge?

In this introduction, we will first look at the different aspects of the ontology, epistemology, and teleology of M&S as they are reflected in current literature, including some of the chapter contributions to this book as well. What will quickly become obvious is that M&S experts are like the United Kingdom and the United States of America as described by George Bernard Shaw: "two nations divided by a common language!" M&S experts use the same ontological means to address methodological ontologies as well as referential ontologies [5]. Methodological ontologies capture knowledge about *"how to model,"* which means they focus on modeling paradigms, techniques, and formalisms. Conversely, referential ontologies capture knowledge about *"what to model,"* which means they model the referents and relations of the application domain. Both are necessary contributions enabling intelligent M&S applications, but both are not sufficient when exclusively applied. So, although the same 'language' is used, the experts are often talking past each other, not recognizing what the other side is saying.

The third section of this chapter will deal with the challenge to find agreement on a *Canon of Research for M&S*. If M&S truly is a scientific discipline, then it must be possible to address the questions typically addressed by the canons applied in other academic disciplines. Again, the focus is on developing a general M&S canon for research, not the application of M&S within the cannon of research within a supported other discipline. The questions to be addressed here are, in particular, how to gain new knowledge about M&S and how to apply it!

The fourth section will examine M&S with regard to the conflicting world views of rationalism and empiricism and will attempt to offer a compromise that is based on research findings so far. This section will also address questions related to validity, specifically under which constraints the question regarding validity of models makes sense at all and what alternatives exist [6]. The questions will be framed by the three topics presented in the title of this introduction, namely: *Truth – What can be considered truth in the context of M&S? Trust – Can we trust results obtained using M&S?* And finally, in the year of the 100th birthday of Alan M. Turing, an appropriate tribute to his work: *Turing – What are the computational constraints that we must take into consideration when considering the application of M&S?* The resulting proposed framework can hopefully support research that will enable an increased understanding of the philosophical foundations for intelligent M&S applications.

Finally, the fifth and last section of this chapter will provide a brief overview of the invited chapters and contributions featured in this book within the light of these introductory remarks. Overall, these contributions are state-of-the-art work of experts in related fields. They purposefully span a wide range, from philosophical foundations and their implications to technical aspects of semantic validations and service-oriented simulation frameworks. Hopefully, this book will spawn more research and increased collaborations, in order to bridge the gaps between the fields of contributing ideas.

2   Ontology, Epistemology, and Teleology

The topic of epistemology has been the subject of research in several recent publications, such as [5, 6, 7, 8, 9]. However, despite these fundamental papers and overviews, in particular the epistemological perspectives documented among others in [8, 9], many M&S researchers are not explicitly aware of their philosophical research assumption and the ontological, epistemological, and teleological implications thereof.

Within this chapter, we define ontology as the 'study of being' or the 'study of what exists.' Epistemology is the 'study of how we come to know,' or how we define knowledge, represent it, and communicate it with others. Teleology is the 'study of action and purpose,' resulting in methods, or how we apply knowledge. A non-exclusive summary of current research in theses domains follows in the subsections of this section and aids in setting the stage for particular discussions conducted in the chapter contributions of invited subject matter experts.

2.1   Ontology: What Exists?

Ontology is one of the oldest philosophical disciplines. Nonetheless, for many M&S experts ontology is practically limited to the artifacts produced using software tools like Protégé [10]. Such artifacts, which capture ontological commitment within a community of interest, have tremendous use and are absolutely necessary for the formal specification of a conceptualization in communicable and reusable form [11]. However, they not only deal with a small subset of ontological means in the technical sense, or as a branch within information science, they also overlook that ontology itself is a branch of metaphysics dealing with the nature of being.

The history of ontology as a branch of metaphysics goes back to the Greek philosophers, including Aristotle, Plato, Heraclitus, and others. There have been few philosophers since that time who did not make attempts to contribute to discussions on what exists, what categories can be defined for things that exist, and how these things are related. A significant accomplishment, generally attributed to Aristotle, was the creation of terminology to deal with such ontological questions. For many, classic ontology is the cradle of scientific work; without its concepts and terminology the canons of research and the foundation for scientific research could not even be formulated.

When experts are building an ontology to capture the knowledge for a given domain, Gruber [11] points out that it is important to understand what the ontology is going to be used for. In the M&S realm, Hofmann et al. [5] identify two main application domains: methodological ontologies to address the question *"How we model?"* and referential ontologies to address the question *"What do we model?"*

Examples of the former are the Discrete-event M&S Ontology (DeMO) [12] or the Component-oriented Simulation and Modeling Ontology (COSMO) [13]. The entities of interest of these ontologies are the entities of simulation systems, federations, etc. They address all simulation implementation details that are necessary to implement an M&S application. The reality captured in methodological ontologies are methods, techniques, and formalisms to address the abstract concepts of M&S applications, such as items, attributes, information exchange requirements, events, event queues, time management objects, etc. While current research focuses on discrete event models, other modeling paradigms are likely to produce similar ontologies soon as well, such as Monte-Carlo simulation, continuous simulation, and agent based simulation [14]. The same problem can be approached using a variety of different paradigms and implementation methods. Methodological ontologies capture the knowledge about the paradigm and methods, not about the problem that needs to be solved in using them.

Consequently, what often is of more interest to those who use simulation mainly to solve problems is an ontology that describes the "real world and the problem" as it is perceived by experts of the supported domain. The United States Department of Homeland Security (DHS) developed an Infrastructure Taxonomy to better address the different concepts and their relations that DHS has to deal with when orchestrating the various branches that support them in protecting the critical infrastructure. In order to be able to protect the critical infrastructure, Adam [15] reported that it is important to facilitate communication and dissemination of information between the various different supporting branches. To this end, infrastructure assets are first grouped onto broad infrastructure sectors (e.g., energy, transportation, etc.) and further categorized in more details of up to five levels, referred to as sector, subsectors, segment, sub segment, and asset type. The Office of Infrastructure Protection (OIP) publishes the DHS-OIP Infrastructure Taxonomy and provides access to organizations that want to support them with research. By doing so, it is ensured that the results of different contributors are referring to the same concepts with the same terms, and all participants are aware of the relations between the concepts. Referential ontologies are therefore pivotal to ensure that all experts within the domain use the same terminology and relations when addressing problems.

As observed by Hofmann et al. [5], both types are necessary to capture all ideas needed to fully capture the epistemological aspects of what we model and how we model it, so that we understand how we must interpret the results accordingly. The idea to clearly differentiate between what we model conceptually does and how we implement these models technically is well known to engineers, in particular software engineers. Tolk et al. [16] use the Part III of the ISO/IEC 11179 Standard for a Metadata Registry (MDR) [17] to motivate model-based data engineering. The following figure shows the four concepts used in ISO/IEC 11179.

- The conceptual domain describes the concepts that are derived in the conceptualization phase of the modeling process. This domain comprises all the concepts that are needed to describe the referent or referents relevant to the information exchange.
- The property domain describes the properties that are used to describe the concept. Concepts are characterized by the defining properties. ISO/IEC 11179 refers to this domain as the data element concept.

- The property value domain comprises the value ranges, enumeration, or other appropriate definition of values that can be assigned to a property. ISO/IEC 11179 refers to this domain as the value domain.
- Property instances capture factual information describing the current value of a property. They minimally comprise the value of one property, which can be interpreted as updating just one value, or they can become an n-tuple of n properties describing a group of associated concepts, which represents complex messages or updates for several objects. ISO/IEC 11179 calls these property instances data elements.

Conceptualization Resulting in Propertied Concepts

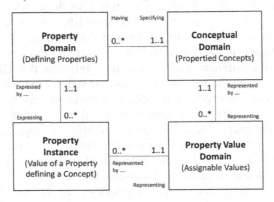

Implementation Resulting in Data Specifications

Fig. 1 Metadatamodel used in ISO/IEC 11179 [16]

The referential ontologies described in [5] can be used by definition and intent to represent the conceptualizations and their relations of the application domain, and with the same argument, the methodological ontologies can be used to represent the implementation and resulting data specifications (plus additional M&S specific information that was not within the realm of the ISO/IEC work).

But why is this so important for the ontology of intelligent M&S applications? Following the arguments presented in [18], the modeling phase conceptualizes the real world referent in form of an executable theory. The conceptual model represents, as a purposeful abstraction and simplification of a perception of reality, everything that according to the world view of the model developers is necessary to address the underlying research questions, but not more. In the simulation phase, these conceptual models are implemented and executed, normally on a digital computer. As a result, the real world referent is replaced in the modeling phase by its representing conceptualization. The conceptualization of the real world referent becomes the reality of the M&S application. If two groups of experts create different conceptualizations of the same real world referent, their solutions are conceptually misaligned. Ontologically, the resulting simulation systems are rooted in different realities, and no technical standard can solve this challenge.

Part of this problem of multiple versions of reality is captured in the M&S literature as multi-resolution modeling as described by Petty [19], and sometimes as multi-models, as described by Winsberg [20]. A respective taxonomy with special view on the applicability for intelligent systems of various current approaches has been compiled by Yilmaz and Tolk [21].

Even if two model development teams agree on the intended use of the model and the real world referents to use, they still may differ in resolution, scope, and structure of the conceptualization. The reasons for this can be manifold. It is possible that the sensors used to observe the system to be modeled differ. Additional data accessible to the model developers may differ as well. Based on the education of the team members, individual as well as group perceptions may be different as well. Cultural and organizational biases may also influence the process of conceptualization. In practice, several of these reasons are often interwoven. For example, one group may have a very fine sensor and may use a high resolution property to represent an attribute of the system, while the second group may be limited to medium or low resolution measures and likely will use a medium or low resolution property as well. Similarly, experience and culture can also lead to conceptual differences.

If you don't believe this, discuss the color of a purse you believe to be "orange" with your wife or girl friend … or likewise try to explain the different important shades of "amber, apricot, carrot, coral, peach, persimmon, rust, or salmon" to your husband or boy friend. The same is true for identifying the characteristic and descriptive properties for a concept. While one group may focus on a minimal set, the other group may take a more inclusive approach resulting in different scopes.

Finally, there is more than one way to group a set of observed or observable attributes of a system into propertied concepts. An abstract example is to group the properties {a1, a2, b1, b2} into sets. Two obvious choices are to group by the letters: {{a1, a2}, {b1, b2}}, or to group by the numbers: {{a1, b1}, {a2, b2}}. Again, the context of the model developers can have a significant influence on what taxonomical structure is chosen.

In all cases, however, what becomes the basis of truth evaluations in the simulation may look very different, even if the model developers are starting from the same common ground. If they start from different schools of thought, such as is often the case in social, organizational and human sciences, the conceptualization results may be even more diverse. Referential ontologies can help to make these differences obvious and hopefully accessible to computers and intelligent systems as well [16]. In any case, even when starting with the same real word referents there is a high likelihood of ending up with a surprising variety of conceptualizations. Each of these conceptualizations can then be implemented in variant ways as well, as different modeling paradigms can be chosen, different computer types can be selected, algorithm implementations can differ, and so forth. Methodological ontologies can help to address these challenges and help to identify interoperability challenges.

More obvious become such challenges when two or more simulation systems are needed to address a problem. Winsberg [20, pp.72-92] uses nanosciences as an example where scientists are interested in how cracks evolved and move through material. To address this problem, three different levels of resolution are necessary. In order to understand how cracks begin, sets of atoms governed by quantum mechanics are modeled in small regions. These regions are embedded into medium scale regions that are governed by molecular dynamics. Finally, most of the material is neither cracking nor close to a developing crack and can be modeled using continuum mechanics based on linear-elastic theory. The challenge is that these three theories cannot be mapped to each other; the result is that coupling heuristics (Winsberg calls them handshakes) need to be developed and applied to allow the parallel execution and mutual influence of more than one theory in more than one region. In the given example, the common expression of energy was utilized to exchange information between the regions and allow for a common model.

Petty [19] uses combat models on different military levels to make similar observations. In his overview chapter, he distinguishes between simulation systems that simulate single weapon systems and those that simulate units that group such weapon systems into one aggregate. While logistics and movement away from potential domains of conflicts can be covered well and without much computational effort using unit level simulations, areas of interest and potential combat zones are evaluated in more detail using the weapon system level resolution. As before, the challenge is to remain overall constant when information between zones of different resolution level is exchanged. Again, common concepts known in both zones – sender and receiver – and representable in common concepts – information exchange elements – need to be identified. Various methods have been applied, from the use to derive all elements from a common taxonomy to ensure consistency [22], over the definition of standardized information exchange elements as Protocol Data Units (PDU) in the IEEE 1278 Standard for Distributed Interactive Simulation (DIS) [23] or common and mandatory Federation Object Models in the IEEE 1516 Standard for the High Level Architecture (HLA) [24], or the application of model-based data engineering to derive such common ground [16].

Ontology answers the question: What exists? For M&S applications, this question has many facets and must be answered on many levels. This subsection merely scratches the surface on this topic and initiates a research agenda. In the following sections, and parts of the following book chapter, the quest to better understand what this common ground means and what is necessary to create consistence representation of truth in all participating components will be a reoccurring topic. The ontological implications are directly connected with these research elements.

2.2 Epistemology: How Do Come to Know?

In the public domain, M&S is celebrated as a new way of science. This can be in particular observed in the United States.

- In their 2006 Report on Simulation-Based Engineering Science: Revolutionizing Engineering Science through Simulation [25], the US National Science Foundation Blue Ribbon Panel introduced simulation as the third column of scientific work, standing as an equal partner beside theory and experimentation.
- In 2007, the US Congress recognized M&S as a National Critical Technology in the House Resolution 487 [26].

Currently, many organizations in government and industry are using M&S to solve complex problems and use new decision support methods that are based on M&S on a regular basis. There is no doubt that M&S engineering delivered many good heuristics allowing addressing problems and challenges that could not be addressed before. M&S is ubiquitous.

However, at several academic conferences and in journal contributions more critical statements emerge regarding the often insufficiently understood use of M&S. Several of the book chapters in this volume also address the question: *"Why and when can we trust simulation results?"* As observed by Winsberg [7], and also addressed later in this chapter, there are many steps that lead to a simulation result. The typical application of M&S to solve a problem follows three steps [27]: First, we develop a design based on the model of an actual or theoretical system to answer the research question, then we implement and execute that model as a simulation on a digital computer, and, finally, we analyze the output results.

Many compromises are made on the journey from the customer's problem to the presentation of results, and the question often arises: how much compromise can be tolerated before the approach is diluted to a degree that the results can no longer be trusted? In practice, we leave this justice too often to subject matter experts and their opinion, but we do not answer the question of what criteria qualify such an expert. Is it the knowledge of the application domain? If so, then how is this expert qualified to judge the choice of the correct or appropriate modeling paradigm? Is it the knowledge of software engineering principles? Then how is this expert qualified to deal with the various steps of M&S' typical conceptualization and its implications? As already noted in the last section, a mix of methodological and referential competency is needed here as well. Dealing with this challenge most likely will require a group of interdisciplinary experts who represent all aspects of contributing areas of expertise that make up the body of knowledge of M&S and the problem domain.

In practice, meeting such challenges most often boils down to the credibility of the results and to what degree we really trust them. Simple Turing-like tests to find out if we really trust the simulation results may include very personal questions, like the following:

- If an airline educates their pilots with very extensive flight simulator training in all possible situations, but they get no real live-flight training hours during their education, will you feel good knowing that your pilot is making his first flight with you on board?

- If a surgeon has extensive credentials from virtual operations using a medical simulation system, but the first human he will conduct the surgery on is you, will you trust him?

These are not rhetorical or academic questions. I use the following example for my students to make them aware of the ethical responsibility they share when they design simulation systems to train and educate people:

> *"In April 2000, during a Marine Corps training mission in Arizona, the pilot of an MV-22 airplane dropped his speed to about 40 knots and experienced "Vortex Ring State (VRS)," a rotor stall that results in a loss of lift. Attempts to recover worsened the situation and the aircraft crashed, killing everybody on board. The pilot had 100 hours in the Osprey simulator and nearly 3,800 hours of total flight time. However, none of his training or experience involved coping with a vortex ring. In January 2001, the General Accounting Office, in a presentation to the V-22 Blue Ribbon Panel, attested that the flight simulator used to train the soldiers in handling this aircraft did not replicate the VRS loss of a controlled flight regime." [1, p. 25]*

Epistemology of M&S can at least start to build a foundation to address these questions scientifically. Addressing the various activities of conceptualization, selecting or developing the simulation, executing the simulation, and analyzing the results to deal with the customer's problem under the lens of epistemology, will at least allow us to identify domains that cannot be addressed appropriately due to limitations of accessible tools and means, e.g.: undecidable problems from computer science remain undecidable for M&S as well. Computational complexity is just as significant an issue for M&S as it is for every algorithm; non-computable functions cannot be used for digital simulations, etc.

In summary, epistemology of M&S applications must address methodological constraints of the M&S domain as well as referential constraints of the application domain. In the context of intelligent M&S application, the methodological parts are essential, as they assist in building the common part for applicable solutions in various domains, not just a limited set of examples. Winsberg observes in his conclusion the following:

> *"Theory can stand in a wide variety of relations to its applications; sometimes theory is applied directly – in a process that is well captured by the idea of derivation – and sometimes the path from theory to application is much more indirect, with theory playing only a contributing role in generating local representations of phenomena. There is, consequently, a whole category of epistemological issues in the sciences that have escaped the attention of philosophers, who have traditionally concerned themselves with the justification of theories and not with their application."* [20, p. 136]

Epistemology of M&S falls exactly into this gap, as *"a simulation system is an executable hypothesis or – once proven to be valid – an executable theory!"* [18, p.16] In particular, when the simulation supports various domains following different and potentially competing theories, it becomes obvious that epistemology of M&S cannot be subsumed by the epistemology of a supported domain. Although some domain-specific heuristics have been developed, the main question regarding the epistemology of M&S has not been answered generally so far. Defining how to cope with this issue remains one of the grand challenges.

2.3 Teleology: How Do We Apply This Knowledge?

In practice, the question of how to apply M&S knowledge seems superfluous, as it is obvious that you apply it by executing the simulation. However, this view may be too naïve. The question regarding applicability of simulation has been often discussed under the question of validity: under which constraints can I use an M&S application and expect valid results?

As Velten points out in the introduction to his book [28], models are simplifications of the real world created in order to answer questions. They are not supposed to become unnecessarily complex duplicates of the reality that can't be used to answer the original question. After all, the model was developed to find an answer that we could not find using the real world referents. He highly recommends applying the principle of Occam's razor to modeling and simulation as well.

Pragmatics of simulation also drives the need for validation. Models are purposeful abstractions and simplifications of reality resulting in a conceptualization that is transformed into an executable simulation system. If we want to apply this simulation, and apply it for a special purpose, which needs to be close to the purpose the developers originally had in mind – captured by the requirements used when the developed the simulation system – it must overall make sense.

In his overview paper on the role of M&S, Reynolds [29] introduced two main categories, namely *models to represent knowledge* in reusable form to solve problems, such as training soldiers for a well known task (often referred to a drill) or provide well understood test data for hardware-in-the-loop testing, and *models to gain understanding* about how a system may work, or how alternative actions in a given situation make work out. When we use M&S to provide knowledge, we have a very clear solution that we want to represent to train people, educate them, or provide realistic and reliable test data for a clear defined problem. When we use M&S to explore options or gain understanding of how things may work together it is the nature of the challenge that we do not have a well understood solution yet. What if analysis with M&S could be cheaper, less dangerous, and easier to configure and access than empirical experiments with the real world referent, or what if the real world referent is not accessible at all, as we want to explore a concept of a future prototype or simulate the conditions for a vehicle landing on Mars. From the pure task, the pragmatics of M&S are very different.

This differentiation directly links back to the question of whether or not it is possible to validate such a simulation system, and if so, how to do this. The traditional view on validation and verification – as captured in the work of Balci [30] and Sargent [31, 32] – mainly focuses on the so called "black box" validation.

This style of validation works well when we reproduce knowledge to solve a problem, as this requires full understanding of the problem to be solved as well as the environment it is solved in. A system is understood to be valid if it produces the same input/output pairings under the same time constraints as the real system does. Such a rigorous testing is not possible for the second category, as we want to use M&S exactly for those cases in which we do not have any empirical data yet. Instead of validating the behavior, such systems can be validated on whether or not they are reasonable. If all parts and their relations and functional transactions are reasonable – which translates to their following an accepted theory that can be used to describe the problem – we assume the model to be reasonable as well. This is often called a "white box" validation. It can be argued that black box validation uses quantitative methods to ensure realism, while white box validation uses qualitative methods to ensure axiomatic consistency. Furthermore, these views are not exclusive but complementary: the more I understand an available black box the better I can reproduce the underlying axioms that rule the internals of the box, leading to white box validation.

An alternative view is to implement a new theory as a simulation and observe the simulated system behavior in comparison with accessible empirical data to evaluate the theory. Simulation also allows for combining theories that were not necessarily developed to support each other using heuristics. We can use the nanosciences example described by Winsberg [20, pp.72-92] again to understand the challenges. If we have two valid theories, is there a way to compose them into one model that is still valid in a general way?

An interesting question for the application of intelligent M&S application to support gaining new knowledge is whether this is computationally possible at all: Can we really find something new with simulations? In particular, the literature on agent-directed simulation is full of examples of emerging behavior creating new insight into how a system works without having to make this explicit in the code. Nonetheless, from the pure mathematic standpoint, simulation systems are nothing but production systems. Whatever they produce is already implicitly available in the axioms and rules. Algorithmic Information Theory [33] actually proves that this is the case.

On the other side, we actually see unexpected behavior produced by agent-based simulations, such as wave patterns in traffic streams flowing through a city based on individual car simulations. What most examples illustrate, however, are behaviors that we normally connect with centralized planning, but the same behavior can be observed in the absence of such planning on the macro level simply by utilizing a set of basic rules on the micro level. But is this new knowledge produced by teleology? These are open questions urgently requiring answers to avoid giving bad advice to the decision makers who use such methods.

In summary, a formal approach to ontology, epistemology, and teleology of M&S will provide a framework to address many fundamental questions systemically and holistically and in a way that the results can easily be transferred across the borders of supported discipline and application domains. They are also essential to evaluate the potential, and limitations, of intelligent M&S systems.

3   Canons for M&S Research

It is generally accepted that science aims to gain understanding and knowledge while engineering aims to apply such knowledge. In order to gain new knowledge, we conduct research. Canons for research have been established to ensure the systematic treatment of subjects of research; they build a guideline for the credible research within the discipline. Brewer and Sousa-Poza [34] evaluate multi-disciplinary research encompassing studies based on different ontological and epistemological underpinnings in order to derive generalized research canons for extended justified true belief (JTB(+)) knowledge. Their proposal extends the review and analysis of research canons provided by Bozkurt [35]. An in depth discussion of JTB(+) goes beyond this chapter, and later contributions will deal with the question of what knowledge is and how intelligent M&S applications can – or cannot – support gaining new knowledge.   But to set the state for the following contributions we shall discuss here what it means to build a guideline for credible research for M&S. In order to apply the research of Brewer, Bozkurt, and Sousa-Poza [34, 35], some definitions are needed that are not generally part of the syllabi for M&S education, namely positivism and constructivism.

Positivism is the philosophical foundation for the scientific method that every student learns in school. It assumes that general laws of understanding can be derived by observation and conducting experiments. We conduct scientific efforts to discover such laws. Applying these laws allows not only understanding of what is going on, but also allowing us to predict what will go on in reality. These scientific results are testable, and the tests must lead to the same result independent of who conducts them, as long as the constraints are the same. Bozkurt [35] observes that canons of research rooted in positivism must therefore by internally valid (as they follow the general law defining the behavior) generalizable (as the law should no be limited to one observation) reliable and objective (as the scientific results are testable and independent from the observer).

Constructivism and naturalistic research do not believe that findings are discovered but that they are created by the experiment and the observers in a joint effort. It allows for multiple versions of truth, depending on who made what observations. As such, validity can no longer be the yardstick for canons, but it is replaced by credibility of the efforts. As no general independent laws are assumed, generalizability needs to be replaced with transferability between observations. Although results in constructivism are not necessarily reliable, they should still be dependable, and although objectivity is not in the scope of naturalistic research, results should be governed by the principle of confirmability. For the constructivist, truth is in the eye of the beholder, and internal consistency and relevance to the real-world challenge are more important than conformability to some general law that may or may not exist.

The following table lists these characteristics for canons of research from the view of positivism and constructivism as compile by Bozkurt [35]:

Table 1 Canons of Research Characteristics

Positivism	Constructivism
Validity	Credibility
Generalizability	Transferability
Reliability	Dependability
Objectivity	Confirmability

Modeling and simulation supports many domains, and therefore also may be applied within various canons of research rooted in different philosophies of science, some of which may even be diametrically opposed. Establishing truth or belief in something to be true can be based on validity or credibility. Justification for such a belief may be given by the characteristics of generalizability or transferability. The methods need to be reliable or at least dependable, and the context that establishes a shared understandings must be objectivity or confirmability accepted throughout the team. Depending on the philosophy, the role of M&S within the research can be very different.

The positivist will likely see the M&S application as an approximation of reality. He will try to make the simulation as realistic as possible, using validation and verification methods as discussed above in order to come as close to the real thing as he can get. As scientific truth is objective and testable, this is a justified approach.

The constructivist will see M&S as a general way to express his believes. The intelligent M&S application can serve as the common denominator in the group; it becomes the common tool that creates the finding by providing the necessary means for truth, justification, method, and context. Inconsistencies become clear, as the logic underlying the M&S application will show them clearly. Reliability of justification, method, and context can be expressed and explicitly used in the naturalistic research. Implicit assumptions become explicit and, as such, negotiable.

In order to take full benefit of these research results and insights, a clearer guideline on how to use M&S in the canons of research – on all scales between positivism and constructivism – is needed. Students, scholars, and practitioners need to be educated on how to apply M&S to best represent their philosophical view when utilizing M&S to conduct research.

Another viewpoint not sufficiently covered in the literature is to apply canons of research towards M&S as the object of study. While at least some guidelines and best practices for the application of M&S to conduct research exist, the task to evaluate intelligent M&S applications using positivistic or naturalistic research is not covered in the current literature. The field is wide open and waits to be exploited. Certain chapters in this book begin to look at this issue, but more research is needed to drive this topic forward.

4 Truth, Trust, and Turing

In this section, the insights so far shall be brought into a framework of three different perspectives regarding ontological, epistemological, and teleological views on intelligent M&S applications. The last section examined the canons of M&S research and provided insight regarding how to answer the three fundamental questions: *what is truth in M&S, how do we access truth in M&S*, and *how do we gain new knowledge using M&S?* We used the concepts proposed by Bozkurt [35] recommending using truth, justification, method, and context as the guiding principles. We realized that in the engineering education of M&S experts we often assume, without further philosophical reflection, a positivistic view, although the view of the constructivist is often as justifiable as well and may even be more appropriate under many circumstances where the constraints of positivism are not met.

We will conduct the same discussion in this chapter again, but we will be focusing more on the questions raised in the beginning of this chapter, namely: *what can be considered to be truth in the context of M&S, can we trust results using M&S*, and *what are the computational constraints that we have to take into consideration?*

4.1 Truth: What Is Truth in M&S?

Schmid [36] explicitly asked the question: "What is the truth of simulation?" He observed that to address the truth of M&S, philosophical truth theories have to be taken into account as well, in particular correspondence, coherence, and consensus theory. It is interesting that Schmid observes that – as discussed in the last section as well – the literature of M&S experts rarely addresses even the existence of such theories and implicitly falls back into the traditional positivistic worldview driven by physics-based modeling of systems governed by the Newtonian laws of physics. Tolk et al. [18] identify this as a main challenge that the community must overcome: to generalize their principles based on philosophy and science instead of naively applying solutions that seemed to work in one domain in another domain as well. Introducing truth theories is a necessary first step.

Correspondence theory of truth assigns truth values to simulations based on their correspondence with a fact of reality. In other words: if a simulation corresponds to the matters of fact in reality, it is true. In a positivistic world, this is at least fine for simulations of real systems, as facts are considered to be generally true. Schmid observes more challenges with this approach. Correspondence theory has disadvantages for (a) experimental simulation applications, for (b) simulations of imaginary systems, for (c) simulations of future systems, and for (d) simulations of systems under future potential conditions. In all these cases, the systems are not yet observed, but without doubt they can be reasonable, possible, or even likely. By definition, however, they cannot have any corresponding matters of fact.

Coherence theory of truth translates these ideas into constructivism. If the simulation belongs in a set of members of a coherent system of beliefs, then it is considered to be true among these members. As such, the simulation must be, in and of itself, consistent and may not contradict any other beliefs in the set of members. The main difference between correspondence and coherence theories is that correspondence only accepts one truth reference, while coherence recognizes that there may be several. However, once the truth reference is understood, coherence truth is as rigidly defined in this system as correspondence truth is defined in the general system.

Consensus theory assigns truth interpretations to something depending on whether it is rationally acceptable under ideal or optimal conditions. These conditions are agreed upon in the participating group. Consensus does not necessarily imply that something is de facto true (assuming de facto truth can be evaluated) but that the group believes it to be true.

The question regarding what is true in M&S remains unanswered by these observations, as long as the M&S expert does not take a stand regarding his philosophical foundation. He may believe in empiricism, where measures and observations provide evidence for underlying laws. Then his simulation must be coherent with these observations and follow such laws. Or he may support rationalism and emphasize the coherence of rules and internal deductions to keep a set of truth statement consistent as the yardstick for his simulation.

The notion of credibility may serve as a connector between truth and trust. As stated by Brade:

> *"The credibility of a model is based on the perceived suitability and the perceived correctness of all intermediate products created during model development. The correctness and suitability of simulation results require correctness and suitability of the model and its embedded data, but also suitable and correct runtime input data and use or operation of the model. Verification and validation aim to increase the credibility of models and simulation results by providing evidence and indication of correctness and suitability."*
> [37, p. 13]

This introduces the concept of suitability, which in turn introduces the intended use of the M&S application. In practice, this often translates into the question of whether an M&S application should be used in support of a given task. This leads to the concept of trust, which will be discussed in the following section.

4.2   Trust: Can We Trust Research That Used M&S?

The famous quote *"All models are wrong, but some are useful!"* is generally attributed to the statistician George E.P. Box. Trust and credibility can be understood to answer the question of if a model is suitable to be applied for research addressing a given research challenge. Even if it cannot be decided whether a model is true – or even if we know that it is wrong – it may still be useful in addressing certain research questions. Winsberg [20] gives several examples in his book as well.

The Guide for Understanding and Implementing Defense Experimentation [38] recommends several sound practices for setting up experiments utilizing M&S to address defense research question, many of which are easily transferable to other application domains. In most defense experiments, the objective is to evaluate a new capability. Four principles for good experimentation are identified:

a) The ability to utilize the new capability must be present.
b) The ability to detect change based on the use of the new capability must be present.
c) The ability to isolate the reason for the effect must be present.
d) The ability to relate the experimentation results to operational scenarios must be present.

While these principles seem to be trivial at first sight, many practical applications of M&S are violating some of these good practices. Depending on the intended use, fidelity and resolution must be defined. *Fidelity* of a simulation is the accuracy of the representation when compared to the real world system represented. A simulation is said to have fidelity if it accurately corresponds to or represents the item or experience it was created to emulate: How realistic does the simulation react? The *resolution* of a model or a simulation is the degree of detail and precision used in the representation of real world aspects in a model or simulation. Resolution means the fineness of detail that can be represented or distinguished in an image: How much detail do I observe? In particular when doing experimentations, we often do not know how the new ability will be used or what the effects, and potentially effects of effects, will be. Also, in complex systems, one observed effect can have various causes. The principle requesting the ability to isolate the reason for the effect requires a reductionist approach where attempting to reduce the chain to a preferably single cause-effect event. The real challenge in complex systems is, however, that we have a multitude of interconnections that are often non-linear. There is no easy solution.

A new research domain of interest – particularly in the light of intelligent M&S applications – is the domain of computational trust [39]. Unfortunately, while some applications for multi-agent systems have been evaluated [40], research evaluating the applicability of such ideas for intelligent M&S applications could not be found. The central idea of this new domain builds a bridge to the next section: trust is understood as a computable measure of soft security that complements traditional hard security like encryption, authorization, and authentication. How much I trust another intelligent agent is captured in a metric that is computationally well defined for use in distributed intelligent agent applications. The semantic interpretation of such a metric is hard, although the application in the context of executing the computer program is easy. Looking at ontology, epistemology, and teleology of intelligent M&S applications through Turing's eyes will result in similar observations: the computational questions are challenging but solvable, however, the interpretation for the real problem – or the ability to relate the experimentation results to operational scenarios – is much harder.

4.3 Turing: What Are the Computational Constraints?

The main interest of this series in general, and this book in particular, are intelligent systems that are executed on a computer. Looking at the computational constraints, we are creating somewhat of a fractal here, as the topics of truth, justification, method, and context come back up as computational constraints again.

The development of computers over the last few years is breathtaking. Today, we have more computing power available in a laptop than NASA had available to bring a man to the moon. Computers enable building cars that can park themselves and warn the driver of dangerous developments, intelligent agents observe the schedule of travelers and rebook connections on the fly if needed, complex equations are solved in minimal times, and computer technology illustrates everyday many more impressive accomplishments that seem to have no limit. But no matter how impressive computers become, they are limited to what a Turing machine can do. Since the Church-Turing thesis, we know that problems that can be solved by an algorithm are equivalent to those that can be executed on Turing machines. The resulting computable functions are similarly limited. Hunter's essay [41] gives a great introduction that does not require being a genius in computer science or mathematics.

This has immediate implications on computer simulations as well: as they are a subset of what can be executed on a computer, they cannot expose anything more than what computer programs can do. This sounds trivial, but the number of papers that claim that certain models provide new insight increases. As Bonder clearly states:

> "Not models produce new insights, but the analysts with ten to fifteen years of experience in the respective domain." [42]

While we can get insight by developing and using models, the simulation itself is a production system. It takes the input data and applies the algorithm to produce the output data. Axiom and rules unambiguously define this production which is limited to what is already in the data and the algorithm. Algorithmic Information Theory [43] has clearly proven that no new knowledge can be produced by such computational efforts. Consequently, we cannot discover something we do not already know and put it into algorithm or data. That does not decrease the value of M&S, but we just must be aware that we are dealing with mathematics and not a magic mirror.

The advantage of this formal approach is, however, that the questions for truth and trust become formal questions as well. In particular, model theory [44] offers a wealth of approaches to deal with different interpretations of truth in models. Using mathematical logic in general and Tarski's truth definition for formal languages in particular, model theory provides a rich set of mathematical methods to evaluate the representation of truth in different formal languages. As simulation systems are closed systems that work on discrete, finite sets described in formal languages, the findings of model theory and algorithmic information theory can both be applied to provide a solid foundation for M&S. This research, however, has just begun and is still in its infancy, although initial results are promising.

Another aspect needing discussion here is the well-known fact that when implementing a simulation, the computational decisions of developers often drive the simulation. Such computational engineering decisions are rarely driven by conceptual constraints but by efficiency implications of possible alternatives. Very often, the reason behind a decision is the need to discretize continuous events in order to bring them onto a computer. Typical examples include using the Runge-Kutta method or the Euler method to solve differential equations. The methods are similar, but Runge-Kutta employs parabolas and quartic curves to better approximate the function, which makes it in general more accurate but also computationally more expensive. It is often a decision of the software designer which method to apply for the implementations while the model developer is sometimes not even aware of such challenges. The simulation results can differ significantly, in particular in complex systems with multiple, non-linear interrelations where small changes in one input parameter can lead to significant changes in the observed outcome. Similar dramatic changes have been observed when the computer systems were upgraded and provided a higher resolution in the computation, such as using a 64-bit processor instead of a 32-bit processor. In particular when such errors are propagated, a conceptually sound approach may lead to inaccurate results. Before deciding to trust simulation results, scrutinizing numerical challenges and how these challenges are resolved is therefore necessary.

We began this chapter with discussing ontology, epistemology, and teleology. Ontology has been understood as the 'study of being' or the 'study of what exists,' and is often captured as a system defined by a finite set of concepts and their relations. Epistemology was described the 'study of how we come to know,' including how we define knowledge, represent it, and communicate it with others. Teleology focuses on the application being the 'study of action and purpose,' resulting in methods. Together, they build the philosophical foundation of a discipline. This led to questions of validation and verification, and what various truth philosophies imply for such questions. Many engineers are still rooted in a positivistic worldview. M&S can support positivism, but the concept of internal consistency of axioms and rules are supportive of naturalistic research and rationalism and constructivism. The variety of canons of research that are supported by M&S needs to be better understood by the M&S community, as canons for M&S-based research are as much needed as a canon for M&S-related research. Finally, focusing on computer simulation allows us to apply the rigor and elegance of mathematics to capture and describe challenges for intelligent M&S applications.

In summary, we are just beginning to agree on a research agenda to which many experts from various domains will need to contribute. However, in order to define and understand Modeling and Simulation as a discipline, and not merely as a set of powerful tools, establishing such a research agenda and working on it is absolutely necessary. If we really want to base out scientific judgment on M&S

result, more philosophical rigor is necessary. If we want to apply intelligent M&S applications in support of better decisions, more research is needed.

The expected benefit is worth such efforts. Following the well-known value chain proposed by Ackoff [45], intelligent M&S applications are helping to improve the value by orders of magnitude:

- Factual *data* alone has only limited value. It can only serve syntactical issues.
- Only when data is embedded into context, does it become *information*. The information addresses semantic issues. Information provides concepts needed to evaluate completeness, correctness, currency, consistency, and precision of the data items and information statements available.
- *Knowledge* is procedural application of information. M&S applications provide the means to put information into procedural applications, to produce output data based on input data plus the M&S application. This addresses pragmatic issues.
- *Awareness* includes the cognitive level, not only understanding observed pragmatics, but being aware of possible future developments, their likelihood, and other insights of interest. This is the domain of intelligent M&S applications. However, we will only be successful if we understand what we represent, why this is valid in the applied context, how we apply it, and all other aspects addressed in this introduction to the book.

The following figure shows the quality improvements in Ackoff's value chain by applying the various components of the spectrum of intelligent M&S applications.

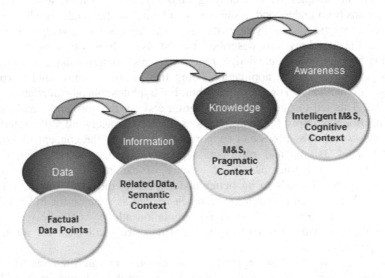

Fig. 2 Ackoff's Value Chain and Intelligent M&S Applications

5   Contributions in the Book

As mentioned in the introduction, this book has not been compiled to become a textbook espousing one consistent theory on a certain problem domain. It purposefully brings together internationally recognized experts from the various different domains of philosophy of science, computer science, and modeling and simulation to highlight important aspects of ontology, epistemology, and teleology from philosophical, computational, and conceptual viewpoints. The objective in selecting the chapters was not to build consensus but to start the discussion and to identify open questions and challenges. And the result is indeed characterized by diversity. As such, whatever domain the reader may be an expert in, he will likely find something truly new, be it a collection of philosophical concepts for the computer scientists, or the computational implications of semantic metadata for science philosophers.

In their chapter, Chris Partridge, Andy Mitchell, and Sergio de Cesare contribute to a better understanding of meta-ontological concerns for M&S systems. They provide a guideline for developing ontological architectures, i.e., an ontology describing the system of concern becomes the conceptual basis for the simulation. They embed their proposal in a rich philosophical discussion that shows how various scientific philosophy concepts are directly connected to architecture choices and vice versa. This is captured in the form of meta-ontological concerns.

In the third chapter, Marko Hofmann defines ontologies as setting the relationship between formal signs used in computer simulations and "meaning" as a notion of human minds. Of particular interest is the ability of ontologies to capture knowledge. He shows the implication of various epistemological interpretations, which adds valuable discussion points to the observations made earlier in this chapter. His primary contributions are his valuable insights into the two-sided nature of ontologies in modeling and simulation. He clearly illustrates that for successful progress in the domain of intelligent M&S application, we need to be addressing both aspects.

Brian Heath and Ross Jackson provide a critical look at the task underlying this book in their discussion on the ontological implications of M&S in Postmodernity. New developments in animation allow the presentation of simulation results to look and feel as real as a movie. Many users enjoy seeing simulations accordingly as "the real thing" and request validity to support their view of using models. The fourth chapter uses the notion of validity to show that we are in danger of forgetting that models are tools to think with and makes a strong statement that we need to engage more in the epistemological discussions of postmodernity.

Chapter 5 makes a similar request, to focus more on the value of models to help solve a problem rather than on reproducing reality. Weirich observes that the value attributed to models is their applicability for analogical explanations, not the capability to mimic exactly what is observed. As such, models offer only partial explanations, but they are well understood. The approach is applicable to descriptive and normative model types allowing composing their individual contributions to overall increase the insight. Explanatory models offer partial explanations which accurately depict the ability of models to elucidate phenomena.

In the sixth chapter, Troitzsch presents 'non-statement view' reconstructions of three versions of the 'garbage can' theory of organizational decision making behavior in order to determine under which conditions these theoretical approaches could be tested against the results of empirical research. Troitzsch shows the similarity between reconstructing a theory and designing a simulation model representing this theory. The approach may be used as a template for intelligent support systems.

Andreas Pyka and Simon Deichsel use Occam's razor to cut back models in chapter seven. Using their experiences with users of agent-based models, they present a plea for discontinuing the philosophical idea of scientific realism. Within their work, they show that particularly for economic models the need for bottom-up realism of observations on the micro-scale and predictions on the system level is not given. Starting with simple and 'unrealistic' models is often preferable to description-rich models that are as complicated as reality and often even more obscured by their implementation decisions.

The philosophical aspects of M&S presented by Tuncer Ören and Levent Yilmaz in chapter eight are based on multiple years of experience in the M&S domain. Their chapter highlights the relationships between reality, representations of reality, and simulation as related concepts but with significant differences. Various views on the reality-model dichotomy are presented, and the implications for experimentation, knowledge generation, and gaining experience using simulation are evaluated. The chapter also addresses ethics in a short section.

Andrew Collins and D'An Knowles Ball introduce a non-traditional topic in chapter nine. In their contribution, they evaluate the philosophical and theoretic underpinnings of simulation visualization rhetoric and its practical implications. As stated before, animation and visualization play an increasingly significant role in simulation credibility, so that understanding the ontology and epistemology behind such visualization is as important as the represented model. Exposing audiences to the rhetorical nature of visualization will ensure that decision makers using intelligent M&S application in support of their process are aware of potentials and limitations. It will ensure that quality of information will be displayed accordingly. In addition, it will ensure that 'smoke and mirrors' cannot be used to conceal inappropriate or inferior M&S applications.

Current modeling and simulation applications focus on simulation as a computational activity that algorithmically produces output data based on valid input data. However, the approach described by Saikou Diallo, Jose Padilla, Ipek Bozkurt, and Andreas Tolk in chapter ten produces knowledge in form of new theory. To this end, theory components are represented as simulation components. This allows composing and recomposing theory components to process new theory out of existing theory.

Chapter eleven introduces a new concept for improving the credibility of simulation-based result, namely the 'e-Portfolio' concept. Levent Yilmaz and Tuncer Ören introduce this concept to capture conceptual model, simulator design, experimental frames, and scientific workflow specifications in addition to assumptions and constraints that influence the results. This schema can also be used to communicate result constraints between M&S application which can become a cornerstone to enable collaborative research between intelligent M&S applications.

The topic presented by John Elias in chapter twelve is immersion. He introduced a new viewpoint on M&S application by showing the intimate relationship of simulation technology and cognitive research. When simulation is truly used as the main technology in support of developing a theory of the mind, this is of direct interest to the intelligent systems audience. Starting from historical considerations of the development of notions of mind, he makes a strong case for intelligent M&S applications to become a backbone for cognitive research in the near future.

Roger Smith observes in chapter thirteen that the current practice of modeling allows almost any approach, while its measure of correctness is determined solely by the usefulness of the resulting product. This freedom is not a bad thing, but it hinders mutual exchange of insights. He therefore proposes a 'Periodic Table' for modeling that captures the various approaches and shows the interrelation and potentials for mediation between different solutions. Following scientific tradition, he motivates a deeper exploration of the nature of modeling through a taxonomy of the scientifically supportable methods and structures that have been created and discovered over many years of research and practice.

By applying a Bayesian approach to the validation of agent-based models, Kevin Korb, Nicholas Geard, and Alan Dorin make their contribution to more formal methods in chapter fourteen. They observe that validation of M&S in general and of intelligent M&S, such as presented by agent-based models in particular, has been neglected by research and has left the theory behind its practice. By combining data-directed and expert validation efforts, they are able to obtain a balanced picture of the empirical merits of a model.

Scott Douglass and Saurabh Mittal present their framework for M&S of the artificial in chapter fifteen. Coming from the background of conducting M&S support for the United States Air Force, their environment was dominated by physics-based models for decades. The introduction of knowledge-based approaches, addressing the cognitive domain as well, required new methods. The resulting Knowledge-based Contingency-driven Generative Systems (KCGS) framework combines philosophical aspects, as discussed in this introduction, with the rigor of the Discrete Event System Specification (DEVS) formalism. A major contribution of this work is illustrating how ontology, epistemology, and teleology play critical roles in the simulation of agents. Agents select other agents based on their behavior to meet situational constraints while achieving goals. By doing so, they provide critical characteristics of intelligent M&S applications and support their collaborative effort in agile and dynamic environments.

Claudia Szabo and Yong Meng Teo are known for their contributions to the body of knowledge of semantic validity. In chapter sixteen, they use their insight to apply it to intelligent M&S applications through evaluating the possibility and constraints of semantic validation of emergent properties in component-based simulation models. This chapter describes and compares grammar-based, variable-based, event-based, and our objective-based approaches for the identification and validation of emergent properties. Although not sufficient, this is a necessary contribution for collaboration of intelligent M&S applications.

The final chapter is a contribution of Wenguang Wang, Weiping Wang, Qun Li, and Feng Yang. Their research focuses on the use of service-oriented architectures and semantic web methods in support of distributed simulation systems. In this chapter, they extend their work by referential and methodological perspectives and build a systematical philosophical foundation of the service-oriented simulation paradigm. This results in a general framework that is not only applicable to intelligent M&S applications but may be extended into a general integration framework for other IT solutions that support the service-oriented paradigm.

All seventeen chapters provide unique, but mutually supportive, insights into philosophical, conceptual, or computational aspects that are all highly relevant. Although the chapters are only loosely coupled, they all contribute to a common view on intelligent M&S systems. Some views presented here are competing, but this is to be expected when entering terra nova. Which views are 'correct' or will survive in practice will be judged later. For now, at least we have made an initial step toward establishing a common agenda in the field.

After completing this book, I personally feel that something has been accomplished. The various chapters can all be seen as starting points for continued collaboration on a common foundation for intelligent M&S applications that comprise the identified philosophical, computational, and conceptual aspects. Such a strong foundation will not only help us to reach the next level of intelligent M&S applications, it will also contribute to the body of knowledge for intelligent systems and modeling and simulation through closing gaps that have been identified in this book.

References

[1]   Tolk, A.: Engineering Principles of Combat Modeling and Distributed Simulation. John Wiley and Sons, Inc., New York (2012)

[2]   Sokolowski, J.A., Banks, C.M. (eds.): Modeling and Simulation in the Medical and Health Sciences. John Wiley and Sons, Inc., New York (2011)

[3]   Ören, T.I.: Modeling and Simulation: A Comprehensive and Integrative View. In: Yilmaz, L., Ören, T.I. (eds.) Agent-Directed Simulation and Systems Engineering, pp. 3–36. John Wiley and Sons, Inc., New York (2009)

[4]   Padilla, J.J., Diallo, S.Y., Tolk, A.: Do we Need M&S Science? SCS M&S Magazine 8, 161–166 (2011)

[5]   Hofmann, M., Palii, J., Mihelcic, G.: Epistemic and normative aspects of ontologies in modelling and simulation. Journal of Simulation 5, 135–146 (2011)

[6]   Schmid, A.: What is the truth of simulation? Journal of Artificial Societies and Social Simulation 8(4) (2005)

[7]   Winsberg, E.: Sanctioning models: The epistemology of simulation. Science. Context 12(2), 275–292 (1999)

[8]   Frank, U., Troitzsch, K.G.: Epistemological Perspectives on Simulation. Journal of Artificial Societies and Social Simulation 8(4) (2005)

[9]   Grune-Yanoff, T., Weirich, P.: The Philosophy and Epistemology of Simulation: A Review. Simulation & Gaming 41(1), 20–50 (2010)

[10]   Noy, N.F., Sintek, M., Decker, S., Crubezy, M., Fergerson, R.W., Musen, M.A.: Creating Semantic Web contents with Protege-2000. IEEE Intelligent Systems 16(2), 60–71 (2000)

[11]   Gruber, T.R.: Toward principles for the design of ontologies used for knowledge sharing. International Journal of Human-Computer Studies 43(4-5), 907–928 (1995)

[12]   Silver, G.A., Miller, J.A., Hybinette, M., Baramidze, G., York, W.S.: DeMO: An Ontology for Discrete-event Modeling and Simulation. Simulation 87, 747–773 (2011)

[13]   Teo, Y.M., Szabo, C.: CODES: An Integrated Approach to Composable Modeling and Simulation. In: Proceedings of the 41st Annual Simulation Symposium, pp. 103–110. IEEE CS Press (2008)

[14]   Guizzardi, G., Wagner, G.: Towards an ontological foundation of agent-based simulation. In: Proceedings of the Winter Simulation Conference, pp. 284–295. IEEE CS Press (2011)

[15]   Adam, N.: Ontology Applications in Homeland Security. In: Proceedings of the Ontology Summit 2011: Panel Session-9: Grand Challenges (2011), http://onolog.cm3.net (last visited April 2012)

[16]   Tolk, A., Diallo, S.Y., King, R.D., Turnitsa, C.D., Padilla, J.J.: Conceptual Modeling for Composition of Model-based Complex Systems. In: Robinson, S., Brooks, R., Kotiadis, K., van der Zee, D.J. (eds.) Conceptual Modeling for Discrete-Event Simulation, pp. 355–381. CRC Press (2010)

[17]   International Organization for Standardization/International Electrotechnical Commission (ISO/IEC). Information technology metadata registries part 3: registry metamodel and basic attributes (2003), Standard ISO/IEC 11179 3

[18]   Tolk, A., Diallo, S.Y., Padilla, J.J., Turnitsa, C.D.: How is M&S Interoperability different from other Interoperability Domains? In: Proceedings of the Spring Simulation Interoperability Workshop, pp. 12–20. SISO, Orlando (2011)

[19]   Petty, M.D., Franceschini, R.W., Panagos, J.: Multi-Resolution Combat Modeling. In: Tolk, A. (ed.) Engineering Principles of Combat Modeling and Distributed Simulation, pp. 607–640. John Wiley and Sons, Inc., New York (2012)

[20]   Winsberg, E.: Science in the Age of Computer Simulation. The University of Chicago Press (2010)

[21]   Yilmaz, L., Tolk, A.: A Unifying Multimodel Taxonomy and Agent-Supported Multisimulation Strategy for Decision-Support. In: Phillips-Wren, G., Ichalkaranje, N., Jain, L.C. (eds.) Intelligent Decision Making: An AI-Based Approach. SCI, vol. 97, pp. 193–226. Springer, Heidelberg (2008)

[22]   Davis, P.K., Bigelow, J.H.: Experiments in Multiresolution Modeling. MR-1004-DARPA, The Rand Corporation/National Defense Research Institute (1998)

[23]   Institute of Electrical and Electronics Engineers. IEEE 1278 Standard for Distributed Interactive Simulation. IEEE publication, Washington, DC

[24]   Institute of Electrical and Electronics Engineers. IEEE 1516 Standard for Modeling and Simulation High Level Architecture. IEEE publication, Washington, DC

[25]   National Science Foundation (NSF) Blue Ribbon Panel, Report on Simulation-Based Engineering Science: Revolutionizing Engineering Science through Simulation. NSF Press, Washington, DC (2006)

[26]   US Congress. House Resolution 487: Recognizing the contribution of modeling and simulation technology to the security and prosperity of the United States, and recognizing modeling and simulation as a National Critical Technology. Introduced 14 June 2007, passed 16 July 2007. US Congress, Washington, DC (2007)

[27] Banks, C.M.: What is Modeling and Simulation? In: Sokolowski, J.A., Banks, C.M. (eds.) Principles of Modeling and Simulation: A Multidisciplinary Approach, pp. 3–24. John Wiley & Sons, Hoboken (2009)

[28] Velten, K.: Mathematical Modeling and Simulation: Introduction for Scientists and Engineers. John Wiley & Sons, Hoboken (2009)

[29] Reynolds, P.F.: The Role of Modeling and Simulation. In: Sokolowski, J.A., Banks, C.M. (eds.) Principles of Modeling and Simulation: A Multidisciplinary Approach, pp. 25–43. John Wiley & Sons, Hoboken (2008)

[30] Balci, O.: Verification, Validation, and Testing. In: Banks, J. (ed.) The Handbook of Simulation, pp. 335–393. John Wiley & Sons, Hoboken (1998)

[31] Sargent, R.G.: Verification, Validation, and Accreditation of Simulation Models. In: Proceedings of the Winter Simulation Conference, pp. 50–59. IEEE CS Press (2000)

[32] Sargent, R.G.: Some Approaches and Paradigms for Verifying and Validating Simulation Models. In: Proceedings of the Winter Simulation Conference, pp. 106–114. IEEE CS Press (2001)

[33] Chaitin, G.J.: Information-Theoretic Limitations of Formal Systems. Journal of the Association for Computing Machinery (ACM) 21, 403–424 (1974)

[34] Brewer, V., Sousa-Poza, A.: Generalized Research Canons for JTB(+) Knowledge. Paper V732ST. In: Proceedings of the International Symposium on Peer Reviewing. International Institute of Informatics and Systemics (IIIS), 6 pages (2009)

[35] Bozkurt, I., Sousa-Poza, A.: A Comparison of Canons of Science used in Positivistic Research and Constructivist/Naturalistic Research. In: Proceedings of the 26th American Society of Engineering Management National Conference, pp. 143–146. ASEM publications (2005)

[36] Schmid, A.: What is Truth of Simulation? Journal of Artificial Societies and Social Simulation 8(4), 5 (2005), online journal accessible via, http://jasss.soc.surrey.ac.uk/8/4/5.html (last visited May 2012)

[37] Brade, D.: A Generalized Process for the Verification and Validation of Models and Simulation Results. Ph.D. Thesis at the University of the Federal Armed Forces of Germany, Munich (2004)

[38] Technical Cooperation Program (TTCP), Guide for Understanding and Implementing Defense Experimentation (GUIDEx). Canadian Forces Experimentation Centre, Ottawa, Canada (2006)

[39] Sabater, J., Sierra, C.: Review on Computational Trust and Reputation Models. Artif. Intell. Rev. 24(1), 33–60 (2005)

[40] Lu, G., Lu, J., Yao, S., Yau, Y.J.: A Review on Computational Trust Models for Multi-agent Systems. The Open Information Science Journal 2, 18–25 (2009) ISSN 1874-947X

[41] Hunter, G.: What Computers Cant't Do. Philosophy 63(244), 175–189 (1988)

[42] Bonder, S.: Military OR, Science, and Models – Some Lessons learned. In: Proceedings of the 68th Military Operations Research Symposium, Colorado Springs, Colorado (2000)

[43] Chaitin, G.J.: Algorithmic Information Theory. Cambridge University Press, Cambridge (1987)

[44] Prestel, A., Delzell, C.N.: Mathematical Logic and Model Theory. Springer, Heidelberg (2011)

[45] Ackoff, R.L.: From Data to Wisdom. Journal of Applied Systems Analysis 16(1), 3–9 (1989)

Guidelines for Developing Ontological Architectures in Modelling and Simulation

Chris Partridge[1], Andy Mitchell[1], and Sergio de Cesare[2]

[1] BORO™ Solutions Limited
  Henley on Thames, Oxfordshire, United Kingdom
[2] Brunel University
  Uxbridge, Middlesex, United Kingdom

1 Introduction

This book is motivated by the belief that "a better understanding of ontology, epistemology, and teleology" is essential for enabling Modelling and Simulation (M&S) systems to reach the next level of 'intelligence'. This chapter focuses on one broad category of M&S systems where the connection is more concrete; ones where building an ontology – and, we shall suggest, an epistemology – as an integrated part of their design will enable them to reach the next level of 'intelligence'.

Within the M&S community, this use of ontology is at an early stage; so there is not yet a clear picture of what this will look like. In particular, there is little or no guidance on the kind of ontological architecture that is needed to bring the expected benefits.

This chapter aims to provide guidance by outlining some major concerns that shape the ontology and the options for resolving them. The hope is that paying attention to these concerns during design will lead to a better quality architecture, and so enable more 'intelligent' systems. It is also hoped that understanding these concerns will lead to a better understanding of the role of ontology in M&S.

1.1 Chapter Structure

The chapter starts with some background and then reviews the selected concerns and associated choices that characterise the meta-ontological landscape. Some concerns relate to the process of producing the ontology-based models; others are more metaphysical and focus on the nature of what is produced. The main sections address these topics:

- The basis for assessing these choices
- The major meta-ontological choice: what kind of ontology to adopt
- Some key methodological and metaphysical choices
- How to approach epistemology

A. Tolk (Ed.): Ontology, Epistemology, & Teleology for Model. & Simulation, ISRL 44, pp. 27–57.
springerlink.com   © Springer-Verlag Berlin Heidelberg 2013

2 Background

The starting point for the analysis in this chapter is the use of ontologies in M&S systems; the first section below clarifies which systems these are.

Much of the discussion in this chapter is of necessity highly abstract and theoretical. However it is also informed by the authors' experience with developing and implementing ontological architectures, particularly the BORO ontology. The second section below introduces this.

2.1 Which Kind of M&S System Benefits from Using an Ontology?

M&S is a broad church with a variety of types of member. It is used in both science and engineering. Well known examples in science are the billiard ball model of a gas and the Bohr model of the atom; in engineering the scale model of a bridge or an airplane wing. Within this broad church, there is one kind of M&S system – large-scale, engineering, computing systems - that has been identified as likely to benefit from ontology-driven design.

What characterises this kind of system? One key underlying factor is the mode of representation. Models have different ways of representing. This is clearly evidenced by examples of the same subject represented in different ways; for example, a scale model of the wing of an airplane represents the wing in a way that is different from how a mathematical model of its shape does.

One of the ways of classifying the different ways of representing is by the nature of the representation. The scale model of the wing is a straightforward physical object ('a material model'), so are analogue models like electric circuit models of neural systems. Other models are conceptual; for example, Bohr's model of the atom. These are located in the minds of scientists or engineers rather than in the laboratory or workshop and they do not have to be physically realised and experimented upon to perform their representational function.

The development of computing led to a new kind of representational mechanism, where descriptions or data can be given behaviour and so simulate. Morgan [1, p. 231] comments on the "degree of materiality" of computer data, though as [2, p. 495] points out the computer system is a "material/physical system". This has been incredibly successful in both engineering and science. Humphrey [3, p. 64] suggests that this computational technique 'constitutes a significant and permanent addition to the methods of science'.

Building large scale computing M&S systems requires careful design. Balci et al. [4, p. 158] identified 'conceptual models' as useful tools and lists four main types and seventeen sub-types of engineering simulation systems where they can be deployed in the design of large scale complex applications. They note "... a simulation conceptual model (CM) as a repository of high-level conceptual constructs and knowledge ... intended to assist in the design of any type of large-scale complex M&S application. ... M&S application designers can be assisted by a CM in the design of large-scale complex M&S applications for solving problems ..." [4, p. 158].

Hofmann et al. [5] are more specific about the possible nature of these conceptual models and states that "... ontologies have been proposed for modelling and simulation (M&S) as well" listing [6] [7] [8] [9] [10] [11] [12] [13] [14] as support. These papers use the term 'ontology' in a variety of related senses; one of the themes of this chapter is that these senses need to be clarified and their use made explicit. With reference to [15], Hofmann et al. note that "Among other advantages, ontology-based simulation is said to support consistent semantic model interchange, which leads to higher quality models, lower costs and a faster development process ... indeed a promising solution for interoperability and composability. These identify the design of large-scale, complex, engineering, computing M&S systems as an area that can benefit from an ontological approach.

2.2 Experience of Developing and Implementing Ontological Architectures

The ultimate requirement here is practical – more intelligent M&S systems. One sensible concern is whether the abstract issues raised are actually practically significant. The authors have developed an appreciation of what features are important for ontological architecture through the development and implementation of systems. In this process, they have contributed to the development of BORO, an approach to building ontological or semantic models for large complex applications. This includes a top ontology and a process for constructing the domain ontology. A top ontology is the upper general layer of the ontology; it is this layer that is shaped by the meta-ontological decisions. The top layer then provides a structure for the lower layer, called the domain ontology. In BORO's case, the top ontology is broken out into a separate component so that it can be shared across the domain ontologies of individual systems. As well as economies of scale, this facilitates re-use and simplifies interoperability.

Partridge [16] [17] describes in detail an early version of BORO. It was originally developed to mine a single coherent ontology from multiple legacy systems – as the first stage in an architectural transformation [18] or software modernization, but has since been used for a variety of purposes. An early version was the basis for much of ISO 15926. It is used in the U.S. Department of Defense Architecture Framework Meta Model and is currently being used to develop a metamodel for the UK Ministry of Defence's Architecture Framework. A core use is enhancing the semantic interoperability of federated systems [19-22].

The authors' practical experience has guided them in their identification of the issues in this chapter. In a later section, how BORO addressed these issues is discussed.

3 Making Good Meta-ontological Choices

The meta-ontological choices highlighted in this chapter can seem esoteric; certainly some of them will seem highly abstract and maybe obscure to many people. In these situations it is helpful to have some explicit criteria for assessing the choices. One helpful resource is Kuhn [23]. He took an empirical approach and studied the characteristics of successful improvements in scientific theories, uncovering this list of six features:

- Generality: where the scope of the improved theory increased.
- Simplicity: where the improved theory is less complicated (it is typically more 'deeply simple' in the complexity theory sense).
- Explanatory power: the ability of the improved theory to give increased meaning.
- Fruitfulness: the ability of the improved theory to meet currently unspecified requirements or to be easily extendable to do so.
- Objectivity: the ability of the improved theory to provide a more objective (shared) understanding of the world.
- Precision: the ability of the improved theory to give a more precise picture of the world.

Making the ontological choices explicit provides an opportunity to take a position that improves on a number of features; explanatory power and objectivity are obvious candidates.

These assessment criteria should be used as a tool to assess the choices made for the issues identified in this chapter. As the focus here is on the architectural choices in M&S system design, these criteria operate at one remove; the goal is to make design choices that lead to artefacts that score well against these criteria.

3.1   The Right Basis for Assessment: Science or Engineering?

Kuhn was considering scientific theories and not engineering theories; and science and engineering have different bases for assessment. While scientific M&S is motivated by a pure search for scientific knowledge, engineering M&S is motivated by more practical, pragmatic engineering concerns. While it is important to ask whether a scientific model is true (this does not mean it has to be 'true' as [24] point out, there are cases where it is false, and known to be false, but still explanatory), an engineering model may be false yet extremely useful. There are many examples of this difference of approach in the wider world; civil engineers will knowingly elect to use Newtonian physics because it is significantly more efficient for them than Einsteinian physics, despite physicists regarding it as 'false'. This distinction is recognised in the M&S community: "The development of ontologies in computer science is motivated not so much by the pure search for knowledge (in contrast to the philosophical endeavour of finding the appropriate universal 'ontology', and also in contrast to enquiries of natural science), but by the urgent need to design, engineer and manage 'knowledge', and, more tangible, complex software systems effectively." [5].

The task of designing engineering M&S systems is an engineering task. Building ontologies for engineering, computing M&S systems is ontological engineering rather than pure ontology. What should concern M&S is the usefulness and effectiveness of the approach, not its truth per se. Hence our discussion is framed by a pragmatic engineering context. It is particularly important to bear this in mind as much of the philosophical ontological content discussed here was developed in the philosophy community where truth is, if anything, a more important concern than in science – and certainly a bigger concern than in engineering.

Kuhn's criteria are sufficiently grounded to be useful for both science and engineering. However, the main sections of this chapter focus on engineering, computing M&S (hence we shall use 'M&S' to mean 'engineering, computing M&S' for the rest of the chapter unless stated otherwise). For these it makes sense to prefix Kuhn's list with some engineering specific criteria, pragmatic criteria such as 'Usefulness' and 'Effectiveness'.

4   What Kind of Ontology to Adopt

One of the challenges holding back the successful deployment of ontology in the M&S community is the use of the term 'ontology' with a number of quite different (though related) senses. There are two intertwined factors at play here. One is the use of the same term to refer to different things (the real world and the model); another, and more important factor, is a different view on what ontology is (realism or idealism).

This section aims to tease apart the two factors and particularly to make clear the choice one has between the different views. We crystallise the views into two broad alternatives; the realist (real world) stance and the idealist (conceptual) stance. It will become clear as we discuss these below how different they are and how important it is that an informed choice between them is made when designing the top ontology. One of the key reasons is that the alternatives have different benefits. Unfortunately the lack of a clear distinction has led to situations where the benefits that accrue to one alternative are claimed for the other. So this chapter aims to clarify what benefits accrue to which alternatives.

A good way to understand the current situation is by putting the term into its historical context, showing how we got to where we are today; we do this below.

4.1 History of the Term

The different senses have emerged in different communities, but they have a common root in philosophy where the term originated and has been significantly researched. Typically the communities into which the term has crossed-over, like M&S, have had little overlap with the philosophy community. One of the interesting questions, relevant to M&S, is how established ideas in one community (ontology within philosophy, in this case) can be fruitfully transplanted into another distant community. In particular, how the ideas should be adapted to the needs of the new community.

4.2 Origin in Philosophy

The original sense comes from philosophy, where ontology is the study of existence. Though the etymology is Greek, the word has its origins in the 17th century (the oldest extant record of the word itself is the Latin form ontologia, which appeared in 1661, in the work Ogdoas Scholastica by Jacob Lorhard and in 1631 in the Lexicon Philosophicum by Rudolph Göckel), where the subject was regarded as one of the major branches of metaphysics. However, the practice is much older and can be traced back to the Ancient Greeks. For example, Aristotle, Metaphysics, IV; "all the species of being *qua* being and the attributes which belong to it *qua* being".

It has over the millennia developed into a significant practice; part of which is an understanding of what is required to produce a general characterisation of reality, known as an ontology. This has led to the modern 'objectification' derivative sense of ontology as "the set of things whose existence is acknowledged by a particular theory or system of thought: it is this sense that one speaks of 'the' ontology of a theory, or of a metaphysical system"(Jonathon Lowe in [25, p. 670]). This sense is the one most relevant to information systems, such as M&S systems. Their information element can be seen as a 'theory' that represents in various ways (explicitly and implicitly, directly and indirectly) the M&S domain [26]; so an M&S system's ontology is "the set of things in the domain whose existence is represented in some way by the information in the M&S system" or more simply, the domain.

4.2.1 Grounding Ontology in Reality

Philosophical ontology's focus is on reality – the 'real' world – and for it to get off the ground one needs to accept that we can know this reality. We do not have to accept this; this is illustrated by a key episode in ontological history, which is briefly outlined below.

In the late 18th century, Kant undermined this acceptance, claiming the idea that we can know reality as a "transcendental illusion (transzendentale Illusion)", a propensity to "take a subjective necessity of a connection of our concepts ... for an objective necessity in the determination of things in themselves" [27, A297/B354] (and in the Analytic "...the proud name of ontology ... must give way to the more modest title of a transcendental analytic" [27, A247/B304]). Kant's position is epistemic – it is not that the world (noumena) does not exist, it is rather that we cannot know it; and if we cannot know it, we cannot ontologise about it.

Kant's claim was largely accepted by the philosophical community and as a result ontology was neglected until the 20th century. In the late 19th century, Frege produced [28] [29] which eventually led to interest in ontology being rekindled. He argued the Kantian outlook led to a kind of psychological logic that conflated 'true' and 'being-taken-to-be-true', that we need to distinguish between psychological 'ideas' (Vorstellungen) and their objects. One outcome of this was the emergence of a clear recognition within the community that there is a choice between adopting a Kantian or an ontological position; where adopting the Kantian position typically means rejecting ontology in the philosophical sense. Another outcome was the development of a large body of analytic tools for detecting which position was being adopted and how it was being deployed.

4.3 Emergence of the Realist Stance in the Information Systems Community

With the development of computing in the second half of the 20th century, a number of related communities emerged. In the information systems community, the need for ontology in the philosophical sense was clearly recognised from the start; Mealy [30, p. 525] quite clearly says "The issue is ontology, or the question of what exists". And what exists was clearly recognised as a 'real world' outside the mind, often reflected in the phrase "real world models" [31]. Within philosophical ontology, this position is known as realism, hence we call it here the 'realist stance'. Hirschheim et al.'s research [32] found that this position was mainstream among practitioners; however, as their book illustrates, academics often adopted a quite different stance, which is described in the next section.

4.4 Emergence of the Idealist Stance in the Informatics Community

In the broader informatics community a different stance emerged. There was a shift from the assumption that we cannot know what objective reality (the Kantian 'noumena') is like to the view that there is no such thing as an objective reality, that all that exists is our ideas and concepts. This leads to the ironic conclusion that we can know 'reality' as it is nothing more than a construction built out of our concepts; where everyone's concept-system constitutes a reality that has in principle an equal claim (indeed, the only claim) to constituting one of the multiplicity of 'realities' - Kusnierczyk [33] and Smith [34] describe this development in more detail. This is taken to imply that the information in systems must reflect our ideas or concepts (though quite a few steps are required to reach this conclusion: for example, while I may see the information in systems reflecting my ideas, how can I be sure that other people see it as reflecting *my* ideas).

Viewing reality as mentally constructed is known in philosophical ontology as idealism, so here we call this conceptual idealism in the informatics community the 'idealist stance'. It contrasts with the realist stance; which accepts that both the 'real' world exists and we can know it. Clearly if one adopts this kind of idealist stance the study of ontology becomes the study of concepts rather than a mind-independent real world.

Smith at al. [35] review the adoption of the idealist stance and note "Sadly, elements … are found mixed up together in almost all terminology-focused work in informatics today." Smith [34] argues that the idealist stance is flawed and notes that this situation "is a matter of considerable astonishment to ontology-minded philosophers".

In our view, what is damaging from an engineering perspective is that there is often a reluctance in the informatics community to face up to the implications of this situation with the result that many of those developing ontologies have no real awareness that they have, in effect, made the choice to adopt the idealist stance and live with its implications. There are many exceptions both at the individual and sub-community levels. Tolk [36] is one example. He makes a distinction between positivism and interpretivism that appears in behavioural research in Information Systems. Though this is not exactly the same as the distinction made here, it has the same broad thrust. Tolk states that "positivism is rooted in the belief that truth exists on its own, it is independent of the observer and reality is separated from the individual who observes it" and "The alternative viewpoint is interpretivism that holds the belief that truth is a construct of the observer. Reality is relative and cannot be separated from the individual who observes it." One difference is that unlike the discussion here, this brings 'truth' into the distinction. This is a live issue, as there seems to be some equivocation; where the idealist stance is adopted, but the benefits of the realist stance are claimed.

4.4.1 Explaining Concepts and Modelling Methodologies

One way of understanding the different implications of the stances is looking at the fundamentally different ways they need to regard models. Models are central to M&S systems and one of the most basic requirements for a model to be of any use, is that people need to be able to agree on what the icons in a model represent.

In the realist stance, this is straightforward. People agree on what an icon represents by agreeing on the thing in the domain it represents. In the idealist stance, things are not so clear-cut. Here concepts have a central role and icons need to reflect (maybe represent) them. The usual explanation is that two people agree on what an icon represents, if they agree it represents the same concept. The problem is that to do this they need to share the same concept and it is not clear that this is even, in principle, possible.

Given the importance that the idealist stance places on concepts, one would expect there to be a reasonably clear picture of what they are in the community. This seems to be missing in the informatics literature, which seems to rely on a naïve folk notion of concept. Looking outside the community, there is one discipline that has researched the topic, philosophy of mind, developing a couple of mainstream possible views. Both of these illustrate the problem of sharing concepts.

One mainstream view is that concepts are psychological entities that are part of an internal system of representation; internal in that they are only visible to the owning mind [37], [38], [39], [40], [41], [42], [43], [44] and [45]. From one perspective, this has a strong immediate attraction to modellers adopting an idealist stance, as then modelling can be regarded as a process of transcribing one's private representations into a public model. Though if one developed this line of thought, one would need to explain how we get conscious access to the private representations.

This view has well-known problems with explaining how people share concepts. If we take it seriously, then the common claim that two people have the same concept, cannot mean that both of them have the identical concept (the usual meaning of 'same') as they cannot literally share their private internal concepts. Without a 'real world' to coordinate their concepts, they have no way to build a shared model.

Another common view is that concepts are Fregean senses (roughly speaking, meanings) [46] [47]. Typically, proponents of this view are realists who see concepts as abstract rather than mental objects that make the connection between thought and (real world) referents. If one adopts an idealist stance, then the problem with shared concepts reappears. How can a concept created by my mind, whether abstract or mental, be the same as a concept created by your mind?

Hopefully the preceding discussion has given some idea of how fundamentally different the implications of the two stances are. And also an appreciation of the need for the idealist stance to clarify what it means by 'concept' and of the hurdles it needs to negotiate to develop a useful approach to modelling.

4.4.2 Intentional Construction: An Argument for the Idealist Stance?

One common misconceived argument for the idealist stance is that the existence of intentionally constructed objects implies it is correct. Clearly these do exist, money and marriage are examples; they depend upon human beings to construct them. But this does not imply all objects are intentionally constructed in this way. There are also natural objects, examples are mountains and rivers; these exist whether we do or not.

Furthermore, this does not mean that intentionally constructed objects must exist as concepts in our minds; that, for example, when I look at a £5 note in my wallet, I perceive a concept in my mind. Searle [48] has explained how a realist stance towards these kinds of objects can work. He notes firstly that intentionally constructed objects are, and need to be, ultimately rooted in natural objects – without the natural objects they could not exist. And secondly that while the intentionally constructed objects are ontically subjective - that is, they depend upon human minds, they are also epistemically objective – so they can be known objectively unlike concepts.

4.5 Emergence in Computer Science

More recently, in the computer science community (particularly the AI community) a new sense for the term 'ontology' has emerged. The earliest documented expression is Gruber's [49]; "a formal explicit specification of a shared conceptualization". It claims that "The term [i.e. ontology] is borrowed from philosophy, where an ontology is a systematic account of Existence" but does not make clear that it is being used in a very different sense.

Guarino [50] [51] clarifies the terminology. In [51] he clarifies the shift in sense by describing the ontology (in the AI sense) as "an engineering artefact" and suggests using "the word *conceptualization* to refer to the philosophical reading" and attempts to relate these. As the earlier discussion should make clear, for people with a philosophical background, it is perverse to call a philosophical ontology a 'conceptualisation'. However, it may also be revealing as the AI community seems to be leaning towards an idealist stance and so a rejection of "the philosophical reading". For example, Guarino's [50] Figure 1 lists the "Possible interpretations of the term ontology"; this contains no mention of the strict philosophical reading and three of the seven entries contain the term 'conceptual'.

There seem to be disagreements on the specifics of what a conceptualization is. Gruber [49], referring to Genesereth et al. [52], says that it is: "the objects, concepts, and other entities that are presumed to exist in some area of interest and the relationships that hold [between] them". Though he muddies the water with the odd claim that "For knowledge-based systems, what "exists" is exactly that which can be represented". Whereas Guarino [50] says it is "an intensional semantic structure which encodes the implicit rules constraining the structure of a piece of reality" having earlier claimed that this is "the philosophical reading". Neither of these are exactly what a philosopher would recognise as an ontology, though the Genesereth/Gruber description seems closer, the reluctance or inability to recognise the philosophical sense noted earlier seems to have been there from the start.

4.5.1 Implications of the Different Choice of Sense

Within AI, one of the things that happened was a shift of the sense from the represented to the representation. This is a natural progression given that the focus of work is on producing the "engineering artefact". However, the utility of the engineering artefact depends upon it characterising the so-called 'conceptualization' – so this is important as well. Giving priority to one or other of these two foci can and has led to different flavours of ontology; Hofmann et al. [5] give examples of "two classes of ontologies in M&S: ontologies defining modelling methods and simulation techniques … and ontologies representing real world systems to be simulated"; they name the former 'methodological ontologies' and the latter 'referential ontologies'.

Clearly Hofmann et al.'s [5] 'methodological ontologies' are only loosely related to philosophical ontologies. However, it appears that despite the name 'referential ontologies' – where 'referential' might be taken to imply the model refers to 'real' things - Hoffman et al. assume these adopt the idealist stance. For example, the paper states that "Models are conceptualizations of (real world) referents and computer simulations are executable expressions of these conceptualizations." Firstly, this identifies the models, that is the representations, as the conceptualisations – unlike Gruber [49] and Guarino [50] [51]. Secondly it sees the relationship between the representation/model and the represented/ referents as one of conceptualisation rather than one of representation or reference. This is made clear in the next sentence "Conceptualization, however, is a cognitive, purpose-driven act that varies from individual to individual and from task to task." Clearly the authors have at least partly adopted the idealist rather than the realist stance.

This is not an isolated example in M&S. Tolk et al. [53] say "The goal of conceptual modeling in Modeling and Simulation (M&S) is not focusing on describing an abstract view of the implementation, but to capture a model of the referent, which is the thing that is modeled, representing a sufficient simplification for the purpose of a given study serving as a *common conceptualization of the referent and its context* within the study." This again identifies the models (the representation) with the conceptualisation unlike Gruber and Guarino. It also suggests both that the model has a referent and that the relation between the model and the referent is conceptualisation – in other words, not reference or representation. This is further confirmed in Figure 14.1 'The semiotic triangle for M&S' where Ogden et al.'s [54] semiotic triangle shows this relation diagrammatically, implying the idealist stance has been adopted.

In part this 'confusion' between the senses is understandable given the lack of an agreed definition for the term 'conceptualization' that would resolve which stance had been adopted.

4.6 Meeting the Requirement for Semantic Interoperability

In information systems in general, and M&S systems in particular, there is also a growing requirement for systems integration which drives a requirement for semantic interoperability (often called composability at the model level). Within this, there is a growing recognition that semantic interoperability is a challenge and that ontology may be the answer [5].

At the heart of this claim is a view of how the semantic mapping between information systems (and models) works. If one has adopted the realist stance, then the method for identifying the correct mapping is simple. Take the simplest case; if given node a in Model A and node b in Model B, then a should map to b if and only if a and b represent the same thing [51]. All one needs to do is identify the 'thing' which will be in both domains; from a realist stance, their ontologies.

However, if one adopts the idealist stance, then there is not an obvious methodologically robust approach. Furthermore, one cannot discount the possibility when faced with exactly the same domain that two systems may have radically different conceptualisations – implying there is no straightforward semantic mapping.

One could argue that it is just the case that there are not always (or indeed often) straightforward semantic mappings; and the challenges people face when trying to map between systems would seem to back this up. On the other hand, this natural result of the idealist stance is at odds with our everyday experience. One can easily imagine a military engagement where one side launches a missile against the other. We might expect that (from an idealist stance) the land and air divisions of the targeted combatants would have very different conceptualisations of the missile, given their different interests. But from a practical perspective, we would resist the idea that these conceptualizations, however different, imply that there are two real missiles. For example, we would have grave doubts about a missile defence system that reported two missiles – presumably in the same portion of airspace – and we cannot conceive how one of these might be shot down without this affecting the other.

There is another explanation for these mapping difficulties, one that is compatible with the realist stance; that difficulties in identification arise when the intuition is inadequate to the task. In most current projects, the identification of the objects in the domain is left to the mappers' untutored naïve (albeit experienced) intuition. Most mappers are unaware of the analytic tools developed in philosophical ontology. If one built an M&S engineering discipline for identifying the objects in the domain based upon these 'industrial strength' tools, then the mapping difficulties might disappear.

The key point here is that the realist stance provides a robust solution to semantic interoperability – as there is a 'real' world to underwrite the semantic mapping. Whereas the idealist stance cannot provide the same simple explanation for the semantic mapping, and indeed may suggest such mappings are difficult if not impossible.

Given this, there is a good case for projects that aim to improve semantic interoperability by using ontology to be clearer about which stance they are adopting. If they adopt the idealist stance, they will need to explain how they see the benefits accruing. If they adopt the realist stance, they have an explanation (given above) for how the benefits should accrue. The real engineering test is whether these benefits can be harvested in practice. We believe they do and have documented some of our experiences [55] [56-58] [22].

From the more general perspective of the nature of M&S's ontology there is probably more useful work to be done exploring how the idealist stance can, at least in principle, support semantic interoperability.

4.7   Generalising to a Requirement for a Canonical Representation

The problem with semantic interoperability arises because currently modellers seem to have an uncanny knack for producing quite different models for the same domain. Though common in practice, if one takes the realist stance it seems slightly counter-intuitive, as the models are representing the same objects in the domain. This suggests a solution to a wider requirement – one for a canonical representation.

If a system already exists one can reconstruct its ontology and use this to drive the semantic mapping. However, if one is starting to build the model, it makes sense to start with the ontology and use this to produce the model. All models of the domain produced this way would have the same structure, as they would be representing the same objects. In this sense the model would be a canonical representation of the domain, though the form of the representation may be different: for example, one model may be textual and another graphical. However, as they have the same structure, there will be an isomorphism between them. The business benefits of this are clear; as well as supporting semantic interoperability from the start, it greatly simplifies re-use.

Canonical representation is also a good way of distinguishing the realist and the idealist stance. The realist stance implies that there is a canonical representation of a domain, whereas the idealist stance suggests there probably is not. Though this is broadly right, there are some further considerations. There are some meta-ontological (metaphysical choices) that shape the ontological architecture and the representation will be canonical within an agreed set of choices; different choices will lead to different representations. This is a good reason to be clear about which choices have been made.

5 Key Methodological and Metaphysical Choices

There are a number of meta-ontological choices that need to be considered when developing a top ontology for M&S systems. Some concerns relate to the process of producing the ontology-based models. Others are more metaphysical and focus on the nature of what is produced [59-62]. Some of the choices are more general, leading to guiding architectural principles for the design. Others are more specific, leading to specific architectural features.

Developing a better understanding of the issues will help to ensure a coherent approach to them. It will (hopefully) lead to a more coherent ontological architecture where the meta-ontological choices are made explicit and so bring engineering benefits. A lack of understanding typically leads to a much less coherent architecture where different choices are made ad hoc across the architecture. This puts at risk the benefits ontology brings, particularly the goal of more intelligent support.

The choices are closely related and some choices naturally fit together. In the sections below, we start by looking at the individual choices and assess how they meet M&S engineering goals. The individual choices are:

- Setting clear expert governance
- Avoiding abstract objects
- Providing ontological completeness
- Providing criteria of identity
- Explaining parallel worlds
- Explaining simulations and time
- Separating the concerns

In a final section, we look at how these individual choices depend upon one another.

5.1 Expert Governance: What Should They Be Responsible for

The design of a top ontology raises some specific governance issues. Typically, when building an M&S system, there will be experts in its domain. It is currently common practice to give these experts responsibility for assessing whether the M&S model is a true picture of the domain. Introducing a top ontology brings out a governance issue; who should be responsible for the way the top ontology shapes the domain ontology. The domain experts are usually not experts in top ontologies; similarly the ontology experts are not usually experts in the domain.

One could take the view that the top ontology deals with general basic things that are common currency for everyone including the domain expert, and that the ontologist's job is restricted to identifying these so that the domain expert can specialise them in her domain. Or one could think that the ontologist needs to be given the freedom and the responsibility to devise the best top ontology possible. This could be either because one believes that there is no real common view or that the common view can and should be improved, maybe substantially. If one takes the latter view, then it is likely that the resultant top ontology will encourage (even enforce) a domain model quite different in structure from that assumed by the experts. However, one of its benefits would be that it could form the basis for a deep common understanding of the domain.

This choice is clearly recognised in philosophical ontology; Strawson [63] coined the terms 'descriptive' and 'revisionary' to distinguish between the two approaches. Strawson says descriptive metaphysics seeks to "lay bare the most general features of our conceptual scheme... a massive central core of human thinking which has no history... the commonplaces of the least refined thinking... the indispensable core of the conceptual equipment of the most sophisticated human being" [63, p. 10]. Whereas, he says, revisionary metaphysics is "concerned to produce a better structure" [63, p. 9]. Strawson gave Aristotle and Kant as examples of descriptive metaphysics and Descartes, Leibniz and Berkeley as examples of revisionary metaphysics. In the descriptive approach (Strawson's preferred approach) the ontologist aims to find a top ontology that preserves as far as possible the accepted picture of the world. In the revisionary approach the ontologist has the responsibility for devising the best top ontology even if this transforms the accepted picture.

The descriptive assumption of a common underlying general picture may be optimistic. A point often made by metaphysicians is that most people unfamiliar with philosophy tend not to be consistent in the way they apply philosophical principles across their picture of a domain (as someone unfamiliar with general accounting theory may not be aware that they are applying different, maybe inconsistent, accounting rules in different situations). If one person is unlikely to have a consistent general picture, then a whole community is even less likely to. Multiple domains are another source of inconsistency – even the most conservative descriptive common picture possible may have significant differences from the individual domain pictures. Cartwright [64,65] strongly argues the case that this situation is commonplace in science, that it has different incompatible theories to model different situations in the world. If it is common in science, which places a high premium on consistency, then it is likely to be common in M&S domains. The requirement for consistency in the top ontology will almost inevitably enforce a degree of change on the domain model. However the conservative descriptive approach aims to minimise these. In so far as this is successful, it has the benefit of producing models that are more likely to be immediately recognised by the domain's community.

Lewis [66, p. 134] points out that when one adopts the revisionary approach typically one is trying to improve the theory, not replace it, and the degree of revision is "a matter of balance and judgement": noting that when "trying to improve the unity and economy of our total theory" ... "I am trying to *improve* that theory, that is to change it. But I am also trying to improve *that* theory, that is to leave it recognisably the same theory we had before." Following Lewis one can make an argument that in an engineering approach one should look to improve the model but take account of any benefits of preserving the domain experts' picture of the domain.

M&S projects that make use of a top ontology will inevitably come across this issue. As with many of these meta-ontological concerns, implementing a consistent approach may be practically impossible once the development project is well underway. So it is more effective to make a clear informed decision on this aspect of governance from the start.

5.2 Tacit Knowledge and the 'Transparent Vision' Fallacy

A lack of understanding of two inter-related topics often leads to uninformed decisions; these are the transparent vision fallacy and tacit knowledge. Both of these deal with the nature of the domain experts' knowledge. The following sections provide an outline of the issues.

5.2.1 'Transparent Vision' Fallacy

When a domain expert builds a model of the domain, a common assumption is that she has a transparent vision of the domain; in other words, she sees the domain's structure intuitively and that her expertise guarantees this vision's correctness. This needs to be distinguished from the weaker 'Transparency of Experience' [67] which is more concerned with our immediate experiences than the way in which we classify the world. Clearly, if one accepts 'transparent vision' as a background assumption, one would be more comfortable with a conservative/descriptive approach.

For an idealist, as introspection is transparent, domain vision is transparent. As we mentally construct the objects in the domain, and we have a transparent vision of these through introspection, we have a transparent vision of the domain.

For a realist, it is difficult to see how this assumption can be maintained with clear evidence that different domain experts build different models. A useful perspective on this assumption is given by the critics of the 'transparent vision' fallacy [68]. This criticism has a long history, going back to Hume [38]. In Hume's time this position was justified using 'The Insight Ideal' [69]; which argues that a good God would give man the ability to see the world he created clearly ("God in his goodness endowed human beings with faculties that enable them, in principle, to gain knowledge of the world he created for them. It is totally taken for granted that 'the universe was in principle intellectually transparent ...'" [69, p. 38]). While this religious argument would have little traction in modern times, something similar to transparent vision is commonly assumed by modellers and their managers.

5.2.2 Tacit Knowledge

Tacit knowledge is knowledge that is difficult to write down or verbalise – a standard example being the ability to ride a bicycle. This can be regarded as nonconceptual mental content [70-73]. Obviously this causes immediate problems for the idealist stance. It also raises issues for the realist stance.

One might think that experts have their vision trained so it becomes more transparent; that while they might disagree between themselves, they have a more accurate conceptual picture than ordinary people. Tacit knowledge raises doubts that this is true for traditional expertise (know-how). The issue can be illustrated by a common problem that occurs when subject matter experts are asked to produce models. It is often assumed that their expertise is a form of mental conceptual representation and that modeling is simply a matter of recreating this in a public model. So it seems odd that, in practice, experts often have great difficulty in articulating their tacit knowledge in a form that can be directly represented. And they have similar difficulties comprehending or agreeing on the representations produced by others. This is a critical issue as the design of a computer system relies on the knowledge it needs being represented in the kind of excruciating level of detail that will enable it to carry out the task.

There has been a reasonable amount of research on the distinction between having expertise as an ability to do something and being able to represent this ability as knowledge. Ryle [74, Chapter 2] and Habermas [75] ("... we can distinguish between know-how, the ability of a competent subject who understands how to produce or accomplish something, and know-that, the explicit knowledge of how it is that he is able to do so") make a clear distinction between know-how and know-that (or know-what): though there have been defences of the position, e.g. [76]. Polanyi [77] provides a description of know-how as tacit knowledge. John Searle [78,48] makes the case for gaining expertise being a move from conscious control to unconscious action or ability, where the more expertise one has, the less one has an internal representation of that expertise (or conscious access to that representation). Collins et al. [79] and Collins [80] provide a detailed analysis of this situation.

If no conceptual picture exists in the expert's mind (or it is not accessible), then a different kind of approach is required. There is a reasonable literature describing candidates for these; for example Carnap's [81] 'rational reconstruction' ("... [in rational reconstruction] the distinction between drawing on a-priori knowledge and drawing on a-posteriori knowledge becomes blurred. On the one hand, the rule consciousness [i.e. intuitive know-how] of competent subjects is for them an a-priori knowledge; on the other hand, the reconstruction of this calls for inquiries undertaken with empirical [methods]".) and Lipton's [82] 'inference to the best explanation'. Earlier Peirce [83] called this abduction – saying that "Long before I first classed abduction as an inference it was recognized by logicians that the operation of adopting an explanatory hypothesis - which is just what abduction is - was subject to certain conditions. Namely, the hypothesis cannot be admitted, even as a hypothesis, unless it be supposed that it would account for the facts or some of them." He also more light-heartedly said "Abduction is no more nor less than guessing".

A further problem arises because experts often feel an obligation to be able to provide a representation. In these cases, they, post hoc, rationalize one, which only needs to be plausible not correct, as it is not involved in the deployment of the expertise. Shaffer et al. [84] provide a good example: in which expert baseball players provide a plausible, but completely false, (post hoc) rationalization of how they catch a fly ball. (There are many examples of this blindness in the literature; see also [85], where chess players falsely claim to be following a new strategy.) However, when this plausible representation is included in an M&S system, it is deployed. In this case, using an expert's representation (or judgment about one) is likely to be misleading.

Clearly, if one regards traditional expertise (know-how) as largely tacit and inaccessible then one would be reluctant to adopt a conservative/descriptive approach which is aimed to preserve the experts' non-existent conceptual picture of the domain. One would be more likely to adopt a revisionary approach provided this offered a way to ensure a better representation.

Ontological expertise should complement tacit expert knowledge. Ontologists (philosophical ontologists, at least) are trained to make and organize an explicit (i.e. not tacit) representation of the ontology so that it can be publically examined. Historically this was done with text, including mathematical logic; now it is being done with computer models.

5.3   Are There Abstract Objects?

When asking about an element of an M&S model, it is reasonably common to get the answer that it is abstract. A common example is 'roles' such as the President of the USA – an example we return to later. This can have implications for the ontology; if the object is accepted as-is into the ontology, then the top ontology needs to accommodate abstract objects. However, philosophers have spent some time clarifying the cost of doing so.

In modern ontology there is a fundamental distinction made between concrete and abstract objects; where abstract objects are defined as those that are not spatial (or spatiotemporal) and have no causal powers. Lewis [66] calls this the Way of Negation, as abstract objects are defined as those which lack certain features possessed by typical concrete objects.

There are several well-known issues with accepting abstract objects into one's ontology, and an extensive literature arguing for (e.g. [86-88]) and against (e.g. [89-92]) – where these arguments are tied into related arguments about realism, materialism and physicalism.

The main challenge supporters of abstract objects face is explaining how we can know anything about them, even know that they exist. These challenges are particularly acute for mathematical objects and were raised in relation to them in [93]. More specifically, how we explain that we know about the existence of things that are not spatial and have no causal powers. As Field [91, pp. 232–3] says "we should view with suspicion any claim to know facts about a certain domain if we believe it impossible to explain the reliability of our beliefs about that domain".

This translates into a serious issue for M&S modellers [94]. If they need to model an abstract object and its characteristics then, as they cannot explain how anyone could know anything about it and why their beliefs about it are reliable, then their resources are severely limited. It is difficult to see what analytic process could be used to determine the characteristics. When some kind of intuition is developed, how do two (or more) modellers reconcile their conflicting intuitions, when they cannot explain their own intuitions? There seems, in principle, no analytic way of resolving this.

For M&S modellers the pragmatic option is to avoid any commitment to abstract objects. This option can be built into the top ontology helping to ensure conformance in the domain ontologies. Note however, that if the domain experts' picture makes use of abstract objects, this may imply the adoption of a revisionary stance.

5.4 Why Sets Are Not Abstract

There is a misconception about sets that sometimes clouds the discussion of abstract objects. Sets are sometimes talked about as abstract, but they are not abstract in the sense outlined above. Consider a set of located objects, this has the location of its members; so, for example, the set of objects located on my desk is also located on my desk [66, p. 83]. If the location is dispersed, it may not be interesting; the set of atoms is dispersed around the whole universe, its location is of no real use. But having an uninteresting location is quite different from not having one at all.

5.5 Finding an Ontologically Complete Framework

When modellers add a new element to the model, they expect that it will fit into the existing framework. One does not want to discover when building a domain ontology for an M&S system something that does not fit. One would then need to revise the framework to accommodate it and rollout this change across all the other systems that use the framework. So, when one is devising a top ontology for M&S systems, it is essential that it provides a framework that covers all the things that might be in a domain.

This boils down to a requirement to list, at some general level, all the types of things that might be in the range of domains that the top ontology is likely to be used for. This is closely related to categories in traditional philosophy, which are a complete list of the highest, most general kinds. In philosophy one considers everything that could exist, within M&S engineering one may wish to tailor a top ontology restricted to a specific range of domains. Interest in categories can be traced back to Aristotle [95, 1b25] who divides the world into the ten most general kinds of entities.

There are a variety of different ways to derive categories, but a common way is ontological; dividing things by how they exist. Lowe [96, p. 5] takes categories to be categories of "what kinds of things can exist and coexist". Such categories, he argues, are to be individuated according to the existence and/or identity criteria for their instances. Johansson [97, p. 1] aims to "develop a coherent system of all the most abstract categories needed to give a true description of the world". There is also the question as to whether there is a single or multiple classifications (e.g. [98, §10] [99, Chapter 2] [100]). And also the question of whether a particular classification provides mutually exclusive categories.

From an M&S perspective, what is required is completeness; so providing a single exhaustive list of categories is sufficient, it does not have to be the only possible list.

5.6   Categories Based Upon Criteria of Identity

When modelling there are many times where it would be useful to know whether two modellers are talking about two different things or the same thing (whether two elements in different models represent the same thing); this is often colloquially expressed as knowing whether the two things are the same. If one had a criterion of identity, a way in principle (though maybe not always in practice) of deciding whether two things are the same or different, then one would have a principled basis for doing this. Without this (or something similar) the goal of a canonical representation would be difficult to achieve.

One might think that devising a criterion is simple, but several puzzles have been devised to illustrate this is not so. The Ship of Theseus and Locke's socks are historically well-known examples of one kind. In these, small parts of an object are replaced until eventually none of the original parts remain. It seems intuitively clear that the thing is the same after each small part is replaced, but it is much less intuitively clear whether it remains the same or not after all the parts are replaced. The challenge is to devise a criterion that sensibly resolves the issue.

There is a link between this criterion and categories. Dummett [101, pp. 73-75], Lowe [102] and Wiggins [103] suggest building a classification based upon criteria of identity – where each category in the classification has its own criterion of identity. This relies on the 'in principle' nature of the criteria; a rough rule of thumb would not provide clear classifications. The attraction of a criterion of identity based set of categories from an M&S perspective is a simple coherent structure.

One mainstream example is the group of extensional criteria of identity. This divides objects into three broad types, each with its own criterion of identity; elements, types and tuples. The criterion of identity for elements is spatiotemporal extension; two things are identical if they have the same spatiotemporal extension. For types the criterion is extension, their members; two types with the same members are identical. For tuples, the criterion is their places; two tuples are the same if they have the same objects in the same places. There is a literature explaining the details of this classification [66,16,62,55,17]. The attraction of this classification is that it is general and simple (almost minimalistic); and that it cleanly resolves a number of identity problems as well as issues on how to deal with representing the past and future. As well as the identity over time problems such as 'The Ship of Theseus' it also addresses problems where there are two things occupying the same space at the same time – see [17, Figure 8.18]. However, adopting this classification in a top ontology is likely to be revisionary, as experts' pictures of their domain are likely to need some kind of transformation to conform to it.

5.7 Representing Parallel Worlds

A standard explanation of the way M&S in general functions is that it creates parallel or imaginary worlds. Weirich [104] says that following Robert Sugden [105], "I assume that a model is an imaginary world. I allow it to be a small world, including only the features under investigation. To underscore this point, I say that a model is an imaginary system, a component of an imaginary complete world." In a review of the current work, Frigg et al. [106, p. 597] find many people making the same point: "Several philosophers, historians and scientists claim that simulations create 'parallel' (or 'virtual' or 'artificial') worlds" and refer to [107-109]. Frigg at al. continue "The most plausible interpretation of this idea, we think, is that the simulant investigates proximate systems that differ more or less radically from the systems she is ultimately interested in. This usually means that inferences about those latter systems, the 'target systems', are made in two analytically separable (though in practice not always thus separated) steps: first, a conclusion is established about the proximate system; second, the claim is exported from the proximate system to the target system."

There are various ways to interpret this. Material models, such as a scale model of an airplane wing, are straightforward physical objects. Because of the way these represent, one could think of the parallel world being physically similar to the model – or even the model being the parallel world. However, for computing models this interpretation is less feasible as the data running in the computer has no physical similarity with what it represents in the parallel world. A more obvious explanation here is semantic, that the M&S system represents a parallel world. In either case, there is a need to explain why we think that what happens in the parallel world tells us something about the actual world where the simulation is carried out. This in turn raises interesting epistemic questions about how we can know anything about the parallel world that is being represented.

From an M&S model development perspective these translate into questions about how this is to be modelled. How do we know how to model the parallel (possibly imaginary) world? Does it have the same kind of top ontology as this actual world? Does it have the same categories and criteria of identity?

Interestingly, the characterisation of what is being simulated as a 'parallel' world is not quite right. It is possible to simulate an event in the past to gain an understanding. Similarly, we may simulate a possible future event and then find that one of the simulations actually happened. Both these cases suggest that in some cases the parallel world is this actual world. One can argue here that the future and, to some extent, the past and future of the actual world is inaccessible in a weaker but similar way to the parallel worlds.

Similar issues to these arise in the study of possibility (also known as alethic modality) and a number of approaches have been developed. These can be regarded as ways for M&S to explain how simulation works.

One approach that has had an influence in computing is 'possible worlds' [51,16,17] where statements about possible objects are taken to refer to objects that exist in possible worlds. This approach was first suggested by Leibnitz; it was developed in modern times by a number of people starting with Kripke (in a series of papers beginning with [110]) and then later Lewis [111,66] – and is known as Kripke semantics or possible world semantics. Lewis took a particularly strong stance with respect to possible worlds, saying that they were real; a position called 'modal realism'. A weaker option is called ersatzism (or actualism, or abstractionism), which does not commit to their reality.

This approach provides a neat semantic consistency for talk about objects in parallel worlds; it can refer to these in the same way as we talk about actual objects. It also avoids complicating the overall ontological framework; the top ontology's categories and criteria of identity span the actual and possible worlds, applying equally to both. This simplification is a benefit from an M&S top ontology development perspective.

If an M&S system is representing a 'parallel' world, it makes sense to have a clear understanding of what this world is. There are clear attractions to adopting a possible world approach in the development of the top ontology. However as has been noted with other options, for most domains this is likely to have a revisionary effect on the domain models and so needs to be undertaken within a revisionary governance.

There are several well-known issues with the possible world approach, but these do not seem relevant here. For example, if one is a modal realist, then for every choice a person makes in this world, there is some possible world where their counterpart makes a different choice; this is thought to raise ethical problems. Engineers tend to ignore these issues. For example, a civil engineer is unlikely to regard the implications of Newton's mechanistic theories on free will to be relevant to using them in the construction of bridges. From an M&S engineering perspective, the relevant issue is whether this makes the development of better M&S systems easier; the choice of approach is methodological.

5.8 Simulations and Time

Simulations can be regarded as dynamic models; that is models that change over time. The changes over time in the simulation are designed to represent the changes over time in the system it is representing; that a simulation imitates a process by another process [112,3].

This suggests two models of representation are in play; one where a static sign is representing an object and the other where a process working on the signs represents a process happening to the objects. This division is familiar in computing, which has a clear delineation between data and process, though there has been much confusion about whether the data-process distinction reflects a similar distinction in the real world [16, Ch. 2.4.1]. This is also a different picture from that given in much early computing system literature (e.g. [113]) – see Figure 1. This assumed that only the data represented, and the processing was manipulating the represented data. Simulation implies that it is the same distinction at work in data and process in the representation and what it represents.

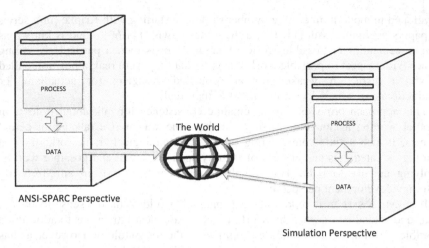

Fig. 1 Data-process to world mapping

This raises the question of how ontologically fundamental this distinction is. As is common in philosophy opinion is divided. In modern philosophy, the debate was started by McTaggart [114], though the issues go back to Heraclitus. Simplifying slightly, one approach favours a fundamental ontological division into continuants and occurrents, where occurrents are changes that happen to continuants. Another approach regards these distinctions as a matter of perspective. For example, a glacier may seem like an object in human time, but a process in geological time – it is not intrinsically one or the other. This approach regards all objects as extended in time and change as just different timeslices of the object having different properties (this is sometimes known as a process ontology, as it can be regarded as treating all objects as processes). For example, if someone grows an inch, then this is seen as two timeslices of the extended object which differ in height by an inch. At the heart of these differences is a choice about the reality of change; as the height example illustrates in the second approach there is no object corresponding directly to the change. These two approaches are sometimes known as 3D and 4D, though this can be misleading as the issue is change rather than a number of dimensions.

From an M&S perspective, a choice between these approaches needs to be made as it will significantly impact the top ontology. A 4D approach is usually simpler (it has less distinctions) but is likely to be revisionary, with all that entails. As the glacier example illustrates, it also downgrades the static-process distinction at the heart of simulation distinction to one of perspective and context rather than ontology.

5.9 Ontology, Semantics and Separation of Concerns

The successful delivery of ontology-based M&S systems depends not just on building a good ontology but also on fitting this into an appropriate development process. The ontological approach makes explicit an endemic development issue and so this requires particular attention. All computing information models, including M&S models, suffer from a semantic schizophrenia. On the one hand, the model represents the domain; on the other hand, it represents the implemented system, which then represents the domain. These different representation requirements place different demands upon its structure. With an ontological approach, one has a far more structured and effective way of representing the domain, making the need to manage the different demands more acute.

One of the common ways to manage this is a separation of concerns, described in many current textbooks; Pressman [115, p. 313] describes a model that is constructed by asking the customers what are "...the "things" that the application or business process addresses", and that "These "things" evolve into a list of input and output data objects as well as external entities that produce or consume data". A more structured example is the Object Management Group's Model Driven Architecture (MDA) where a model is built for each concern and this is transformed into a different model for a different concern.

With the introduction of ontology, particularly a top ontology, the process of building a domain model becomes an engineering task, much more than "asking the customers". The ontological demands on the structure will be much clearer. Hence its separation also needs to be clearer, probably much clearer than in most M&S projects undertaken at the moment. The benefits of semantic interoperability and those arising from canonical representation rely on this; that the ontology is developed independently of the particular system's implementation requirements.

5.10 Managing Dependencies across the Framework - Between the Choices

There is also a close link between the different choices we have been exploring [116]. For example, if one wishes to adopt an extensional approach to the criterion of identity for physical objects, then one is obliged to go 4D. 3D is not sufficiently fine grained to distinguish between objects that are in the same location at the same time. Barack Obama and the President of the USA are a common example, as is shown in the 4D space-time map below.

Possible worlds are also attractive to an extensionalist, as then the extension of types across possible worlds is able to capture nuances of meaning it could not do otherwise.

5.11 BORO Meta-ontological Choices

The BORO top ontology is an example of an ontology that has explicitly addressed these choices in its ontological architecture. It has adopted a realist stance towards ontology (it takes for granted a mind-independent real world). It has adopted a revisionary stance – accepting that if we want better models, we need to change the ways we look at the world. It has explicitly adopted completeness categories based upon extensional criteria of identity.

It chooses a 4D and possible worlds approach as these fit best with its commitment to extensionalism. Figure 2 shows how it approaches the Obama

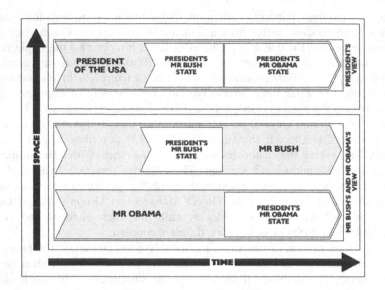

Fig. 2 4D space-time map

– President of the USA issue. The individuals, 'President of the USA' and 'Mr Obama', are both extended in space and time. At some points in time they overlap – where they are 'in the same place at the same time'.

Note that this explains away the same place, same time concerns – ticking the 'increase explanatory power' box in the six assessment features identified at the beginning of the chapter.

One can see further kinds of revisionary transformations in the various BORO models [16,55-58,17,22]. In each case, the transformation is justified by Lewisian arguments that it "improves the unity and economy" of the M&S system.

6 How to Approach Epistemology

The literature contains a reasonable amount of discussion of how ontologies can, in the concrete way described in this chapter, be integrated into the design of computer systems. As we mentioned earlier, there developed within philosophy an objectification of ontology, where as well as the sense of ontology as a practice or discipline there emerged the sense of ontology as the set of objects being studied – "the set of things …".

Epistemology is the study of knowledge and justified belief, and is given the same high ranking as ontology in the philosophical literature. However, there has not been the same objectification of it; a shift to a sense of epistemology as the sets of objects that are known. From an M&S modelling perspective, this would be useful, recognising an epistemology as "the set of things whose existence is *known* by a particular system" [117,55,118,58]. This is particularly useful when allied with ontology; "the set of things in its ontology that are *known* by a particular system".

To understand why this is useful, one first needs to recognise that systems have an epistemology and that this diverges from its ontology [119]. Ironically, it turns out that many system models are for good practical reasons epistemologies, models of what it can know rather than what is. Here is a simple example. An insurance system may wish to represent whether a person is married and, in cases where it has the information, represent the 'spouse of' relation. The system's model will represent what it can know (its epistemology), that married persons optionally have a spouse – whereas ontologically, being married is the same thing as having a spouse; ontologically if one is married the spouse relation is mandatory.

This suggests a need for a finer separation of concerns, one which recognises the distinction between an ontology and an epistemology. It also reveals an interesting relation between the two – that the epistemology can be regarded as a view of the ontology. This has positive implications for semantic interoperability between systems with different epistemologies but the same underlying ontology.

7 Summary

This chapter has concerned itself with the category of M&S systems that can make concrete use of ontology in their design; large-scale engineering computing M&S systems. Currently the community is exploring how to do this and a clear understanding has yet to emerge. One area where it is useful to develop a clearer understanding is the meta-ontological concerns that shape the ontology and the options for resolving them; the ontological architecture. These have been outlined in this chapter.

The major meta-ontological concern that needs to be addressed is what kind of ontology will be adopted, or in the terminology of this chapter, whether to adopt the idealist or the realist stance. This is an engineering decision and needs to be justified in this context, particularly in terms of the engineering benefits. Where semantic interoperability is a major requirement, it needs to be recognised that the realist stance has major advantages.

Another meta-ontological concern is top ontology governance; whether this is descriptive or revisionary. This again needs to be driven by engineering concerns; choosing the approach that delivers better overall results. The benefits of having a familiar looking model need to be balanced with the benefits arising from potential improvements in the model.

There are other detailed meta-ontological concerns, such as alethic modality and the reality of change, which need to be considered and a way of handling them decided upon. It makes sense to apply the decision consistently across the ontology, and the top ontology can be used to manage this.

Addressing these concerns directly is likely to lead to a significant improvement in the design of ontologies and the intelligence of the implemented M&S system.

References

1. Morgan, M.S.: Experiments without material intervention: Model experiments, virtual experiments and virtually experiments. In: Radder, H. (ed.) The Philosophy of Scientific Experimentation, pp. 216–235. University of Pittsburgh Press, Pittsburgh (2003)
2. Parker, W.: Does matter really matter? Computer simulations, experiments, and materiality. Synthese 169(3), 483–496 (2009), doi:10.1007/s11229-008-9434-3
3. Humphreys, P.: Extending ourselves: computational science, empiricism, and scientific method. Oxford University Press, New York (2004)
4. Balci, O., Arthur, J.D., Ormsby, W.F.: Achieving reusability and composability with a simulation conceptual model. Journal of Simulation 5, 157–165 (2011), doi:10.1057/jos.2011.7
5. Hofmann, M., Palii, J., Mihelcic, G.: Epistemic and normative aspects of ontologies in modelling and simulation. J. of Simulation 5(3), 135–146 (2011)
6. Christley, S., Xiang, X., Madey, G.: An ontology for agent-based modeling and simulation. In: Macal, C.M., Sallach, D., North, M.J. (eds.) Agent 2004 Conference on Social Dynamics: Interaction, Reflexivity and Emergence, October 7-9 (2004)
7. Miller, J.A., Baramidze, G.T., Sheth, A.P., Fishwick, P.A.: Investigating Ontologies for Simulation Modeling. In: 37th Annual Simulation Symposium, ANSS, Arlington, VA, pp. 55–71. IEEE Computer Society (2004)
8. Benjamin, P., Patki, M., Mayer, R.: Using ontologies for simulation modeling. In: Perrone, L.F., Wieland, F.P., Liu, J., Lawson, B.G., Nicol, D.M., Fujimoto, R.M. (eds.) 38th Conference on Winter Simulation, Monterey, California, pp. 1151–1159. IEEE (2006)
9. Bell, D., Mustafee, N., de Cesare, S., Taylor, S.J.E., Lycett, M., Fishwick, P.A.: Ontology Engineering for Simulation Component Reuse. Int. J. Enterprise Information Systems 4(4), 47–61 (2008), doi:10.4018/jeis.2008100104
10. Guizzardi, G., Wagner, G.: Using the Unified Foundational Ontology (UFO) as a Foundation for General Conceptual Modeling Languages. In: Poli, R., Healy, M., Kameas, A. (eds.) Theory and Application of Ontologies, pp. 175–196. Springer, Netherlands (2010)
11. Guizzardi, G., Wagner, G.: Towards an ontological foundation of discrete event simulation. In: Johanson, B., Jain, S., Montoya-Torres, J., Hugan, J., Yücesan, E. (eds.) 2010 Winter Simulation Conference, Baltimore, pp. 652–664. IEEE (2010)
12. Livet, P., Muller, J.P., Phan, D., Sanders, L.: Ontology, a mediator for agent-based modeling in social science. Journal of Artificial Societies and Social Simulation 13(1), 3 (2010)
13. Taylor, S.J.E., Bell, D., Mustafee, N., de Cesare, S., Lycett, M., Fishwick, P.A.: Semantic web services for simulation component reuse and interoperability: An ontology approach. In: Gunasekaran, A., Shea, T. (eds.) Organizational Advancements Through Enterprise Information Systems: Emerging Applications and Developments. IGI Gobal Inc., Hershey (2010)
14. Ezzell, Z., Fishwick, P.A., Lok, B., Pitkin, A., Lampotang, S.: An ontology-enabled user interface for simulation model construction and visualization. J. of Simulation 5(3), 147–156 (2011)

15. Lacy, L., Gerber, W.: Potential modeling and simulation applications of the web ontology language-OWL. In: 2004 Winter Simulation Conference, Piscataway. IEEE (2004)
16. Partridge, C.: Business Objects: Re - Engineering for re - use. Butterworth Heinemann, Oxford (1996)
17. Partridge, C.: Business Objects: Re - Engineering for re - use, 2nd edn. BORO Centre (2005)
18. Booch, G.: Nine Things You Can Do with Old Software. IEEE Software 25(5), 93–94 (2008), doi:10.1109/MS.2008.139
19. Bailey, I.: A Forensic Approach to Information Systems Development. Cutter Consortium (2008)
20. Bailey, I.: Enterprise Ontologies – Better Models of Business. In: Tolk, A., Jain, L.C. (eds.) Intelligence-Based Systems Engineering. ISRL, vol. 10, pp. 327–342. Springer, Heidelberg (2011)
21. Daga, A., de Cesare, S., Lycett, M., Partridge, C.: An ontological approach for recovering legacy business content, p. 224a. IEEE (2005)
22. Partridge, C., Lambert, M., Loneragan, M., Mitchell, A., Garbacz, P.: A novel ontological approach to semantic interoperability between legacy air defence command and control systems. International Journal of Intelligent Defence Support Systems 4(3), 232–262 (2011), doi:10.1504/ijidss.2011.043357
23. Kuhn, T.S.: Objectivity, value judgment, and theory choice. In: The Essential Tension: Selected Studies in Scientific Tradition and Change, pp. 320–339. University of Chicago Press (1977)
24. Morgan, M.S., Morrison, M.: Models as mediators: perspectives on natural and social science. Cambridge University Press, Cambridge (1999)
25. Honderich, T. (ed.): The Oxford Companion to Philosophy. Cambridge University Press, Cambridge (2005)
26. Naur, P.: Programming as theory building. Microprocessing and Microprogramming 15(5), 253–261 (1985), doi:10.1016/0165-6074(85)90032-8
27. Kant, I.: Immanuel Kant's Critique of pure reason. Macmillan, London (1929)
28. Frege, G.: Die Grundlagen der Arithmetik. Breslau (1884)
29. Frege, G.: Grundgesetze der arithmetik. H. Pohle, Jena (1893)
30. Mealy, G.H.: Another look at data. In: Fall Joint Computer Conference, Anaheim, California, November 14-16, pp. 525–534. ACM (1967), doi:10.1145/1465611.1465682
31. Kent, W.: Data and reality: basic assumptions in data processing reconsidered. North-Holland Pub. Co., Amsterdam (1978)
32. Hirschheim, R.A., Klein, H.K., Lyytinen, K.: Information systems development and data modeling: conceptual and philosophical foundations. Cambridge University Press, Cambridge (1995)
33. Kusnierczyk, W.: Nontological Engineering. In: Formal Ontology in Information System, FOIS 2006, pp. 39–50. IOS Press (2006)
34. Smith, B.: Beyond Concepts: Ontology as Reality Representation. In: Formal Ontology and Information Systems, FOIS, Amsterdam, pp. 73–84. IOS Press (2004)
35. Smith, B., Ceusters, W., Temmerman, R.: Wusteria. Studies in Health Technology and Informatics 116, 647–652 (2005)
36. Tolk, A.: M&S Body of Knowledge: Progress Report and Look Ahead. M&S Magazine 4 (October 2010)

37. Locke, J.: An essay concerning human understanding. The Clarendon edition of the works of John Locke. Clarendon Press, Oxford (1979)
38. Hume, D.: A treatise of human nature, 2nd edn. Clarendon Press, Oxford (1978)
39. Pinker, S.: The Language Instinct: The New Science of Language and Mind, 1st edn. W. Morrow and Co., New York (1994)
40. Carruthers, P.: Phenomenal consciousness: a naturalistic theory. Cambridge University Press, Cambridge (2000)
41. Millikan, R.G.: On clear and confused ideas: an essay about substance concepts. Cambridge Studies in Philosophy. Cambridge University Press, Cambridge (2000)
42. Fodor, J.A.: Hume variations. Lines of thought. Clarendon Press, Oxford (2003)
43. Harman, G. (Non-Solipsistic) Conceptual Role Semantics. In: Lepore, E. (ed.) New Directions in Semantics. Academic Press, London (1987)
44. Margolis, E., Laurence, S.: The Ontology of Concepts - Abstract Objects or Mental Representations? Noûs 41(4), 561–593 (2007), doi:10.1111/j.1468-0068.2007.00663.x
45. Fodor, J.A.: Psychosemantics: the problem of meaning in the philosophy of mind. Explorations in cognitive science, vol. 2. MIT Press, Cambridge (1987)
46. Peacocke, C.: A study of concepts. Representation and mind. MIT Press, Cambridge (1992)
47. Zalta, E.N.: Fregean senses, modes of presentation, and concepts. Noûs 35, 335–359 (2001)
48. Searle, J.R.: The construction of social reality. Free Press, New York (1995)
49. Gruber, T.R.: A translation approach to portable ontology specifications. Knowledge Acquisition 5(2), 199–220 (1993)
50. Guarino, N., Giaretta, P.: Ontologies and Knowledge Bases: Towards a Terminological Clarification. In: Mars, N. (ed.) Towards Very Large Knowledge Bases, vol. 1, pp. 25–32. IOS Press (1995)
51. Guarino, N.: Formal ontology in information systems. In: Guarino, N. (ed.) Formal Ontology in Information Systems, FOIS 1998, Trento, Italy, June 6-8. Ios Pr. Inc. (1998)
52. Genesereth, M.R., Nilsson, N.J.: Logical foundations of artificial intelligence. Morgan Kaufmann, Los Altos (1987)
53. Tolk, A., Diallo, S., King, R., Turnitsa, C., Padilla, J.J.: Conceptual modeling for composition of model-based complex systems. In: Robinson, S., Brooks, R., Kotiadis, K., van der Zee, D.-J. (eds.) Conceptual Modeling for Discrete-Event Simulation, pp. 355–381. CRC Press (2010)
54. Ogden, C.K., Richards, I.A.: The meaning of meaning; a study of the influence of language upon thought and of the science of symbolism. International library of psychology, philosophy, and scientific method. K. Paul, Trench, Trubner & co., London (1923)
55. Partridge, C.: What is pump facility PF101? LADSEB CNR, Padova, Italy (2002)
56. Partridge, C.: What is a customer? The beginnings of a reference ontology for customer. In: Kilov, H., Baclawski, K. (eds.) 11th OOPSLA Workshop on Behavioral Semantics, Seattle, Washington, Northeastern (2002)
57. Partridge, C.: A new foundation for accounting: Steps towards the development of a reference ontology for accounting. LADSEB CNR, Padova, Italy (2002)
58. Partridge, C.: A new foundation for accounting: Steps towards the development of a reference ontology for accounting. In: ECAIS 2003 - European Conference on Accounting Information Systems, Seville (2003)

59. Masolo, C., Borgo, S., Gangemi, A., Guarino, N., Oltramari, A.: WonderWeb Deliverable D18: Ontology Library. Ontology Infrastructure for the Semantic Web. Laboratory For Applied Ontology - ISTC-CNR, Trento (2003)

60. Borgo, S., Gangemi, A., Guarino, N., Masolo, C., Oltramari, A.: WonderWeb Deliverable D15: Ontology RoadMap. Ontology Infrastructure for the Semantic Web. Laboratory For Applied Ontology - ISTC-CNR, Trento (2002)

61. Semy, S.K., Pulvermacher, M.K., Obrst, L.J.: Toward the Use of an Upper Ontology for U.S. Government and U.S. Military Domains: An Evaluation. The MITRE Corporation, Bedford (2004)

62. Partridge, C.: Note: A Couple of Meta-Ontological Choices for Ontological Architectures. LADSEB CNR, Padova, Italy (2002)

63. Strawson, P.F.: Individuals, an essay in descriptive metaphysics. Methuen, London (1959)

64. Cartwright, N.: How the laws of physics lie. Clarendon Press, Oxford (1983)

65. Cartwright, N.: The dappled world: a study of the boundaries of science. Cambridge University Press, Cambridge (1999)

66. Lewis, D.K.: On the plurality of worlds. Basil Blackwell, Oxford (1986)

67. Tye, M.: Representationalism and the Transparency of Experience. Noûs 36, 137–151 (2002), doi:10.1111/1468-0068.00365

68. Goodwin, C.: Transparent vision. In: Ochs, E., Schegloff, E.A., Thompson, S.A. (eds.) Interaction and Grammar, vol. 13, p. 370. Cambridge University Press (1996)

69. Craig, E.: The mind of God and the works of man. Clarendon, Oxford (1987)

70. Dretske, F.I.: Seeing and knowing. University of Chicago Press, Chicago (1969)

71. Dretske, F.I.: Knowledge & the flow of information, 1st edn. MIT Press, Cambridge (1981)

72. Evans, G.: The varieties of reference. Clarendon Press, Oxford (1982)

73. Stich, S.P.: Beliefs and Subdoxastic States. Philosophy of Science 45(4) (1978)

74. Ryle, G.: The concept of mind. Hutchinson, London (1949)

75. Habermas, J.: On the pragmatics of communication (trans: Cooke M). Studies in contemporary German social thought. MIT Press, Cambridge (1998)

76. Stanley, J., Williamson, T.: Knowing How. Journal of Philosophy 98(8), 411–444 (2001), doi:10.2307/2678403

77. Polanyi, M.: The tacit dimension. Terry Lectures, 1st edn. Doubleday, Garden City (1966)

78. Searle, J.R.: Intentionality, an essay in the philosophy of mind. Cambridge University Press, Cambridge (1983)

79. Collins, H.M., Evans, R.: Rethinking expertise. University of Chicago Press, Chicago (2007)

80. Collins, H.M.: Tacit and explicit knowledge. The University of Chicago Press, Chicago (2010)

81. Carnap, R.: Der logische Aufbau der Welt. Weltkreis-verlag, Berlin-Schlachtensee (1928)

82. Lipton, P.: Inference to the best explanation. Routledge, London (1991)

83. Peirce, C.S.S.: Prolegomena to an Apology for Pragmaticism. Monist 16, 492–546 (1906)

84. Shaffer, D.M., McBeath, M.K.: Naive beliefs in baseball: systematic distortion in perceived time of apex for fly balls. J. of Experimental Psychology Learning Memory Cognition 31(6), 1492–1501 (2005), doi:10.1037/0278-7393.31.6.1492

85. Bilalić, M., McLeod, P., Gobet, F.: Why good thoughts block better ones: the mechanism of the pernicious Einstellung (set) effect. Cognition 108, 652–661 (2008), doi:10.1016/j.cognition.2008.05.005

86. Bealer, G.: Universals. The Journal of Philosophy 90(1), 5–32 (1993)

87. Putnam, H.: Philosophy of Logic. In: Mathematics, Matter, and Method, pp. 323–357. Cambridge University Press, Cambridge (1975)

88. Tennant, N.: On the Necessary Existence of Numbers. Noûs 31(3), 307–336 (1997), doi:10.1111/0029-4624.00048

89. Carnap, R.: Psychology in Physical Language. Erkenntnis 3, 104–142 (1932/1933)

90. Field, H.H.: Science without numbers: a defence of nominalism. Library of philosophy and logic. Blackwell, Oxford (1980)

91. Field, H.H.: Realism, mathematics and modality. Basil Blackwell, Oxford (1989)

92. Neurath, O.: Physicalism: The Philosophy of the Vienna Circle. The Monist 41(4), 618–623 (1931)

93. Benacerraf, P.: Mathematical Truth. The Journal of Philosophy 70(19), 661–679 (1973)

94. Partridge, C.: Modelling the real world: Are classes abstractions or objects? JOOP 7(7) (1994)

95. Aristotle: Categories, and De interpretatione (trans: Ackrill JL). Clarendon Aristotle series. Clarendon Press, Oxford (1963)

96. Lowe, E.J.: The four-category ontology: A metaphysical foundation for natural science. Oxford University Press, USA (2006)

97. Johansson, I.: Ontological investigations: an inquiry into the categories of nature, man, and society. Routledge, London (1989)

98. Husserl, E.: Ideas; General Introduction to Pure Phenomenology (trans: Gibson WRB). The Muirhead Library of Philosophy. Allen & Unwin, London (1967)

99. Ingarden, R.: Time and modes of being (trans: Michejda HR). American lecture series. publication no. 558. American lectures in philosophy. Charles C. Thomas, Springfield, Illinois (1964)

100. Thomasson, A.L.: Fiction and metaphysics. Cambridge studies in philosophy. Cambridge University Press, Cambridge (1999)

101. Dummett, M.A.E.: Frege: Philosophy of language, vol. 1. Cambridge University Press (1981)

102. Lowe, E.J.: Kinds of being: A study of individuation, identity and the logic of sortal terms. Blackwell, Oxford (1989)

103. Wiggins, D.: Sameness and substance renewed. Cambridge University Press, Cambridge (2001)

104. Weirich, P.: The Explanatory Power of Models and Simulations: A Philosophical Exploration. Simulation & Gaming 42, 155–176 (2011), doi:10.1177/1046878108319639

105. Sugden, R.: Credible worlds: the status of theoretical models in economics. In: Mäki, U. (ed.) Fact and Fiction in Economics: Models, Realism and Social Construction, pp. 107–136. Cambridge University Press, Cambridge (2002)

106. Frigg, R., Reiss, J.: The philosophy of simulation: hot new issues or same old stew? Synthese 169(3), 593–613 (2009), doi:10.1007/s11229-008-9438-z

107. Edwards, P.N.: Representing the global atmosphere: computer models, data, and knowledge about climate change. In: Miller, C., Edwards, P. (eds.) Changing the Atmosphere: Expert Knowledge and Environmental Governance, pp. 31–65. MIT Press, Cambridge (2001)

108. Galison, P.: Computer simulations and the trading zone. In: Galison, P., Stump, D. (eds.) The Disunity of Science: Boundaries, Contexts, and Power, pp. 118–157. Stanford University Press, California (1996)

109. Sterman, J.D.: Learning from evidence in a complex world. Am. J. Public Health 96(3), 505–514 (2006), doi:10.2105/AJPH.2005.066043

110. Kripke, S.A.: A completeness theorem in modal logic. The Journal of Symbolic Logic 24(1), 1–14 (1959)

111. Lewis, D.K.: Counterpart theory and quantified modal logic. The Journal of Philosophy 65(5), 113–126 (1968)

112. Hartmann, S.: The World as a Process: Simulations in the Natural and Social Sciences. In: Hegselmann, R., Mueller, U., Troitzsch, K.G. (eds.) Modelling and Simulation in the Social Sciences from the Philosophy of Science Point of View. Theory and Decision Library, pp. 77–100. Kluwer, Dordrecht (1996)

113. The ANSI/SPARC DBMS Model. In: Jardine, D.A. (ed.) Second Share Working Conference on Data Base Management Systems, Montreal, Canada, April 26-30, pp. vii, 225. North-Holland Publishing Company, Amsterdam (1976)

114. McTaggart, J.E.: The Unreality of Time. Mind 17(4), 457–474 (1908)

115. Pressman, R.S.: Software Engineering, 5th edn. McGraw-Hill (2001)

116. Sider, T.: Four-dimensionalism: an ontology of persistence and time. Clarendon Press, Oxford (2001)

117. Partridge, C.: The Role of Ontology in Integrating Semantically Heterogeneous Databases. LADSEB CNR, Italy, Padova (2002)

118. Partridge, C.: The Role of Ontology in Semantic Integration. In: Second International Workshop on Semantics of Enterprise Integration at OOPSLA 2002, Seattle (2002)

119. Lycett, M., Partridge, C.: The challenge of epistemic divergence in IS development. Communications of the ACM 52(6), 127–131 (2009)

Ontologies in Modeling and Simulation: An Epistemological Perspective

Marko Hofmann

IT IS, University of the Federal Armed Forces Munich
Neutraubling, Germany

Abstract. Ontologies are formal specifications of concepts. They represent entities of a specific knowledge domain and the relationships that can hold between the entities. Ontologies are formal descriptions of the so called "body of knowledge" that composes a domain. Regardless of being implicitly or explicitly applied during the modeling, ontologies set the relation between formal signs used in computer simulations and "meaning" as a notion of human minds. Unfortunately, the essence of this relation is disputed, especially in modern epistemology, which deals with the "nature of knowledge" and the methods and limitations of gaining knowledge. Therefore, the chapter introduces first the debate which epistemological view is most appropriate for modeling and simulation. On the basis of this introduction ontologies are scrutinized with respect to their ability to capture knowledge. As a consequence of this analysis two main classes of ontologies for M&S are distinguished: Methodological and referential ontologies. Their values and limits are discussed in detail.

1 Introduction and Motivation

Why should anyone designing, programming or simply using serious computer simulations (The term „*serious* computer simulation" is used to differentiate simulations used for analysis, decision making, training and education from simulation intended to entertain.) be familiar with such sophisticated philosophical concepts like "ontology, epistemology, and teleology"? The long answer to this question is given on the next pages; a short answer can be sketched within a short paragraph: The most prominent attribute of a *serious* simulation is its validity. Whereas validation is a relatively straightforward activity for many technical computer simulations, it is a much more complicated endeavor when dealing with complex (social) systems simulation. In general, in such simulations it is inevitable to make some basic assumptions about what exists (ontology), about the right approach to come to know about reality (epistemology) and about the appropriate methods to achieve a given purpose (teleology). Unfortunately, there is no commonly accepted set of assumptions usable for all complex system simulations. Moreover, many standard assumptions used in modeling and

A. Tolk (Ed.): Ontology, Epistemology, & Teleology for Model. & Simulation, ISRL 44, pp. 59–87.
springerlink.com

simulation are disputed among experts. Consequently, a lot of (self-)critical reflection about the assumptions being used is unavoidable in order to assess the validity (and thereby the practical value) of a complex system simulation model.

The broader context of such reflections is the need to bridge the "semantic gap" between any formal computer system (necessarily a completely syntactical structure) and its real world counterpart full of "meaning" which has been attributed to it from human users. Consequently, ontology, epistemology, and teleology provide basic philosophical foundations for all kinds of "intelligence-based system engineering" (Tolk and Jain 2011), including the design and development of simulations and agent-based systems.

The chapter is divided into two sections: The first section provides basic definitions for ontology, epistemology, and teleology, discusses some key concepts given in landmark and recent academic papers, and illustrates the variety of different epistemological perspectives applicable for modeling and simulation. The second section, a concrete application of these philosophical foundations, explores the intricacies of epistemic and normative aspects of ontologies used in modeling and simulation – including practical questions of interoperability and composability of simulations.

2   Basic Definitions

"Ontology", "epistemology" and "teleology" (sometimes called the "triangle of knowledge" (see Figure 1; the underlined questions give a short description of the scope of the three concepts, in italics are advanced questions that reflect the dependencies of the answers from special perspectives or communities)) are terms originally coined in philosophy. Although the traditional meanings of these terms are still visible in the modern and more concrete meaning in modeling and simulation, important differences exist:

In philosophy **"ontology"** (a composition from the Greek words "onto" (*being; that which is*) and "logia" (*study, theory*)) is the study of existence and reality as such, as well as the categorization of beings and their relations. Practically, it deals with questions like: What entities exist, and how such entities can be classified according to similarities and differences. In computer science (and modeling and simulation) **"ontologies"** (mind the plural!) are formal specifications of concepts representing entities of a specific knowledge domain and the relationships that can hold between the entities. They intended to be formal descriptions of the so called "body of knowledge" that composes a domain. The definition of ontologies can become a controversial issue because of two problems: First, in some domains (especially domains including human behavior) there is no commonly accepted "body of knowledge" and secondly, there are some fundamental problems of capturing knowledge in words, at least from the perspectives of modern epistemologies. These challenges are discussed in detail in the second part of the paper.

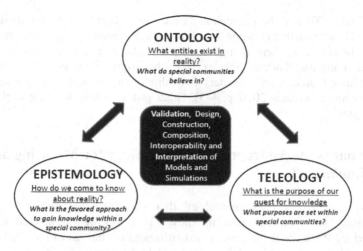

Fig. 1 The triangle of knowledge and its importance for several aspects of modeling and simulation

In traditional <u>philosophy</u> **"epistemology"** (the Greek work "episteme" means "knowledge") deals with the "nature of knowledge" (how it relates to notions such as truth and belief), and the methods and limitations of gaining knowledge. The many different approaches that have been proposed for the appropriate way to gain knowledge are called **epistemologies**. Among them are positivism, constructivism, hermeneutics, and relativism (to name just a few). The <u>central question for</u> <u>modeling and simulation</u> is which of these theories best fits as a general foundation for gaining knowledge with computer simulations. This is an intensively disputed question presented in the first section of this chapter.

In <u>philosophy</u> **teleology** (from Greek *"tele"*: far) is the supposition that final causes exist in nature, meaning that purpose such as found in human actions is inherent in the rest of nature, too. Teleology in this sense is refuted by most scientists. In <u>modeling and simulation</u>, teleology designates the study of appropriate purposes for simulation models as such, as well as actions and methods which are used to achieve a specific purpose (see Turnitsa , Padilla , and Tolk (2010)). Teleology in that sense gives modeling its direction, and it is obvious, that the setting of a purpose dominates any (simulation-based) analysis. Furthermore, the objectives, for which a model has been developed and is used, are essential for what kind of results may be obtained by testing, verification and validation. However, it is no general solution to define weak objectives – such as getting "unspecific insights". Whenever empirical confirmation is difficult some advocates of validation argue that with respect to a modest purpose of the model, for example, getting insights into courses of actions, any simulation model can be validated. However, by rephrasing the model purpose in this kind of manner the model is too easy to validate (and too difficult to falsify). Any model building and execution process will yield *some* insights into the model specific logical consequences of the assumptions put into the model. The term "validation" suffers from such a weak interpretation, which has gained widespread acceptance over the past decades. To use the expression "valid simulation model" for models that fulfill such weak-defined purposes is almost a self-deceiving illusion of the model quality. In any case, there is a severe limit to how far this perspective can be justified (Batty and Torrens 2005, p.762). The origin of the problem is the need for confirmed prediction in order to evaluate any model quantitatively (see Bharathy and Silverman 2010, p. 443). Other issues of teleology are neglected in this chapter.

3   What Is the Appropriate Epistemology for Modeling and Simulation?

The basic questions of epistemology that directly affect serious computer simulation: "How do we come to know? How do we define and represent knowledge? What are the necessary and sufficient conditions of knowledge?" are much too broad in scope and deep in detail to be captured in a few pages (for overviews focused on M&S see Frank and Troitzsch 2005, and Grüne–Yanoff and Weirich 2010). Hence, the following sections focus on the single core problem raised above: *Which of the various epistemologies suggested from philosophers is most appropriate for modeling and simulation?*

3.1   *The Limits of Naïve Realism and Empirical Validation*

Before going into detail it seems necessary to explain, why controversial issues exist at all. Some practitioners of modeling and simulation believe that input and output comparisons between simulation models and their real world reference system ("empirical validation") are sufficient to test model validity. As sound as this comparison-based approach may seem, a lot of serious epistemological qualifications have been made concerning the universality of the concept, its applicability (Some arguments against the applicability of "empirical validation" in natural sciences can be found in: Konikov and Bredehoeft (1992), Oreskes et al. (1994), Dessai et al. (2009), and Ruphy (2011)), and its underlying philosophical perspective, which is called "common sense realism" from its advocates and "naïve realism" from its opponents. Naïve realism or common sense realism is the most "natural" of all epistemologies. It claims that our senses provide us with direct, objective knowledge of a reality, which exists without any dependence from human perspectives. By observing nature meticulously we can establish solid truths. Naïve realism stands in sharp contrast to idealism and skepticism. Among the qualifications against naïve realism and empirical validation are:

- Any model is limited in its validity because of "underdetermination" or "equifinality" i.e. given any finite amount of evidence, there is at least on rival (contrary) model which equally fits with the data. In other words, the evidence cannot by itself determine that some one of the host of competing theories is the correct one (see Quine (1977) and Fraassen (1980) for the philosophical background of this argument, Oreskes et. al. (1994) for a discussion from the perspective of natural science, and Richardson (2003) for a view from computer simulation).

- One reason to build simulations is to explore phenomena for which no or only insufficient empirical data exists. So called "data-poor environments" are widespread through all sciences. In general, these simulations cannot be judged on the basis of comparisons with real data, because data from reality is scarce, uncertain or completely missing (van Horn 1971, Zeigler et al. 2000, Hofmann 2002b).

- Measurement of empiric processes is limited in precision. Unfortunately, minor differences in initial conditions can cause completely different system state trajectories in nonlinear systems. If the sensitivity of the output data is high within the range of the input measurement uncertainty, the correspondence of specific system and model trajectories is of little value (Byrne 1997, Goldspink 2002).

- For huge simulations or simulation federations it is impossible to guarantee that the adjustments of the simulation model to the given system input/output are achievable in reasonable time- for reason of computational restrictions (Hofmann 2005).

- All observation is theory-laden and since all experimentation involves observation, experimental data has to be theory-laden, too. Experiments, according to this view, make sense only in relation to some theoretical background. Consequently, observations are not "bed rock elements" on which theories and models can safely rely (Ahrweiler and Bilbert 2005, Carrier 1994).

Furthermore, "objective observation" is a concept that is only applicable in traditional epistemologies (realism, positivism, empiricism etc.). It is almost irrelevant in others (all kinds of constructivism, for example). Thus, to a certain extent, it is a matter of one's philosophical attitude whether objective empirical validation is possible or not.

3.2 On the Variety of Possible Epistemologies Applicable for Modeling and Simulation

The issue of model and simulation validity as seen from different epistemological perspectives has been addressed first by Naylor and Finger (1967 !) and received thorough investigation by Stanislaw 1986, Barlas and Carpenter 1990, Dery et al. 1993, Miser 1993, Kleindorfer et al. 1998, and Klein and Herskovitz 2005. These papers contrast simulation validation from the epistemological perspectives of empiricism and relativism, logical empiricism (positivism), critical-rationalism, pragmatism, historico-critical relativism, different sociological perspectives, classical rationalism, empiricism, Bayesianism, Lakatos' methodology of scientific research programs and hermeneutics. Not treated in these publications are Paul Feyerabend's anarchistic view of science (Feyerabend 1988) and Ernst von Glasersfeld's radical constructivism (Glasersfeld 1995, 1997), although they are of eminent importance in contemporary philosophy of science. In response to this variety Niehaves et al. (2005) have even constructed a framework for comparing epistemological perspectives on simulation. It is impossible to discuss all of these positions in this paper, and there is no common agreement about which of these epistemologies is most appropriate for modeling and simulation. Fortunately, the point the author wants to make, can also be discussed based on a course distinction between two main classes of epistemology: objectivism and relativism (Kleindorfer et. al (1998) use the alternative expressions *foundationalism* and *anti-foundationalism*), from which two (rather extreme) proxies are taken (which are by no means representative for all positions, but highlight the essential differences): Positivism and constructivism.

The positivistic paradigm of knowledge gaining (a methodological advancement of common sense realism) is based on the belief that reality exists independently from the human observer and is totally governed by laws of nature. It is founded on the notion, that humans can fully understand realty, *and that carefully planned, executed and described experiments can reveal the "true" nature of a phenomenon*. A human observer is seen as an instance of a stimulus-reaction-mechanism. All open questions are formulated as hypothesis which are corroborated or refuted on the basis of experiments. Knowledge is consequently the correspondence between reality and the mental or formal representation (correspondence theory of truth). (Today, in practice, this position is often attenuated to a kind of "pragmatic realism", which means that scientist have the aim of developing and using models that are as "realistic as possible", given the constraints of current knowledge, skills, computing power and available time (see Beven (2002) for a critical discussion of this seemingly impeccable philosophy).)

The positivistic epistemology has been and still is successful (Chen and Hirscheim 2004) in the so called exact sciences. However, in almost every phenomena affecting decisions in complex social systems (in the broad sense of the word "social", comprising all systems with human communication and interaction, e.g. economic, military, political, social in the narrow sense) there is so much inherent uncertainty that trying to follow a *purely rational analytical process*, hoping to find exact solutions is often downright *irrational* and self-deceiving (Hofmann and Hahn 2007). The Nobel laureate Herbert Simon (1997) used the term "bounded rationality" to indicate this limit of rationality and decision making, in general. Some other authors have even argued that the future of social systems is completely unpredictable (see Luhmann, 1991, p.21, for a view form a sociologist, and (Hemez 2004) for a "technical" point of view). The problem is a timeless challenge for scientist working with social simulations (see Churchman 1973 to understand its historic dimension, and Küppers and Lenhard 2005, Windrum, Fagiolo and Moneta 2007, Moss 2008, Garson 2009, Brenner and Werker 2009, Rossiter, Noble and Bell 2010, Tolk et al. 2010, and Neumann et al. 2011 for some recent publications.).

An epistemology reflecting the problems of social sciences is constructivism (in any variant). The ontological foundation of constructivism is idealism: Different subjective realities coexist as mental constructs. The observer is not neutral with respect to the observation but part of it. Each observation is the result of an interaction between observer and observed situation, thus the results are strongly influenced by the observers beliefs, attitudes, and values. To a certain extent humans are the creators of their own reality. The most important methodological technique of the constructivist paradigm is interpretation. The goal of knowledge seeking is not absolute truth but relative "viability" (see Glasersfeld 1997). A mental construct of reality is viable if it helps a subject in its struggle for life.

The paramount idea of constructivism has been described by the philosopher Ludwig Wittgenstein long before the term constructivism has been introduced. In his "Philosophical remarks (1953) he stated that:

"Our expectation anticipates the event. In this sense, it makes a model of the event. But we can only make a model of a fact in the world we live in, i.e. the model must be essentially related to the world we live in and what's more, independently of whether it's true or false."

According to the constructivist view the validation of simulation against empirical data sets *"is not about comparing the real world and the simulation output; it is comparing what you observe as the real world with what you observe as the output. Both are constructions of an observer and his views concerning relevant agents and their attributes. Constructing reality and constructing simulation are just two ways of an observer seeing the world* (Ahrweiler and Gilbert 2005)". Clearly, such an observer-oriented view of the world is an exaggeration to many scientists, and in order to avoid both solipsism and indiscriminate relativism it is indeed necessary to explain, how individual perceptions and constructions of the world converge to common pictures of reality, which are shared and trusted within scientific communities.

Not with standing many clarifying discussions in the papers mentioned in this section the role of simulation as a method of knowledge gain (the "epistemological value" of simulation) is still controversial. Moreover, recently the dispute has even gained new momentum: see Frigg and Reiss 2009, Humpreys 2009, Grüne-Yanoff 2009, Frank, Squazzoni and Troitzsch 2009, Dessai et al. 2009, Phan and Varenne 2010, Ruphy 2011, and Simpson 2011 for exemplary discussions from a variety of different, partially critical perspectives, as well as Varenne 2001 and Grüne-Yanoff and Weirich 2010 for overviews. Despite the many different aspects addressed in these papers, the fundamental question of serious computer simulations still seems to be: *To what extent are simulations valid representations of (behavioral) systems (Hermann 1967, p. 216)*? The history of epistemology in modeling and simulation has shown that this question has to be (and can only be) answered for each single simulation anew. Getting familiar with the epistemological debate is a prerequisite for defending own models more profoundly.

4   The Two-Sided Nature of Ontologies in Modeling and Simulation

This second section of the chapter is based on the   article "Epistemic and normative aspects of ontologies in modeling and simulation, *Journal of*

Simulation, Volume 5, Number 3, 2011, pp. 135-146(12); Hofmann, Palii, and Mihelcic. The most important "semantic" difference to this publication is the new summary and conclusion.

The section illustrates why ontological and epistemological considerations are essential even for practical purposes like model interoperability and composability.

4.1   Introduction to the Use of Ontologies in Computer Science

In computer science ontologies have been defined as formal specifications of concepts representing entities of a specific knowledge domain and the relationships that can hold between the entities. Since the nineties of the last century ontologies have become a promising approach to overcome semantic problems of knowledge sharing and reuse (Gruber 1993, Gruber 1995, Benjamin et al. 1995.) The development of ontologies in computer science is motivated not so much by the pure search for knowledge (in contrast to the philosophical endeavor of finding the appropriate universal "*ontology*", and also in contrast to inquiries of natural science), but by the urgent need *to design, engineer and manage "knowledge"*, and, more tangible, complex, "intelligent" software systems effectively.

Considering the numerous challenges in simulation composability and interoperability (Hofmann 2004) it is therefore consequent that ontologies have been proposed for modeling and simulation, too (Miller et al. 2004; Christley et al. 2004; Benjamin et al. 2006; Bell et al. 2008; Guizzardi and Wagner 2010(a,b); Livet et al. 2010; Taylor et al. 2010; Ezzell et al. 2011). Among other advantages, ontology-based simulation is said to support *consistent semantic model interchange* which leads to higher quality models, lower costs, and a faster development-process (Lacy and Gerber 2004).

Models are conceptualizations of (real world) referents and computer simulations are executable expressions of these conceptualizations. Conceptualization, however, is a cognitive, purpose-driven act, that varies from individual to individual and from task to task.

"Because of this, we can see the implications that there are an uncountable number of possible conceptualizations arising from one referent, and an equally uncountable number of possible symbols that can represent one conceptualization (Turnsitsa *et al.* 2010, p. 645)."

Computer simulations are therefore (1) manipulations of arbitrarily chosen symbols referring to objects that are (2) conceptualized from a specific point of view for a specific purpose (in scientific context: a research question). While the first choice can be easily restricted by strict definitions the second degree of freedom is essential for successful modeling and simulation. It is also a major distinction from other software engineering disciplines, in which a product has to be developed, that supports a real world referent directly (as an example think of the software for anti-blocking systems (ABS) in cars) – instead of conceptualizing a referent within a model that acts like a substitute for reality (for a more detailed illustration of this difference see Turnsitsa et al. 2010). In the following pages the consequences of this fundamental distinction are investigated with respect to the construction of ontologies for modeling and simulation.

For the further considerations it is crucial to realize, that there are two classes of ontologies in modeling and simulation: ontologies defining modeling methods and simulation techniques (for which the term *"methodological ontologies"* seems appropriate), and ontologies representing real world systems to be simulated (coined *"referential ontologies"*) (see Table 1). The expression "domain ontology" is, in general, a synonym for referential ontology. We prefer the former term since many authors use "domain" for methodological domains, too.

Table 1 Methodological versus Referential Ontologies

	Methodological Ontologies	**Referential Ontologies**
Body of knowledge	Methods, techniques, formalisms	Real world systems, domains
Entities	Abstract concepts (event, item, activity, time, scheduling, etc.)	Real world objects (tanks, aircrafts, harbor, terrorist, etc.)
Relations	Logical, syntactic, simple semantic ("is a", "has parts" etc.)	Logical, syntactic, semantic, pragmatic ("depends on", "cooperates with" etc.)
Languages	Formal languages, OWL, RDF, XML	Formal and natural languages
Examples	DeMO and DESO, see Figures 2 and 3	Military relevant actors, tanks, see Figures 4 and 5
Epistemic nature	Reflect current "state of the art" in modeling (abstraction and idealization); certain	Reflect what is currently known or supposed to be known about the real world domain; uncertain

Examples for methodological ontologies in modeling and simulation are:

- the Discrete-event Modeling Ontology (DeMO) described in (Miller et al. 2004, Miller et al. 2008, and most recently in Silver et al. 2009); one part of it, a process interaction world view, is illustrated *informally*, but instructively, in Figure 2 (taken from Lacy 2006, p. 87, his dissertation describes thoroughly the development and application of an ontology in modeling and simulation), and
- the Discrete Event Simulation Ontology (DESO) an extension of the Unified Foundational Ontology (UFO) described in (Guizzardi and Wagner 2010(a,b)); a part of it, the "ontology of events" is illustrated *formally* in Figure 3 (taken from Guizzardi and Wagner 2010b, p. 657)

An example for a referential ontology is depicted in Figure 4. It shows actors in a modern military environment and was developed in cooperation with the German Armed Forces (Bundeswehr). Very roughly, methodological ontologies define

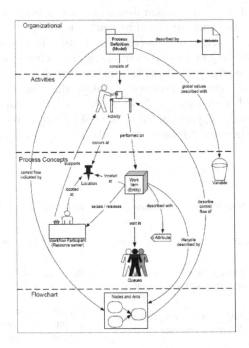

Fig. 2 Example of a methodological ontology (informal description) - Ontological concepts in a process interaction methodology (taken from Silver et al. 2006, p. 1172)

how the keywords of a method, modeling and simulation (M&S) for example, should be defined and how this method has to be transformed into software in order to be *formally* compatible and technically efficient, whereas referential ontologies try to provide a standard description of *what* is essential for modeling a certain part of the reality – given a specific purpose.

In general, in modeling and simulation, a referential ontology is both a *conceptual representation* of a reference system (a part of the real world, like a manufacturing system, a business, or a military conflict area) and a *conceptual specification* for simulation software and evaluation tools. These two dimensions of ontologies are essential but nevertheless often underestimated in their consequences. As representations ontologies are descriptive. They are models *of* a reference system. They describe system behavior from a purpose-oriented perspective. They are the result of a profound system analysis of a domain. As specifications, ontologies are prescriptive; they are models *for* a (simulation) system to be built. They prescribe functionality and formats (syntax). They are the crucial reference for software engineers. They define a formal semantics for automated information processing.

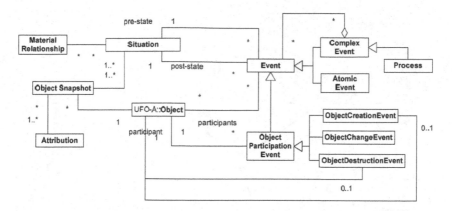

Fig. 3 Example (No. 2) of a methodological ontology (formal description) – Ontology of events in DESO (taken from Guizzardi and Wagner 2010b, p. 661)

The fundamental difference between these two dimensions is that the descriptive dimension reflects epistemic issues whereas the prescriptive dimension has a completely normative nature. (A similar view is expressed in Hesse (2008) using the terms "after-image" and "pre-image"). In other words, as conceptual (meta-)models of systems ontologies incorporate and exclude certain aspects *of reference systems*. The choice of depth and breadth of this modeling is not only a question of consistent definitions but also a question of appropriateness with regard to reality and objectives. Hence, as all scientific models, ontologies are hypothetical, transient representations of current inter-subjective knowledge, *open* with respect to system borders, resolution, and change. As specifications, in sharp contrast, they are *closed* definitions of what constitutes the components of a model and of what the symbols – used to represent real objects – signify *within the model.*

The normative nature of ontologies is dominant in methodological ontologies although they also have an epistemic part: They refer to the state of the art of current praxis – with respect to a specific methodology. However, this reference is, in general, well-established (certain) within the community using that methodology. In discrete event simulation, for example, the match between concept definitions and practical use is very strong.

The normative nature of (methodological and referential) ontologies implies a great enhancement for composability and interoperability, since it ensures, as far as possible, a common, consistent and well-defined world view among the developers of software or simulation components within a specific domain. However, it is by no means guaranteed that the epistemic nature of referential ontologies is always appropriate for the purpose of the simulation. This is especially true, if the real world referent system is messy, the problems to be solved are ill-defined and the appropriate perspective of modeling controversial – like in almost all social, economic or military systems. The basic question posed by this double-edged nature of ontologies in modeling and simulation is, how these different dimensions should be merged within a single linguistic entity, the formal realization of the ontology.

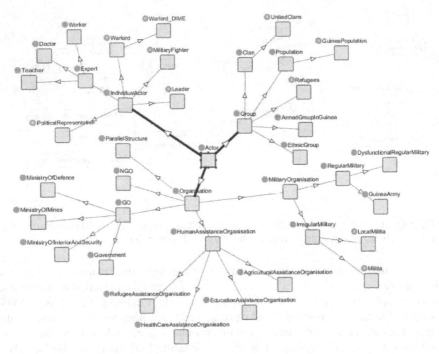

Fig. 4 Example of a referential ontology - Actors in a modern military environment

This paper first discusses the main aspects of the normative and epistemic nature of ontologies, digging down into the possibilities and limits of languages as means of reality representations and as conveyors of meanings. These findings are then discussed with respect to the composability and interoperability of simulation components within ontological frames. A short example will illustrate the theoretical considerations. Finally, some guidelines for the balancing of epistemic and normative aspects of ontologies are given.

4.2   Ontologies as Formal Specifications

As specifications, ontologies are formal, unambiguous descriptions used to define and categorize concepts and the relationships among concepts within a particular knowledge domain. Being formal and unambiguous they can be processed by computers, in order to, for example, logically deduce higher order relationships between concepts or even classes of concepts. Ontologies provide a common formal semantics of information sources that can be communicated between software and human agents. For that purpose they are written in formal machine-processable languages such as XML (Extensible Markup Language), RDF (Resource Description Framework) or OWL (Web Ontology Language). These languages are very powerful for the description of objects, structures and concepts, and their logical relations. Hence, the strength of ontologies as specifications lies in their precision: Precision for a common terminology, for a common logical structure of conceptual relations, precision even for the denotation of concepts as far as possible by definitions.

Any methodological ontology is primarily such a precise specification – in the case of DeMO and DESO specifications of the elements of discrete event simulation methods. Both include an unambiguous terminology, an exact taxonomy, and a detailed interrelation of the concepts used in these simulation techniques (see Miller et al. 2004 and 2008, Guizzardi and Wagner 2010 (a,b)). Such methodological precision is essential for knowledge exchange and reuse, for compatibility with other ontologies and powerful querying capabilities.

The advantages of formally specified ontologies for software engineering in general, and for modeling and simulation, in particular, are obvious:

- First, they can be used to provide a common framework for concepts shared by researchers within a certain domain – in a way readable both for humans *and* computers. Simplifying matters extremely, ontologies are "machine readable dictionaries". As such they *can* reduce conceptual misunderstandings among modelers, and formal inconsistencies between software. Thus, ontologies *can* increase the potential for interoperability and reuse of (simulation) artifacts. Figure 3 illustrates this advantage: DESO defines unambiguously the relation between events, situations, and (general) objects.
- Second, the formal languages mentioned above, especially OWL, the Web Ontology Language, share much in common with the basic ideas of Entity-Relationship Modeling (ERM), and the Unified Modeling Language (UML), the new "lingua franca" of computer scientists (for a profound comparison of UML and OWL see Kiko and Atkinson 2008). The transition from ontologies to models and backwards is thereby at least simplified, since the general notion of thinking in objects (or entities), their attributes and relations *almost* stays the same. (Figure 3 is an excellent example for this similarity).
- Third, for practical reasons, it seems inevitable that simulation models are developed with different modeling techniques and tools. Within a common general methodological ontology it is much easier to describe general transformations between them. The realizability of such transformations (Arena, AnyLogic, ProModel) has been shown by Lacy (2006) using an ontology called "Process Interaction Modeling Ontology for Discrete Event Simulation" (PIMODES). In addition, the DEVS framework (Zeigler et al. 2000) is a theoretical confirmation of the possibilities of such transformations among very different techniques (Vangheluwe 2000).

As an example (No. 1) for the first benefit of precision in referential ontologies, consider the concept "tank" that is ambiguous in natural language, since it denotes a special class of armored vehicles as well as a variety of containers. Within a military domain ontology, for example, a tank could be specified clear-cut as a tracked, armored fighting vehicle, and put into various relation to its weapons, armor, motor, to soldiers and echelons, even to its task and targets (see Figure 5, which is only intended as an illustration and necessarily incomplete). Within such an ontology any aspect of tanks could be specified in accordance with domain experts.

The applicability of ontologies for the specification of technical terms has been demonstrated for a variety of very different domains, such as air battle (Fishwick and Miller, 2004), financial risk management (Cuske et al. 2005), hydrodynamic systems (Isam and Piasecki 2008), agricultural systems (Beck et al. 2010), and cardiovascular physiology (Ezzell et al. 2011).

In order to take maximum benefit from the major advantage of ontologies – their precision as a tool of defining concepts and semantics with a formal frame – however, one has to accept an important restriction: Formal systems must be logically consistent. Otherwise, contradictions lead to arbitrary conclusions. As a formal specification the ideal ontology – even if defined by consensus – is therefore a strict and compulsive consensus. The next section will demonstrate, that ideal ontologies as means of knowledge representation – at least in some domains – should (and try to) reflect ambiguity and contradictory conceptualizations. Unfortunately, there is no quick solution for this dilemma; moreover, it is directly connected to one of the most challenging problems in modern philosophy.

4.3   Ontologies as Means of Knowledge Representation

This section illustrates some limits of ontologies as semantic domain models: The limits of computers in interpreting meanings (semantics), and the limits of language as a means of reality representation.

Referential ontologies can be used to store information about real objects, their attributes and interrelations by means of symbolic representation. They are more powerful than taxonomies or glossaries as they do not only collect and categorize concepts but also relate them to each other from a certain perspective. It is, for example, possible to describe manufacturing from the word view of discrete events. Such a preselected perspective gives the system analysts, contributing to the manufacturing domain ontology, a common world view which allows much more consistency, and consequently, a much closer logical interrelation of the concepts. Ontologies are therefore kinds of metamodels, enabling modelers to define the denotation of their conceptual representations within networks of related concepts linked to reality.

Ontologies are *semantic domain models* in the sense that the linguistic (symbolic) representations of objects should be *meaningfully* related as their real world referents are. *The "semantic relations of the world" should correspond to the "formal semantics" defined in the ontology.* The word "semantics" is set into quotation marks as its meaning is not thoroughly clear in both cases: The "semantic relations of the world", in the literal sense of the expression, are very often beyond human cognitive capacities, and the "formal semantics" of a machine-processable language can only define syntactical relations between signs, since no computer is – up to now – able to attribute meaning to signs in the way humans do. Consequently, not all relations sense-making to humans can be sufficiently described for computers by logical expressions. The relations "fights" and "needs" of Figure 5, for example, are semantically clear for human experts but underspecified in logical terms, in contrast to the relations "is (a)" and "has

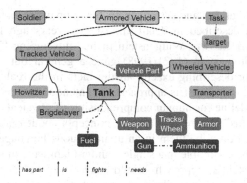

Fig. 5 Part of an ontological description of a tank

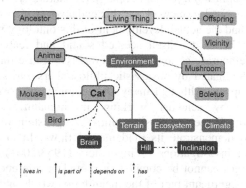

Fig. 6 Part of an ontological description of a cat

part" which have an exact meaning in logical terms. This dilemma becomes obvious when comparing Figure 5 with Figure 6, which is part of an ontology describing the essential relations of the concept "cat" using the same graphical network of relations but with different "names" (identifiers) for the relations.

Again, only the "is (a)" and the "has part" relations are sufficiently defined for automated processing. The other two relations "lives in" and "depends on" are underspecified for computers. Moreover, the – intentionally chosen – exact graphical isomorphism between Figure 5 and Figure 6 illustrate that "fights", "needs", "lives in" and "depends on" are in fact only syntactical terms for computers. With other words, for a computer, there are no differences between both figures except from different identifiers. Consequently, semantic plausibility checks of humans go far beyond the possibilities of automated reasoning in computer ontologies.

Furthermore, the concept of ontologies presupposes an epistemic link between symbols and real objects, between language and reality which has been seriously questioned by philosophers of science (Janich 2001) and linguist (Pinker 2007) as well.

Most computer scientists and system analysts are unaware that their ontological endeavor has been described in detail almost 100 years ago from the famous Austrian philosopher Ludwig Wittgenstein in his outstanding "Tractatus logico-philosophicus". Wittgenstein (1921/1998) gave a general but nevertheless very precise guideline for describing exactly all the facts in the real world by using a formal system of description. His approach was intended to be the final linguistic framework and guideline for human comprehension of physical reality. The basic idea of the "Tractatus" was to find all true concepts about reality using empiric results (facts) and to describe them in an unambiguous language using logic. All knowledge should be codifiable in a single standard language of science. The final result of Wittgenstein's approach would have been a constantly growing "comprehensive ontology", describing exactly what is true in terms of sentences about reality. *With other words he tried to capture the semantic relations of the world within the semantics of a formal language (!).* Naturally, Wittgenstein could only give the philosophical framework for this endeavor, which should have been itemized by others, and the actual development of this ontology would have taken centuries – but the basic idea itself seemed sound and realizable to most of Wittgenstein's contemporary colleagues, presumably because the first half of the 20th century was dominated by positivistic philosophers and scientists (see the first part of this chapter). Although many shortcomings of positivism had been revealed especially by scientist working in humanities (Weber 1949), Wittgenstein's approach seemed feasible until Wittgenstein himself and W.O. Quine (1960) could demonstrate its fundamental flaws. In his second seminal work "Philosophical Investigations" Wittgenstein (1953/2001) showed that the semantics of languages cannot be completely rooted in references to reality and fixed definitions. A significant part of the meaning of words and sentences – and the concepts related to them – is based on their ever changing usage within a community, reflecting the heterogeneity of thoughts, beliefs and emotions. Language is at least as much a group-specific, transient construction than a representation of reality. An "objective ontology", in the ideal sense formulated in the Tractatus is therefore infeasible.

Especially, abstract linguistic concepts like "democracy"," justice", or "progress" are constructions within a community, or society, which becomes immediately evident when transferring them to other cultures.

In this chapter, it is impossible to redraw the philosophical discussions about the power (and impotence) of language as a precise means of reality depiction, however, a "perfect image" of the world as a symbolic representation is judged today as unrealistic. Wolfgang Hesse, a leading computer scientist in Germany, formulated this view in his article on *ontology-based software engineering* (Hesse 2008):

"But whatever philosophical position we might adopt, it is quite obvious that the mental images of the world "out there" we are dealing with are not identical with that world. Often we cannot precisely delimit and describe – much less completely grasp the material objects around us – not to speak about all the immaterial ones like friendship, work, peace, war, etc. Even if we observe that large groups of

humans can arrive at a broad consensus, about certain objects and domains and therefore might be inclined to assume the existence of objective knowledge about such things, we never can be sure about it..." (p. 142).

Consequently, modern ontologies cannot and therefore should not be judged by objective truth but by group specific utility. However, this implies that ontologies as knowledge representations are necessarily subjective, transient and subject to failure.

Moreover, even within a user group and their restricted view on the world it is impossible to define totally unambiguous ontologies, since some concepts, like the expressions "terrorist", "insurgent" or "Taliban", for example (No. 2), simply do not have a precise, objective extension (or perspective-independent reference) to real persons: Is every person who is sympathetic to the ideas of the Taliban necessarily a Taliban? Is someone forced into the Taliban units a Taliban? Is someone without comparable economic opportunities a "real" Taliban? Do you have to call yourself a "Taliban" in order to be one? When exactly does a former neutral civilian become a Taliban? Is mere benefitting from Taliban operations or logistic help enough for being accused? In short, what exactly are the attributes of being a Taliban? An ontological definition of the word "Taliban" might avoid all ambiguity, but this will not make reality itself less ambiguous. In our studies for the German Bundeswehr, we furthermore realized that the interrelation of such concepts is even more critical: the "meanings" of concepts like *soldier, combatant, insurgent, terrorist, non-combatant, civilian, auxiliary*, and the relation of these meanings to each other are extremely difficult. Even within a very special user group, e.g. NATO officers, any agreement of their semantic relation is contended.

Referential ontologies are like maps which refer to real terrain, but which are always imperfect representations of that terrain. The usability of a map and its resolution strongly depends on the purpose of the application. Ontologies, too, have to adapt their resolution, scope and abstraction level to the objectives of their user group. From an optimist's point of view, ontologies are therefore community-mediated and accepted descriptions of the kinds of entities and their interrelations within a certain domain of discourse. From a pessimist's point of view, they also reflect preconceptions, biases, information deficits and group-specific world views.

Regardless of the perspective, the ultimate consequence of the representational nature of ontologies is that they are collections of conceptual descriptions and logical relations that cannot, a priori, assure their correspondence to reality. Although they can demonstrate their appropriateness by successful application, induction to future applications is always uncertain. Ontologies necessarily are mirrors of what is called explicit, language-based knowledge about reality. As such, the degree of congruence between representation and reality depends on the specific domain. Technical domains and pure systems of natural science are much more thoroughly understood than socio-technical systems like traffic or social systems, e.g. the people of Afghanistan. It is a huge gap between selling software on the Web (an artificial system based on a predefined and world-wide accepted ontology) and modeling military conflict scenarios, for example. If a domain and our knowledge about it are volatile, uncertain, subjective and transient, then the ontologies describing them must reflect and not erase this calamity. For such domains heterogeneity of ontologies is therefore almost inevitable. Heterogeneity, however, implies that conflicting – and even contradictory – conceptualizations may overlap. Unfortunately, contradictions are nightmarish for those designing and creating formal systems and using automated logical deduction.

Summarizing so far, the dilemma with referential ontologies is that due to their double edged nature two different ideals exist for their construction: Precision and compulsory strictness as a pre-image and flexible heterogeneity as an after-image. Such ontologies have to evolve between these two poles. In addition, language is always an imperfect means of reality representation, forcing to compromise. These two tradeoffs have consequences for all (referential) ontological approaches, including the composition and interoperation of simulation models.

4.4 Ontological Support for the Composition and Interoperability of Simulation Models

Two perspectives for such an ontological support are addressed in this section: The use of ontologies to increase interoperability on several distinct levels of interoperability and ontology driven modeling itself.

4.4.1 Levels of Interoperability

Many papers have been written in recent years about the levels of interoperability which must be taken into consideration for a successful composition of simulation models. A very comprehensive approach is the Level of Conceptual Inter-operability Model (LCIM) (Wang et al. 2009) differentiating 7 levels of interoperability, starting with a level 0 indicating *no interoperation* at all (see Figure 7).

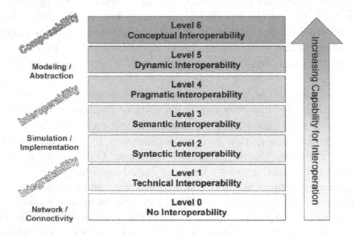

Fig. 7 The Levels of Interoperability Model (LCIM) (taken from Wang et al. 2009)

The indispensable first two levels are called *technical* interoperability (1) and *syntactical* interoperability (2). Technical interoperability is established by a physical connection which ensures, at all, the exchange of data. Syntactical interoperability is reached by the exclusive use of well-defined signs, terms and higher order grammatical expressions. It ensures the sharing of data in the "right" format. The top four levels, called *semantic, pragmatic, dynamic,* and *conceptual* interoperability deal with "meanings", intentions (Hofmann 2002), changing contexts, and concepts, respectively. They ensure that the aspects of knowledge addressed by the adjectives in italics are shared as intended.

Referential ontologies are presumably the only methodological approach capable of dealing with all levels above the technical level simultaneously. Due to their double-edged nature they can ensure syntactical interoperability by formal definitions, semantic interoperability by limiting the scope of the "meanings" of symbols by agreement, pragmatic interoperability by including specific research questions, and even dynamic and conceptual interoperability by a shared understanding of concepts and their evolution of time.

For many technical domains and artificial systems, ontologies will be able to ensure the interoperability of simulation components developed for a similar purpose under a consensual point of view of the world. Especially, if based on a common methodological ontology, e.g. DeMO, such applications could reduce heterogeneity of modeling to a minimum. Hence, composition of simulation models from components developed by different teams' seams realizable without tremendous adjustment efforts.

It is obvious, however, that *methodological* ontologies cannot solve challenges of pragmatic interoperability, because they do not deal with the purposes of the modeler with regard to content. Moreover, teleology, the philosophical study of purposeful human action, struggles with problems like infinite recursion and contradictions that humans can handle by flexible bypassing but formal logic cannot (Turnitsa et al. (2010), p. 648). Hence, it might be difficult if not impossible to capture many purposeful human actions in formal systems – including *referential* ontologies.

In addition, any reduction of the heterogeneity of conceptualizations requires unequivocal "natural" representations or consensus in case of equivocation. This problem is realized by (Benjamin et al. 2006), too. They state that *"the real problem is how to determine the presence of ambiguity and redundancy in the first place. That is, more generally, how is it possible to access the semantics of process-oriented data across different contexts? How is it possible to fix their semantics objectively in a way that permits accurate interpretation by agents outside the immediate context of this data?"* (p. 1156).

Personally, from the experience of 15 years in military modeling and simulation (M&S), we are convinced that there is no *easy* fix for these problems of interoperability using ontologies, for the linguistic reasons given above, and because the modeling of complex adaptive systems is a time-, context- and purpose-specific task that *precisely does not permit* accurate interpretation by human or software agents *outside the immediate context of the scenario.* With other words, pragmatic and conceptual interoperability are not hard to achieve because the right conceptualizations have not been standardized yet, but because there are no standard conceptualizations for many referents – even if they are viewed from the perspective of a given research question (purpose).

Nevertheless, creating ontologies might at least help to find common aspects of referents and purposes that defy easy composability and interoperability.

4.4.2   Ontology Driven Modeling

Miller et al. (2004 and 2008) and Silver et al. (2009) present a framework for ontology driven modeling based on the notion of mapping conceptual models and referential domain ontologies on the methodological Discrete-event Modeling Ontology DeMO (see Figure 8). The idea is that *"the knowledge expert in some particular field may construct a logical model of the relevant system. The user can then map this logical model to one or several formalisms in DeMO ontology (e.g. Stochastic Petri Nets, or Markov Chains) ... In addition, we can argue that the modeler essentially 'maps' her conceptual model on the portion of the ontology."* (Miller 2008 p.27).

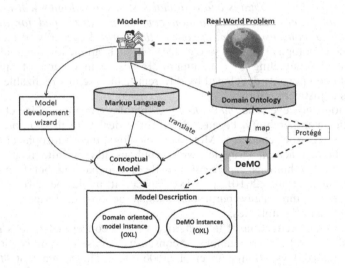

Fig. 8 Detail of ontology driven model development according to Miller et al. 2008

In ontology driven modeling, modelers would (ideally) design conceptual models of real world referents using only concepts defined in a fitting referential ontology (to which the modeler commits himself). Subsequently, they would formalize this unequivocal model using a methodological ontology assuring the exclusive use of well-defined methods. In principle, all models developed with the same referential and methodological ontologies should be interoperable on all levels of interoperability (or at least it should be easy to render them interoperable on all levels).

This approach is indeed promising *"if the knowledge domain is well defined (here this is equivalent to the existence of a good domain ontology)"* (Miller 2008 p. 29). We assume that "good" respectively "well defined" in this sentence means precise, complete, and unequivocal. For the majority of such knowledge domains methods of formalization already exist. A comprehensive methodological ontology can significantly reduce the time spent for finding the most appropriate formalism. A tremendous amount of queuing problems of all kinds, for example, should be perfectly transferable into one or several formalisms in DeMO.

However, in many socio-technical and most social domains the specification of such "well defined" domain ontologies (referential ontologies) will be impossible – for the reasons mentioned in the previous section. Hence, in these cases, there is no easy mapping possible between referential and methodological ontologies (in Figure 8 between the domain ontology and DeMO). This mapping, if possible at all (see next paragraph), would not be a technical matter, but a challenging and subjective task of selection.

In addition, simulation is a technique often applied on *ill-defined problems* (Rittle 1972). The solution to such problems depends on how the problem is framed and the problem definition depends on what is called a solution, since stakeholders have radically different world views and different frames for understanding and evaluating the problem. In order to create an appropriate model for such problems content and method of representation have to be closely interrelated and balanced. Thus, there is no independence between domain knowledge and methodological issues, no single "final match" between a conceptual model and a methodological ontology. The process of non-trivial conceptual modeling is a complicated entanglement of the two questions "what to model?" and "how to model it?" Hence, for such problems, model design will almost completely remain an art of system analysis, creative thinking and specific modeling.

In spite of these limits, ontologies can significantly reduce methodological uncertainty and domain ambiguity, not only in technical and natural scientific applications. The exact range of usefulness of referential ontologies is not determinable, a priori. In our studies for the German Armed Forces (Bundeswehr) we created and implemented referential ontologies for military applications (especially in the context of the "Effects Based Approach to Operation", EBAO). During one of these studies (example (No. 3)), an ontology has been developed that describes "Actors" within a modern military environment (see Figure 4, which is only a partial graphical depiction, also hiding the essential attributes of these actors). As long as this "ontological tree" was used for human expert discussions and regarded to be a dynamic provisional collection of things to be considered when dealing with a specific actor (often but not always reflecting a current state of agreement) all participants praised the utility of the ontology. The problems arose when we started to use it as a defining foundation of objects in a discrete event combat simulation system. In order to be used in a computer simulation one has to exactly define each simulation object (entity) in terms of attributes and operations affecting these attributes. Every event is such an operation on the attributes of entities. However, the experts of the Bundeswehr could not agree on a single authoritative conceptualizations (attributes and operations) of terms like, for example, *Dysfunctional Regular Military* or *Irregular Military*. We tried to circumvent this problem of uncertainty using the definitions of different experts in otherwise identical simulation runs (reflecting different ontologies). Unfortunately, combat simulation is (not very surprisingly) extremely sensitive to the conceptualization of actors. Hence, even minor changes in attributes and operations led to dramatic changes of simulation outcomes. Such high levels of sensitivity with respect to inevitable uncertainties inhibit the drawing of robust conclusions from simulation models. Only exploratory analysis is feasible. For exploratory analysis only, however, ontologies seemed be a lot of work – for all participants of the study.

Thus, the results of these efforts can be seen as successes and failures depending on the perspective. Most computer scientists regarded them as proofs of concept (that ontologies are useful) whereas most practitioners primarily saw a huge amount of work with little return of investment. Regardless of the personal attitude, it has become clear, that ontologies are extremely challenging in ill-defined domains.

Ontology driven modeling is an ambitious endeavor. Some practitioners might even see it as too ambitious: Salt (2008) has described seven habits of highly defective simulation projects, mainly with the focus on decision support projects. Without going into details, it seems to us that at least two of these defective habits lurk in the vicinity of ontology driven modeling: *Methodolatry* (If we only follow the method assiduously enough, everything will be all right) and the *Jehovah problem* (I can, as a perfectly accurate external observer, write a single functional decomposition of the entire system under study). With regard to methodolatry Salt states that: *"Given the essentially creative nature of decision-support modeling, the obvious mistake here is to imagine that there can be a fixed drill for thinking about new things. Related to this is perhaps the belief that the people who wrote the method are more knowledgeable than you are – which they might be in general, but almost certainly not about the situation you are in right now"* (Salt 2008 p 159). With regard to the Jehovah problem Salt writes that *"There is no such thing as perfectly accurate external observer...It is therefore likely that any system obvious enough for an observer to possess the whole truth about it is not a system interesting enough to be worth modeling. Nonetheless, people persist in believing that such accuracy is obtainable"* (Salt 2008, p. 160). Awareness of these (epistemologically) defective habits should help to avoid them when using ontologies.

Despite these critical remarks, ontology driven modeling might be efficient in surprisingly many domains. The crucial challenge for all of these domains, however, will be a domain specific balance between precision in specification and descriptiveness in representation.

4.5   Balancing the Epistemic and Normative Nature of Ontologies in Modeling and Simulation

The following itemization gives some ideas (not all of them are new) which should be considered when working with ontologies in the field of simulation model interoperability.

- *Methodological ontologies* should become *standards*. Modelers using a certain methodology defined by a methodological ontology must be encouraged to commit themselves to the ontological fixing. Such a commitment should be tolerable for most modelers since simulation methods, like the process view in discrete event simulation, for example, have so many degrees of freedom that modeling a referent within a special ontology is not a severe limitation.
- In order to compensate the restriction of fixed specifications in referential ontologies a kind of *sensitivity analysis* (Saltelli et al. 2008) should be compulsory. This "sensitivity analysis" would reflect the heterogeneity of opinions and world-views as well as other discrepancies. Keep in mind that, if the outcome of the simulation model is highly sensitive to conceptualizations, it is going to be hard to establish and justify a common ontology.
- The modifications necessary for the sensitivity analysis can, for example, be based on marks (*annotations*) put from the users of an ontology on concepts to which they do not completely agree; vice versa, each concept in the ontology should have such a disagreement counter. Thus, *ontology versioning systems* are obligatory (Klein and Fensel 2001 and Fensel 2004).
- The ontology should include further annotations, especially *links to successful applications*. The description of these applications has to contain, at least, the purpose of the modeling and simulation project, its scope and its results.
- Militaries all over the world have often been accused of preparing for the last war, not the next. Similar criticism could be and has been formulated for economics, politics in general and social welfare in particular. In such cases, ontologies cannot be limited to the already existing; they must allow the *integration of creative, new concepts* (of what might be).
- Automated information processing algorithms working on ontologies perform deductive reasoning on the current state of these ontologies. Any changes to the ontology might cause a different deduction, for pure syntactical reasons. A single new item might change, for example, the sequence of logical processing. Such *arbitrary effects* must be studied in detail within this context, as they are already well known from inference machines in AI.
- It might be advisable to renounce ontologies in some knowledge *domains that defy easy consensual representation*. A cultural ontology, for example, might easily become absurd. Social-science models often benefit from epistemological modesty (Davis 2009).

5   Summary and Conclusion

Most modeling and simulation (M&S) experts are not explicitly aware of their philosophical research assumptions and the ontological epistemological, and teleological implications thereof. However, in order to realistically assess the validity and value of a complex system computer simulation, a profound understanding of the philosophical triangle of knowledge is mandatory. Ontology is the study of what exists, in M&S often captured as a formal system defined by a finite set of concepts and their relations. Epistemology is the study of how we come to know, how we define knowledge, represent it, and communicate it with others. Various, significantly differing epistemological positions have been proposed for M&S. Teleology is the study of purposes, and actions and methods used to achieve these purposes. In that sense it gives modeling its direction, and it is obvious, that the setting of a purpose dominates any (simulation-based) analysis.

Together these branches of philosophy are indispensable to answer the crucial question: why and when can we rely on the recommendations generated via M&S. In the first section of this chapter a summary of landmark historical and recent academic papers (focusing on epistemology) has been given, trying to illustrate that this question has to be answered for each single simulation anew. However, getting familiar with the debate on M&S is a prerequisite for defending own models more profoundly.

In the second section of the chapter such philosophical considerations are used to show in detail that the two-edged nature of ontologies as both specifications and representations will limit the usefulness of so called referential ontologies (modeling real world domains), whereas no such problems exist with methodological ontologies (modeling methods).

In modeling and simulation ontologies can be used for the formal definition of methods and techniques (methodological ontologies) as well as for the representation of parts of reality (referential ontologies), like manufacturing or military systems, for example. Such ontologies are two-sided: they are both models of a certain body of knowledge and models for automated information processing and further implementation. The first function of ontologies as pre-images (models of) has a strong epistemic nature especially for referential ontologies since they try to capture pieces of the "semantic relations of the real world". The second function as models for further processing, in contrast, is completely normative in nature – it is a specification of a "formal semantics". Unfortunately, the ideal realization of ontologies as epistemic models differs from the normative ideal. As specifications, ontologies have to be as precise (unequivocal) as possible; as representations of reality, in contrast, they have to be as descriptive as possible, which may imply ambiguity and even inconsistency in some domains. Ontology processing is particularly challenging since balancing these ideals is a domain specific task.

Although a major part of scientific progress – especially in computer science (including computer based modeling and simulation) – is done ignoring philosophical foundations, finding and pushing the limits of any scientific branch seems unfeasible without knowing about the philosophical triangle of ontology, epistemology, and teleology.

References

Ahrweiler, P., Gilbert, N.: Café Nero: the evaluation of social simulation. Journal of Artificial Societies and Social Simulation 8(4) (2005)

Barlas, Y., Carpenter, S.: Philosophical roots of model validation: Two paradgms. System Dynamics Review 6, 148–166 (1990)

Batty, M., Torrens, P.M.: Modelling and prediction in a complex world. Futures 37, 745–766 (2005)

Beck, H., Morgan, K., Jung, Y., Grunwald, S., Kwon, H.Y., Wu, J.: Ontology-based simulation in agricultural systems modeling. Agricultural Systems 103(7), 463–477 (2010)

Bell, D., Mustafee, N., de Cesare, S., Taylor, S.J.E., Lycett, M., Fishwick, P.A.: Ontology engineering for simulation component reuse. International Journal of Enterprise Information Systems 4(4), 47–61 (2008)

Benjamin, P., Menzel, C., Mayer, R.J.: Towards a method for acquiring CIM ontologies. International Journal of Computer Integrated Manufacturing 8(3), 225–234 (1995)

Benjamin, P., Patki, M., Mayer, R.: Using ontologies for simulation modeling. In: Perrone, L.F., Wieland, F.P., Liu, J., Lawson, B.G., Nicol, D.M., Fujimoto, R.M. (eds.) Proceedings of the 2006 Winter Simulation Conference, Monterey, pp. 1151–1159. IEEE (2006)

Beven, K.: Towards a coherent philosophy for modelling the environment. Proceedings of the Royal Society London 458, 1–20 (2002)

Bharathy, G.K., Silverman, B.: Validating agent based social systems models. In: Proceedings of the 2010 Winter Simulation Conference, Piscataway, New Jersey, USA (2010)

Brenner, T., Werker, C.: Policy Advice Derived from Simulation Models. Journal of Artificial Societies and Social Simulation 12(4) (2009)

Byrne, D.: Simulation – A way forward? Sociological Research Online 2(2) (1997)

Carrier, M.: The completeness of scientific theories. Kluwer, Dordrecht (1994)

Chen, W., Hirschheim, R.: A paradigmatic and methodological examination of information systems research from 1991 to 2001. Information Systems Journal 14(3), 197–235 (2004)

Christley, S., Xiang, X., Madey, G.: Ontology for agent-based modeling and simulation. In: Macal, C.M., Sallach, D., North, M.J. (eds.) Proceedings of the Agent 2004 Conference on Social Dynamics: Interaction, Reflexivity and Emergence, co-sponsored by Argonne National Laboratory and The University of Chicago, October 7-9 (2004), http://www.agent2005.anl.gov/Agent2004.pdf (accessed March 29, 2011)

Churchman, C.W.: Reliability of Models in the Social Sciences. Interfaces 4(1) (1973)

Cuske, C., Dickopp, T., Seedorf, S.: JOntoRisk: an ontology –based platform for knowledge-based simulation modeling in financial risk management. In: Proceedings of the European Simulation and Modeling Conference, ESM, Porto, Portugal, October 24-26 (2005)

Davis, P.K.: Specifying the Content of Humble Social-Science Models, RAND-RP-1408-1 The Society for Modeling and Simulation International, San Diego, CA (2009), http://www.rand.org/pubs/reprints/RP1408-1.html (accessed March 29, 2011)

Dery, R., Landry, M., Banville, C.: Revisiting the issue of model validation in OR: An epistemological view. European Journal of Operational Research, Special Issue on Model Validation 66, 168–183 (1993)

Dessai, S., Hulme, M., Lempert, R., Pielke, R.: Climate prediction: a limit to adaption? In: Adger, W.N., Lorenzoni, I., O'Brian, K.L. (eds.) Adapting to Climate Change: Thresholds, Values, Governance. Cambridge University Press (2009)

Ezzell, Z., Fishwick, P.A., Lok, B., Pitkin, A., Lampotang, S.: An ontology-enabled user interface for simulation model construction and visualization. Journal of Simulation (February 25, 2011) (advance online publication), doi: 10.1057/jos.2011.5 (accessed March 31, 2011)

Fishwick, P.A., Miller, J.A.: Ontologies for Modeling and Simulation: Issues and Approaches. In: Ingalls, R.G., Rossetti, M.D., Smith, J.S., Peters, B.A. (eds.) Proceedings of the 2004 Winter Simulation Conference, pp. 259–264. IEEE, Washington (2004)

Fensel, D.: Ontologies: A silver bullet for knowledge management and electronic commerce. Springer, Berlin (2004)

Feyerabend, P.K.: Against Method: Outline of an Anarchistic Theory of Knowledge. Humanities Press, London (1975); (reprinted, Verso, London, UK 1978)

Fraassen, B.C.: The Scientific Image. Oxford University Press (1980)

Frank, U., Squazzoni, F., Troitzsch, K.G.: EPOS-Epistemological Perspectives on Simulation: An Introduction. In: Squazzoni, F. (ed.) EPOS 2006. LNCS, vol. 5466, pp. 1–11. Springer, Heidelberg (2009)

Frank, U., Troitzsch, K.G.: Epistemological Perspectives on Simulation. Journal of Artificial Societies and Social Simulation 8(4) (2005)

Frigg, R., Reiss, J.: The philosophy of simulation: hot new issue or same old stew? Synthese 169, 593–613 (2009)

Garson, D.: Computerized Simulation in the Social Sciences: A Survey and Evaluation. Simulation Gaming 40 (2009)

von Glasersfeld, E.: Radical constructivism: A way of knowing and learning. The Falmer Press (1995)

von Glasersfeld, E.: The construction of knowledge: Contributions to conceptual Semantics. Intersystem Puplications (1997)

Goldspink, C.: Methodological Implications of complex System Approaches to Sociality: Simulation as a foundation of knowledge. Journal of Artificial Societies and Social Simulation 5(1) (2002)

Gruber, T.: A Translation Approach to Portable Ontology Specifications. Knowledge Acquisition - Special Issue: Current Issues in Knowledge Modeling 5(2), 199–220 (1993)

Gruber, T.: Toward Principles for the Design of Ontologies Used for Knowledge Sharing. Int. Journal of Human-Computer Studies 43(5-6), 907–928 (1995)

Grüne-Yanoff, T.: The explanatory potential of artificial societies. Synthese 169 (2009)

Grüne-Yanoff, T., Weirich, P.: The philosophy and epistemology of simulation: A review. Simulation & Gaming 41(1) (2010)

Guizzardi, G., Wagner, G.: Using the Unified Foundational Ontology (UFO) as a Foundation for General Conceptual Modeling Languages. In: Poli, R. (ed.) Theory and Application of Ontologies. Springer, Heidelberg (2010a)

Guizzardi, G., Wagner, G.: Towards an Ontological Foundation of Discrete Event Simulation. In: Johanson, B., Jain, S., Montoya-Torres, J., Hugan, J., Yücesan, E. (eds.) Proceedings of the 2010 Winter Simulation Conference, pp. 652–664. IEEE (2010b)

Hermann, C.F.: Validation problems in games and simulations with special reference to models of international politics. Behavioral Science 12, 216–231 (1967)

Hesse, W.: Engineers discovering the "real world" – from model-driven to ontology based software engineering. In: Kaschek, R., Kop, C., Steinberger, C., Fliedl, G. (eds.) Proceedings of the 2nd International United Information Systems Conference on Information Systems and E-Business Technology, vol. 5(2), pp. 136–147. Springer (2008)

Hofmann, M.: Introducing Pragmatics into Verification, Validation and Accreditation. In: Proceedings of the 2nd European Simulation Interoperability Workshop, Harrows. GB (2002a)

Hofmann, M.: Validation: Real world system knowledge, types of validity and credibility levels. In: Proceedings of the 16 European Simulation Multiconference, Darmstadt, DE (2002b)

Hofmann, M.: Challenges of Model Interoperability in Military Simulations. Simulation 80(12), 659–667 (2004)

Hofmann, M.: On the Complexity of Parameter Calibration in Simulation Models. Journal of Defense Modeling and Simulation 2(4) (2005)

Hofmann, M., Hahn, H.: Is it appropriate to use the objective, rational decision-making framework as a foundation for the modelling of social systems? Information and Security 22 (2007)

Hofmann, M., Palii, J., Mihelcic, G.: Epistemic and normative aspects of ontologies in modeling and simulation. Journal of Simulation 5(3), 135–146 (2011)

Hemez, F.M.: The myth of science-based predictive modeling. In: Foundations 2004 Workshop for Verification, Validation, and Accreditation in the 21st Century. Arizona State University, Tempe (2004)

Humphreys, P.: The philosophical novelty of computer simulation methods. Synthese 169 (2009)

Islam, A.S., Piasecki, M.: Ontology based web simulation system for hydrodynamic modeling. Simulation Modeling Practice and Theory 16(7), 754–767 (2008)

Janich, P.: Wozu Ontologie für Informatiker? Objektbezug durch Sprachkritik. In: Bauknecht, K., et al. (eds.) Informatik 2001 - Tagungsband der GI/OCG-Jahrestagung, Bd. II, pp. 765–769 (2001)

Kiko, K., Atkinson, C.: A Detailed Comparison of UML and OWL, University of Mannheim, Germany (2008), http://madoc.bib.uni-mannheim.de/madoc/volltexte/2008/1898/pdf/TR2008_004.pdf (accessed March 28, 2011)

Klein, M., Fensel, F.: Ontology versioning on the Semantic Web. In: Proceedings of the International Semantic Web Working Symposium, SWWS. Stanford University (2001)

Klein, E.E., Herskovitz, P.J.: Philosophical Foundations of Computer Simulation Validation. Simulation & Gaming 36(3), 303–329 (2005)

Kleindorfer, G.B., O'Neill, L., Ganeshan, R.: Validation in simulation: Various positions in the philosophy of science. Management Science 44(8) (1998)

Konikov, L.F., Bredehoeft, J.D.: Groundwater models cannot be validated. Adv. Water Resour. 15(75-83) (1992)

Küppers, G., Lenhard, J.: Validation of simulation: Pattern in the social and natural sciences. Journal of Artificial Societies and Social Simulation 8(4) (2005)

Lacy, L.W.: Interchanging discrete event simulation process-interaction models using the web ontology language – OWL, Dissertation. University of Central Florida (2006), http://etd.fcla.edu/CF/CFE0001353/Lacy_Lee_W_200612_PhD.pdf (accessed March 30, 2011)

Lacy, L.W., Gerber, W.J.: Potential Modeling and Simulation Applications of the Web Ontology Language - OWL. In: Proceedings of the 2004 Winter Simulation Conference. IEEE, Piscataway (2004)

Livet, P., Müller, J.P., Phan, D., Sanders, L.: Ontology, a Mediator for Agent-Based Modeling in Social Science. Journal of Artificial Societies and Social Simulation 13(1), 3 (2010), http://jasss.soc.surrey.ac.uk/13/1/3.html (accessed March 29, 2011)

Luhmann, N.: Soziologie des Risikos, de Gruyter (1991)

Miller, J.A., Baramidze, G.T., Sheth, A.P., Silver, G.A., Fishwick, P.A.: Investigating ontologies for Simulation Modeling. In: Proceedings of the 37th Annual Simulation Symposium, pp. 55–63 (2004)

Miller, J.A., Baramidze, G.T., Sheth, A.P., Silver, G.A., Fishwick, P.A.: Ontologies for Modeling and Simulation: An Initial Framework (2008), http://www.cs.uga.edu/~jam/jsim/DeMO/paper/journal/final4/demojournal_B_1_7.pdf (accessed March 27, 2011)

Miser, H.J.: A foundational concept of science appropriate for validation in operational research. European Journal of Operational Research, Special Issue on Model Validation 66, 204–215 (1993)

Moss, S.: Alternative Approaches to the Empirical Validation of Agent-Based Models. Journal of Artificial Societies and Social Simulation 11(15) (2008)

Naylor, T.H., Finger, J.M.: Verification of computer simulation models. Management Science 14(2) (1967)

Neumann, M., Braun, A., Heinke, E.M., Saqalli, M., Srbljinovic, A.: Challenges in Modelling Social Conflicts: Grappling with Polysemy. Journal of Artificial Societies and Social Simulation 14(3) (2011)

Niehaves, B., Becker, J., Klose, K.: A framework for epistemological persperctives on simulation. Journal of Artificial Societies and Social Simulation 8(4) (2005)

Oreskes, N., Shrader-Frechette, K., Belitz, K.: Verification, validation, and confirmation of numerical models in the earth sciences. Science 263(4), 641–646 (1994)

Phan, D., Varenne, F.: Agent-Based Models and Simulations in Economics and Social Scienes. Journal of Artificial Societies and Social Simulation 13(1), 5 (2010)

Pinker, S.: The stuff of thought: Language as a window into human nature. Viking (2007)

Quine, W.O.: Word and Object. MIT Press (1960)

Quine, W.O.: Ontological Relativity. Columbia University Press (1977)

Richardson, K.A.: On the limits of bottom-up computer simulation: Towards a nonlinear modeling culture. In: Proceedings of the 36th Hawaii International Conference on System Science (2003)

Rittle, A.: On the planning crisis: System analysis of the first and second generation. Bedriftsokonomen 8, 390–396 (1972)

Rossiter, S., Noble, J., Bell, K.R.W.: Social Simulations: Improving Interdisciplinary Understanding of Scientific Positioning and Validity. Journal of Artificial Societies and Social Simulation 13(1) (2010)

Ruphy, S.: Limits to Modeling: Balancing Ambition and Outcome in Astrophysics and Cosmology. Simulation Gaming 42(2), 177–194 (2011)

Salt, J.D.: The seven habits of highly defective simulation projects. Journal of Simulation 2(3), 155–161 (2008)

Saltelli, A., Ratto, M., Andres, T., Campolongo, F., Cariboni, J., Gatelli, D., Saisana, S., Tarantola, S.S.: Global Sensitivity Analysis. The Primer. John Wiley and Sons, New Jersey (2008)

Silver, G.A., Lacy, L.W., Miller, J.A.: Ontology based representations of simulation models following the process interaction world view. In: Perrone, L.F., Wieland, F.P., Liu, J., Lawson, B.G., Nicol, D.M., Fujimoto, R.M. (eds.) Proceedings of the 2006 Winter Simulation Conference, Monterey, pp. 1168–1176. IEEE (2006)

Silver, G.A., Bellipady, K.R., Miller, J.A., York, W.S., Kochut, K.J.: Supporting Interoperability Using the Discrete-Event Modeling Ontology (DeMO). In: Rossetti, M.D., Hill, R.R., Johansson, B., Dunkin, A., Ingalls, R.G. (eds.) Proceedings of the 2009 Winter Simulation Conference, Piscataway, New Jersey, pp. 1399–1410. IEEE (2009)

Simon, H.: Models of Bounded Rationality, vol. 3. MIT Press (1997)

Simpson, J.: Identity Crisis: Simulations and Models. Simulation & Gaming 42(195) (2011)

Stanislaw, H.: Tests of computer simulation validity: What do they measure? Simulation & Games 17 (1986)

Taylor, S.J.E., Bell, D., Mustafee, N., de Cesare, S., Lycett, M., Fishwick, P.: Semantic web services for simulation component reuse and interoperability: An ontology approach. In: Gunasekaran, A., Shea, T. (eds.) Organizational Advancements through Enterprise Information Systems: Emerging Applications and Developments. IGI Gobal, Inc., Hershey (2010)

Tolk, A., Huiskamp, W., Schaub, H., Davis, P.K., Klein, G.L., Wall, J.A.: Towards methodological approaches to meet the challenges of human, social, cultural, and behavioral (HSCB) modeling. In: Proc. 2010 Winter Simulation Conference, pp. 912–924 (2010)

Tolk, A., Jain, L.C.: Intelligent-based system engineering. ISRL, vol. 10. Springer (2011)

Turnitsa, C., Padilla, J.J., Tolk, A.: Ontology for Modeling and Simulation. In: Johanson, B., Jain, S., Montoya-Torres, J., Hugan, J., Yücesan, E. (eds.) Proceedings of the 2010 Winter Simulation Conference, pp. 643–651. IEEE (2010)

Vangheluwe, H.L.M.: DEVS as a Common Denominator for Multi-formalism Hybrid Systems Modelling. In: Proceedings of the 2000 IEEE International Symposium on Computer-Aided Control System Design, Anchorage, Alaska, USA, September 25-27 (2000)

van Horn, R.L.: Validation of Simulation Results. Management Science 17(5), 247–258 (1971)

Varenne, F.: What does a computer simulation prove? Simulation in Industry. In: Proceedings of the 13th European Simulation Symposium, vol. 13, pp. 549–554 (2001)

Wang, W.G., Tolk, A., Wang, W.P.: The levels of conceptual interoperability model: Applying systems engineering principles to M&S. In: Proceedings of the Spring Simulation Multiconference, SpringSim 2009, San Diego (2009)

Weber, M.: The methodology of the social sciences (translated and edited by Shils, E.A., Finch, H.A.), New York (1949)

Windrum, P., Fagiolo, G., Moneta, A.: Empirical validation of agent-based models: Alternatives and prospects. Journal of Artificial Societies & Social Simulation 10(2) (2008)

Wittgenstein, L.: Logisch-philosophische Abhandlung, Tractatus logico-philosophicus. Suhrkamp, Frankfurt am Main (1921/1998)

Wittgenstein, L.: Philosophical Investigations. Blackwell Publishing (1953/2001)

Zeigler, B.P., Praehofer, B., Kim, T.G.: Theory of Modelling and Simulation: Integrating Discrete Event and Continuous Complex Dynamic Systems, 2nd edn. Academic Press (2000)

Ontological Implications of Modeling and Simulation in Postmodernity

Brian L. Heath[1] and Ross A. Jackson[2]

[1] Wright State University
Dayton, OH, United States
[2] University of Phoenix
Dayton, OH, United States

Abstract. Models and simulations are immediately obfuscated by being what they are, abstract representations of reality. With reductionist parameters and defined algorithms, models and simulations obtain a definitiveness lacking in the reality they explain. Increased computational power has enabled the production of complex representations. This increased complexity makes understanding what is happening "behind the scenes" almost entirely unintelligible to consumers. At the same time, advancements in Animation enable practitioners to present the results at almost movie-like levels of production. This subtly transforms the ontological status of the results, making them appear as something one should view rather than something about which one should think. What happens when producers and consumers of models and simulations lose the self-certainty associated with their project? Such a situation calls into question the performative aspects of both groups' maneuvers. We situate the locus of this discussion around the notion of validity. Once considered essential, the quest for validity perhaps increasingly reveals a form of existential absurdity and, in a nihilistic twist of postmodern thought, the radical devaluation of one of the ideals of the philosophy of science.

1 Introduction

Modernity's hold on Modeling and Simulation (M&S) is both extensive and oppressive. Suggestively, practitioners and theorists situate the majority of literature on and practice of M&S within the modern scientific paradigm (Heath & Hill, 2009). Primarily, this means that those familiar with M&S view it as a numerical technique that is decidedly not reality but can none-the-less provide insight into reality through rigorous application of the scientific method. One is able to observe evidence of the popularity of the scientific paradigm in M&S in many different forms. Any textbook or basic literature on M&S will describe the process of M&S that very closely mimics the basic tenets of the scientific method. Many M&S philosophical works (including this book) closely follow and extend fundamental concepts found in the philosophy of science such as Falsificationism and Empiricism. Pragmatically, one can likely attribute the success of M&S to the

A. Tolk (Ed.): Ontology, Epistemology, & Teleology for Model. & Simulation, ISRL 44, pp. 89–103.
springerlink.com   © Springer-Verlag Berlin Heidelberg 2013

pragmatic success of the scientific paradigm and the close following of simulationists to this paradigm.

As a result of the prominence of the scientific paradigm in M&S, a rich rhetorical and philosophical foundation has been created that researchers and practitioners use to continually develop M&S. However, by primarily focusing on scientific philosophy and rhetoric, the foundation of M&S, and its application, has glossed over other philosophical paradigms that can provide key insights into the field, its application, and its implications. In fact, many of these paradigms may be better suited to explain behaviors or phenomena that the scientific paradigm simply struggles to explain consistently or meaningfully. Furthermore, understanding other paradigms is all the more critical when developing philosophical foundations for intelligent M&S application. Since creating intelligent M&S systems further blurs the artificial, understandings of reality, and the competing scientific structures of our world. Simply put, the scientific paradigm is not the only, or even a complete and internally consistent, way of understanding M&S.

In this chapter, we depart from the traditional scientific paradigm and explore M&S within a postmodern paradigm. From this postmodern perspective, we develop some ontological implications for developing and consuming intelligent M&S applications as they relate to the foundational gaps created by viewing M&S from a strictly scientific paradigm. In the next section, we describe postmodernity, its view of reality, and the ambiguity of meaning within the scientific paradigm. Next, we describe the implications of a postmodern paradigm for M&S by examining an existential ontology and the human influence on M&S. Finally, we work toward a postmodern rhetoric of M&S as it relates to the practical application of M&S today.

2 Postmodernity

Situating M&S in postmodernity necessitates at least a cursory development as to what we mean in this particular context by the term *postmodern*. This concept is in some respects "tortured" (Jackson, 2011a) through the multiple attempts to define it. The contours of the concept provided here are not meant to contort postmodernity into a homogeneous body of knowledge, but rather indicate how the concept might be understood and applied in some of its heterogeneous potential. One is able to observe this heterogeneity in the varied applications of postmodern thought to topics as diverse as rhetoric (Brummett, 1976; Gozzi, 1993), leadership and strategy (Grandy & Mills, 2004; Schreiber & Carley, 2006), bureaucracy (Hummel, 2008), historical reenactment (Radtchenko, 2006), and hydroinformatics (Abbott & Vojinovic, 2009). Unifying these studies is a view that situating these phenomena in postmodernity reveals inherent paradoxes. Confronting such paradoxes requires one to interrogate one's presuppositions as to what is *real*. Far from being an academic pursuit, postmodernism deals with the very nature of reality, our understanding(s) of it, and resulting consequences. This is relevant to M&S as it blurs and obfuscates the lines between the real and the represented.

The works of Lyotard (1984) and Baudrillard (1994) provide complementary perspectives on postmodernity. In the view of Abbott and Vojinovic (2006), they are "pioneers in postmodernism" (p. 308). The views developed by Lyotard and Baudrillard are augmented by Debord's (2005) notion of the "spectacle" to understand the "unrealism of the real society" (section 6). Debord's work is in some respects a precursor of postmodernism. The orientations explored in these texts are extended and critiqued by supplementary works, most notably that of Doyle (1992), to provide a slightly more nuanced and poly-vocal perspective on postmodernity. Once formed, this groundwork provides a basis to integrate M&S with postmodernity, thus moving toward a postmodern rhetoric of M&S. Through this extension, we reveal ontological implications.

Lyotard's (1984) work is regarded as "the text that inaugurated the discussion of postmodernity" (Doyle, 1992, p. 113). Given this position of prominence, it is important to address both Lyotard's own construction of postmodernity, and provide some views of his critics. Perhaps Lyotard's most accessible definition of postmodernity is "incredulity toward metanarratives" (p. xxiv). This definition lacks the type of understanding derived from a more rigorous comparison between competing forms of knowledge. Lyotard provided such a comparison by positioning postmodern knowledge vis-à-vis scientific knowledge. This comparison is essential, as scientific knowledge forms the core of what one might call modern knowledge. Through this comparison, one gains greater insight in to the "incredulity" postmodernists hold toward metanarratives.

For Lyotard (1984), "scientific knowledge does not represent the totality of knowledge;" it exists "in competition with, another kind of knowledge," which he called "narrative" (p. 7). This narrative knowledge is, according to Lyotard, "the quintessential form of customary knowledge" (p. 19). One such narrative deals with the objectivity of scientific knowledge. Lyotard noted, "Scientific knowledge cannot know and make known that it is the true knowledge without resorting to the other, narrative, kind of knowledge" (p. 29). Citing Giddens, Grandy and Mills (2004) contended that postmodernism "challenges the foundations of knowledge and the myth that history reflects constant progress" (p. 1154). This is consequential for the scientifically trained.

Given the computational and coding requirements associated with M&S, it is understandable that those engaged in this activity receive their training in technical fields. One might situate the intellectual pedigree of this within the modern, scientific paradigm. In some respects, M&S occurs within a Kuhnian (1996) paradigm. While the modern, scientific paradigm is pragmatically useful, it potentially produces the very cracks in which the postmodern perspective takes root. Lyotard (1984) explained that, "science, far from successfully obscuring the problem of its legitimacy, cannot avoid raising it with all of its implications, which are no less sociopolitical than epistemological" (p. 18), and we are in "a process of delegitimation fueled by the demand of legitimation itself" (p. 39). In some respects, the Kuhnian (1996) paradigm of modern science is a closed, hermeneutic cycle of self-reference. Debord's (2005) "spectacle" understood as the "existing order's uninterrupted discourse about itself" engaged in a peculiar "laudatory monologue" (section 24) about itself, to itself, is applicable here. This view of science is provocative, but not above critique. Lyotard's critics took aim at how he substantiated his claims.

Doyle (1992) explained that many view Lyotard's work as "almost universally, as a seriously flawed work" (p. 114). Doyle's first critique pointed out Lyotard's lack of evidence to support his claims. Admittedly, this is a significant limitation to the work, but one that is predominately within modern notions of the requirements of legitimation. While certainly limiting, one is able to move around this critique through a postmodern turn. It is Doyle's second critique that more directly hits at the core of postmodernity. Doyle explained, "while claiming to be anti-totalizing himself, Lyotard offers his own totalized account of the postmodern condition, his own metanarrative" (p. 115). While it is accurate to make this claim against Lyotard, this limitation is perhaps better understood as being part of postmodernity. Gozzi (1993) explained that it was characteristically postmodern that "in denying something, you must affirm it" (p. 377). Again, one is unable to travel along postmodernity without quickly encountering paradox.

Since models and simulations are often constructed within institutions, for decision-making, it is useful to develop an appreciation for how postmodernists critique the performative quality of maneuvers in this context. Lyotard's (1984) insight draws an important distinction between the self-concepts of analysts, and the actual function they provide in institutions. Lyotard explained, "Scientists, technicians, and instruments are purchased not to find truth, but to augment power" (p. 46). This view is shared in part by Foucault (1980), and is consistent with his notion of "power/knowledge." But what is not altogether clear is what is actually happening through the dialogue between analysts and decision-makers, and how these scripts are created, enacted, and consumed. Within the "spectacle," Debord (2005) explained, "the real consumer becomes a consumer of illusions" (section 47). As indicated earlier, models and simulations are immediately obfuscated by being so obviously what they are, abstract representations of reality. Following Baudrillard's (1994) notion of simulacra, it might be more informative to refer to them as images bearing no relation to reality. This disjointed dialogue between analysts and decision-makers is further constrained by normative expectations as to how these parties should engage in discourse. Lyotard explained:

> An institution differs from a conversation in that it always requires supplementary constraints for statements to be declared admissible within its bounds. The constraints function to filter discursive potentials, interrupting possible connections in the communication networks...there are things that should be said, and there are ways of saying them. (p. 17)

Postmodern understandings produce a gap in which analysts can more effectively maneuver around these institutional scripts.

What is perhaps more interesting is that those engaged in the technical aspects of production (i.e., analysts) and those engaged in consumption (i.e., decision-makers) have different interests, but possibly come to tacitly agree on concealing the implications of operating in postmodernity (Jackson, 2011a). Gozzi (1993) explained that those that benefit from the *status quo* operate just as they would under modern conventions "to preserve their privileged positions while in reality the basis for their elitism had disappeared" (p. 376). Behind these potentially pragmatic performances lurk deeper ontological implications of individual and collective significance. We more fully develop these in the ontological implications section of this chapter. Gozzi's work, while informative about postmodernity in general, was specifically written in response to Baudrillard's notion of the collapse of metaphor. This provides a useful transition point to move from Lyotard's postmodernity to that of Baudrillard.

In employing the work of Baudrillard (1994), it is important to note that Baudrillard's discourse pivots around notions like the "hyperreal" and "simulacra," and not specifically postmodernity. Doyle (1992) noted that to his knowledge, Baudrillard "has never used the term 'postmodern'" (p. 117). Even so, one may consider Baudrillard's writings as part of the postmodern perspective even if Baudrillard did not.

Baudrillard's (1994) notion of the "precession of simulacra" is essential for both an understanding of what is meant here by postmodernity, and later in this chapter what will be offered as a postmodern rhetoric of M&S. In describing the current situation, Baudrillard explained, "today abstraction is no longer that of the map, the double, the mirror, or the concept. Simulation is no longer that of a territory, a referential being, or a substance. It is the generation by models of a real without origin or reality: a hyperreal" (p. 1). Jackson and Dafler (2008) applied a similar construct to explain how cost estimates shape the outcomes estimates are rhetorically attempting to forecast. Similarly, general semanticists make frequent use of map metaphors (Bois, 1975; Hayakawa, 1964; Korzybski, 2000) to distinguish between representations (i.e., maps) and reality (i.e., territory). We further unravel these intertwined threads in the postmodern rhetoric section of this chapter. Looking at the notion of simulation, it is important to understand how Baudrillard (1994) constructed this pivotal concept in his work, when he noted:

> To dissimulate is to pretend not to have what one has. To simulate is to feign to have what one doesn't have. One implies a presence, the other an absence…pretending, or dissimulating, leaves the principle of reality intact…simulation threatens the difference between…the "real" and the "imaginary." (p. 3)

Through simulations, the reality principle is immediately and surreptitiously undermined. However, this occurs largely behind the scenes. In such postmodern movements, a notion of reality collapses in upon itself. Baudrillard (1994) explained this pivotal turning point as the "transition from signs that dissimulate something to signs that dissimulate that there is nothing" (p. 6). This occurrence not only deconstructs the notion of reality, but also the very notion of illusion. For Baudrillard, "illusion is no longer possible, because the real is no longer possible" (p. 19). Because they operate beyond the real, what becomes essential are the meanings humans create and invest in their actions, observations, and communications. Frankl (1992) and Ford (2008) positioned the search for meaning as a central part of human activity. While important, postmodernists tend to consider these meanings as problematic, ambiguous, and contestable.

In examining meaning in postmodernity, Baudrillard (1994) explained that "we think that information produces meaning, the opposite occurs. Information devours its own content. It devours communication and the social" (p. 80). This is increasingly problematic as with advances in information and communication technology, we inhabit "a world where there is more and more information, and less and less meaning" (Baudrillard, p. 79). We develop the existential-ontological implications of this situation in the next section of this chapter. For now, one might ask: What are the consequences of such a notion of postmodernity? For Baudrillard, "the hyperreality of communication and of meaning" produces a situation that is "more real than the real," one where "the real is abolished" (p. 81). It is here that the thoughts of Baudrillard and Lyotard (1984) coincide.

The "incredulity toward metanarratives" produces, by extension, a postmodern focus on localized narratives. Along with paradox, this localized perspective is part of the contours developed here to contextualize postmodernity. Baudrillard (1994) explained, "meaning, truth, the real cannot appear except locally, in a restricted horizon, they are partial objects, partial effects of the mirror and of equivalence" (p. 80). This holds critical implications for the notion of validity developed later in this chapter. For postmodernists this gap is essential and consequential. The closing of such gaps forecloses important spaces for individual and organizational "sensemaking" (Weick, 1979, 1995). Baudrillard asked, "What happens when this distance, including that between the real and the imaginary, tends to abolish itself, to be reabsorbed on behalf of the model?" (p. 121). Gozzi's (1993) response suggests, "in this age of simulation, the distance required by metaphor is closed. For metaphor is a bridge between different domains, asserting a structural similarity between areas normally thought to be separate" (p. 376). The importance of such a stance is not immediately clear, as metaphor is often viewed as not much more than a rhetorical flourish. However, Lakoff and Johnson (2003) developed a construct of human understanding in which metaphors occupy a privileged position.

While not defining the term postmodern, concepts emerging within a postmodern perspective were developed. Among the most central concepts developed were paradox, the ambiguity of meaning, and localized narratives. Provisional links between these concepts and M&S were developed. Through these links and implications, we started to articulate the need for a postmodern rhetoric of M&S. While certainly understandable, it is perhaps the case that those involved in M&S give too much focus on the rudiments of technique, neglecting the consequential exploration of how members of institutions consume, and talk about, models and simulations as part of the decision-making process. We explore these notions, and some of the consequences, in the following section.

3   Ontological Implications

Consistent with our view of postmodernity developed in the previous section, there is a certain ambiguity inherent in employing an ontological perspective to

M&S. If expanded with requisite tension, these complementary views of ontology hold the potential for greater insight. Further, such a focus could lead to greater autonomy and authenticity for both producers and consumers of models and simulations. Ontological perspectives have been applied to an increasingly wide spectrum of topics, to include biomedicine (Rzhetsky & Evans, 2011), occupational therapy (Wilcock, 1999), politics (Petković, 2010), public administration (Raadschelders, 2011), leadership (Jackson, 2008), news media (Ostertag, 2010), artifacts (Kassel, 2010), knowledge management (Bera, Burton-Jones, & Wand, 2011), and M&S (Turnitsa & Tolk, 2006; Turnitsa, Padilla, & Tolk, 2010). These last two works are especially useful here as they provide essential insights into the primary concern of ontological implications of M&S, and provide our starting point for further ontological exploration in this area.

Turnitsa and Tolk (2006) presented a clear rationale for the application of ontology in M&S by noting that ontology produces a representation of reality capable of being processed by computers. It is important to note, such a processing is not merely an algorithmic distillation of an external reality; rather it simultaneously reveals one's internal representation of reality and generates, through an act of social construction (Berger & Luckmann, 1966), an emergent external reality. Turnitsa and Tolk explained, "In the field of simulations, the synthetic environment of simulation has reality *only within* the system" (section 1). Such a perspective is consistent with our reading of Baudrillard (1994) presented in the previous section. Turnitsa and Tolk's "within the system" is not entirely unproblematic. This "within the system" *itself* can take on different meanings depending on one's level of abstraction (Bois, 1975; Hayakawa, 1964; Korzybski, 2000), ranging from a micro-level focus on the model, through a meso-level focus on institutional/organizational dynamics, to a macro-level focus on societal influences. Ontological understandings hinge on subjective selections of focus.

One's aperture of focus ultimately shapes and constrains understandings. In regards to the application of ontology, a narrow construction could limit the focus to methods like Web Ontology Language and Resource Description Framework. Turnitsa and Tolk (2006) consider such a reductionist perspective a "great mistake" (section 2). To prevent such a thin understanding, they employed the ontological spectrum as a means to widen the focus to different types of ontologies available in M&S. In a subsequent work, Turnitsa, Padilla, and Tolk (2010) used a construction of ontology similar to that of Raubal and Kuhn (2004). Such a construction distinguished between an ontology "of philosophy dealing with the appreciation of existence," which we call here existential ontology, and an ontology "of system representation that deals with codifying the contents and knowledge of a system" (Turnitsa, Padilla, & Tolk, p. 644), which we refer to as system ontologies. Understandably, Turnitsa, Padilla, and Tolk focused attention on these system ontologies, as they are more directly applicable to M&S.

While there is a certain logic tilting the primacy of focus toward the type of system ontologies analyzed by Turnitsa and Tolk (2006) and Turnitsa, Padilla, and Tolk (2010), it does not follow that one should neglect the other, existential ontology in M&S. This philosophical notion of ontology provides insight into the possible performative aspects of M&S, and the potential existential impacts on both producers and consumers of models and simulations. The system ontologies are likely more critical in the development of M&S technique and praxis; the existential ontology is likely more essential for understanding the potential human consequences of such an endeavor. To address a small part of this expanded focus, we developed a brief sketch of existential ontology as articulated by Heidegger (1962) and Sartre (1960, 2007).

As already indicated, the system ontologies are of primary concern from an M&S perspective. However, it is possible that even with such a concession one will consider the existential ontology ultimately more consequential. Heidegger (1962) explained:

> Basically, all ontology, no matter how rich and firmly compacted a system of categories it has at its disposal, remains blind and perverted from its ownmost aim, if it has not first adequately clarified the meaning of Being, and conceived this clarification as its fundamental task. (p. 31)

As such, even if there is a pressing, pragmatic concern, an exigency of sorts, necessitating an initial focus on system ontologies, such an approach will be partially inadequate until it clarifies the meaning of *Being*. This places us squarely in the realm of existential philosophy. Heidegger variously described ontological investigation as a "kind of interpreting" as a "working-out" and "appropriation of an understanding," in which the interpretation "has its fore-having, its fore-sight, and its fore-conception" (p. 275). This focus on types of foreknowledge importantly, and correctly, places the ontological investigation within an approach consistent with the semiotic triangle applied to M&S by Turnitsa, Padilla, and Tolk (2010). This understanding leads to a concept much more central to existentialism, a concept that forms the core of Heidegger's ontology. Heidegger explained "*fundamental ontology*, from which alone all other ontologies can take their rise, must be sought in the *existential analytic of Dasein*" (p. 34). While it is possible to translate *Dasein* as something akin to "There-Being," such a translation, while close in a literal sense, inadequately captures the fullness of this deep philosophical concept. Jackson (2008) described Dasein as the type of Being that is "present in one's situation" (p. 26). In a view consistent with both postmodernism and important aspects of Dasein, Turnitsa, Padilla, and Tolk (2010) developed a significant implication to M&S when they noted, "one referent can be represented…as a near infinite number of different models, and how each model could be represented by a near infinite number of simulations" (p. 645). One is able to derive further insights into the existential-ontological implications of M&S in postmodernity by turning to Sartre.

Adequately dealing with existential ontology requires one to distinguish between action and the conscious awareness of one's engagement in some sort of action. Sartre (1960) explained, "The consciousness which says *I Think* is precisely not the consciousness which thinks. Or rather it is not its *own* thought which posits by this thetic act" (p. 45). More directly, as we embark on the production and consumption of models and simulations, we are selecting a way of *being-in-the-world* along with some perceived notion of the value of such a type of being. In other words, "choosing to be this or that is to affirm at the same time the value of what we choose" (Sartre, 2007, p. 24). Without a close interrogation of this dynamic, it is possible that those engaged in M&S not only conform to dominant standards of technique and praxis, but also to implied expectations as to *who* they should be while engaged in the activity.

In an institutional sense, there are further implications associated with enacting models and simulations as part of a decision-making process. For Sartre (2007), "to choose one's advisor is only another way to commit oneself" (p. 33). When a decision-maker directs an analyst to create a simulation, with the intent to inform the decision-making process, that leader has always already made a commitment to the types of decisions to be made under such an approach. This not only directs action (e.g., create a model), but also assigns roles (e.g., be a subordinated technocrat). This nexus of producer and consumer could form an essential part of one's understanding of *self*. Sartre (2007) explained "I cannot discover any truth whatsoever about myself except through the mediation of another. The other is essential to my existence, as well as to the knowledge I have of myself" (p. 41). Kearney (2003) further analyzed the essential role of the *other* in developing an understanding self. Even when engaged in potentially serious and consequential endeavors, this dynamic can take on a performative quality in which the participants subtly conform to roles, which appear to be nowhere explicitly scripted but everywhere implicitly understood. Jackson (2011b) applied such a performance art motif to describe defense acquisition reform. This existential-ontological perspective brings much needed attention to neglected aspects of M&S.

It is perhaps easier to understand the existential-ontological implications resident in M&S, if one first examines how a change in ontological status potentially occurs for objects within a simulated environment. Turnitsa and Tolk (2006) explained through their illustrative example of a truck shifting from a vehicle to an obstacle that, "a simulated environment is a dynamic world. As the entities and processes within that environment exist and interact with each other, the nature of ontological meaning shifts" (section 5.2). This ontological shift occurs "within the system" at the micro-level. Another, more existential shift, potentially occurs at the meso-level, "within the system" of the institution. At this level, it is not just the ontological status of objects that is potentially transformed, but the very nature and understanding of *self*.

Having extended the ontological perspective used in M&S to include its more traditional existential component (Rzhetsky & Evans, 2011) it is now possible to reassess some implications for M&S associated with system ontologies as viewed through an existential-ontological lens. Turnitsa, Padilla, and Tolk (2010) noted that the notion of "a shared conceptualization assumes a common frame of reference, or lens, which can vary depending on the modeler" (p. 643). Further, Turnitsa, Padilla, and Tolk explained, "In order to be able to model, we furthermore need to know why we model (modelers intent), which is defined by requirements derived from the experimental frame or the context of the model," and the "reality" captured in models and simulations is "contingent on the modeler and on a research question" (p. 644). Nevertheless, the intent of the modeler could very well be an amalgam of conflicting, and only partially acknowledged, desires. This is perhaps even more pressing when a senior leader directs an analyst to develop a model or simulation within an institutional context. The work of Foucault (1980) is particularly useful in understanding sublimated, power dynamics within institutions.

Using the phrase modeler's intent perhaps conveys a singular, unified, goal which is too reductionist to be useful. A modeler could hold many, potentially conflicting, intentions to be derived through the creation of a given simulation. A partial list could include, fulfill the direction of a senior leader to create a simulation, demonstrate one's worth to the organization, express one's creativity and intelligence, break out from under institutional subjugation through technical acumen, etc. In a similar fashion, the senior leader directing the action could hold conflicting ambitions as well. A senior leader might be trying to, inform the decision-making process, impress upon others one's up-to-date approach to decision-making, create a dynamic in which one is able to reiterate the subordinated relationship of the analyst to the decision-maker, etc. One might too easily miss such a multitude of intents when a singular modeler's intent is referenced. Foucault's (1980) notion of power/knowledge is particularly useful here. Foucault explained, "'truth' is to be understood as a system of ordered procedures for the production, regulation, distribution, circulation and operation of statements. 'Truth' is linked in a circular relation with systems of power which produce and sustain it" (p. 133). Understood from this position, both the producers and consumers of models and simulations become inextricably linked in a highly specialized form of power contestation regarding how models and simulations are to be used and understood within an institutional context.

Using the ontological approach articulated by Turnitsa and Tolk (2006) and Turnitsa, Padilla, and Tolk (2010) as a foundation, we extended the ontological perspective into the existentialism of Heidegger (1962) and Sartre (1960, 2007). From this, we focused attention on existential implications of M&S in postmodernity. Revealingly, it is possible that how one enacts M&S in an institutional context provokes a situation in which the ontological status of decision-makers are subtly transformed into consumers. With this in mind, it is now possible to move toward a postmodern rhetoric of modeling and simulation. We advance this movement in the following section.

4   Toward a Postmodern Rhetoric of Modeling and Simulation

Through openly questioning what is real, postmodernity posits that the objectivity claimed by scientific knowledge is a façade articulated through and resting upon on a subjective narrative. From this paradox we are free to uncover the frequently ignored and decidedly unscientific component of science: our humanity. It is through our humanity that the postmodern paradigm demonstrates that the scientific endeavor and by extension M&S, is not above social influence and comes to reflect a peculiar aspect of the human condition. In some scientific and M&S communities, practitioners may view this aspect as an unfortunate externality. However, a postmodern paradigm allows one to fully embrace it and openly explore the human dynamics of M&S. In this section, we explore these dynamics and work towards a postmodern rhetoric of M&S by exploring the process of validation in M&S.

There is perhaps no area of M&S more critical or more frequently discussed than validation. This results, in part, from the scientific conjecture that one can say nothing about reality with a model unless one first shows the model to be an accurate representation of reality. In practice, this is roughly how major works in M&S define validation, the process of ensuring that the simulation model is an accurate representation of reality (Balci, 1995; Banks et al, 2001; Davis, 1992; Fishman & Kiviat, 1968; Law, 2007; Sargent, 2005). In essence, simulationists treat their models as miniature scientific theories and through a series of tests attempt to show that their models are valid. However, by treating their models as scientific theories, simulationists fall into the same conundrum that is openly acknowledged in the philosophy of science (Heath, 2010): a model can only be shown to be not true. The fact that models cannot be proven true, or valid, and can only shown to be not yet untrue, or not yet invalid, is an important rhetorical distinction in attempting to understand a deeper meaning of M&S validation.

In practice, and in the eyes of the simulationist, a validated model does not mean that is it necessarily *true* but that it has been deemed an appropriate representation of reality for a given purpose (Heath & Hill, 2009). There are several important aspects of this definition of validity. First, it is clearly not the same definition of validity that implies that *truth* is being conveyed. Instead it effectively avoids the problem of never being able to prove that the model is true while maintaining a rhetorical tone implying that it is indeed a true representation of reality. Second, it is clearly subjective to the person performing the evaluation, since any approach to determine appropriateness resides nowhere outside the mind of the evaluator (Heath, 2010).

This dependence upon a human evaluator to approve the model's appropriateness supports the postmodern view that indeed M&S is not above social influences or beyond the idiosyncrasies produced when one confronts the human condition. Furthermore, it creates a situation where the simulationist is no longer attempting to show that their model is not invalid but is instead attempting to sell the evaluator on their model's abilities as well as their own. While not necessarily a popular perspective, there are several texts referencing the phenomenon of selling the simulation model as a key part of simulation validation (Banks et al, 2001; Feinstein & Cannot, 2002; Law, 2007; Pidd, 2003). However, these analyses miss a key postmodern extension that perhaps selling the simulationist's own capability is as important as selling the simulation's capability. Some may even argue that selling one's own capability is in fact more important than selling the model's capability. Such a radical reversal in values is a somewhat nihilistic twist (Nietzsche, 1968) consistent with postmodernism (Lyotard, 1984).

The act of selling simulation models transforms the process of validation into the game of validation where strategy, sportsmanship, power, and skill overshadows the original intent of demonstrating that the model represents something that is supposedly real. In the game of validation, self-interests, socio-economic status, and relationship dynamics are all in play and can drastically influence the "validity" of the model regardless of how not yet invalid the model may appear to be. Thus, by more thoroughly understanding these dynamics one can learn how to play the game more effectively. Although simulationists may appear to be ill equipped to handle these dynamics, the success of their craft indicates that they are truly creative and skilled artisans with at least an intuitive grasp of these performative aspects of the game.

To this point, our discussion of the validation of M&S has closely mirrored that of other scientific based pursuits. That is say that many of the same postmodern conclusions made about M&S validation could also be said about scientific validation more generally. However, where M&S begins to distinguish itself from more traditional scientific disciplines is the expansive and often untraditional set of techniques that they have developed to help simulationists validate a model (Sargent, 2005). While there are some similar approaches when compared to traditional scientific techniques of validation such as statistical testing, some well-accepted techniques almost completely rely on subjective human judgment. Some of these methods include Face Validation (i.e., asking experts to determine whether the model behavior seems reasonable) and the Turing Test (i.e., asking if the evaluator can tell the difference between real output and the model's output). However, probably the most interesting and disorienting tactic is the use of Animation to demonstrate the model's validity. Animation both pleases and obfuscates as it simultaneously appeals to our human desire for visual stimulation while creating an experience that disconnects us from the reality we are supposed to be evaluating.

At its core, the use of Animation transforms the simulationist into an artist that creates alternative realities and environments that are judged as much on artistic value than any scientific representation of reality. Furthermore, Animation adds no real scientific value to the model and appears to be used purely for the sake of validation. In essence, Animation becomes nothing more than validation performance art. Despite being an emerging art form, one likely far removed from a scientific approach to validation, there is clearly significant value in Animation. Why else would simulationists invest the considerable amount of additional work required to incorporate Animation in their models? With increasing computing power and every major simulation software vendor featuring advanced animation capabilities, one can only expect that Animation will continue to grow in popularity, especially as model complexity begins to exceed the technical abilities of the majority of evaluators. This use of Animation, as well as an extensive and unconventional set of validation techniques and praxis that are well accepted in the community, suggests that simulationists are mastering the game of validation even if they are currently unaware they are playing in the game. Perhaps they are already remarkably postmodern in perspective and practice.

5   Conclusion

As the development of intelligent M&S systems further blurs the artificial, understandings of reality, and the competing scientific structures of our world, it is increasingly valuable to extend the philosophical foundations of intelligent M&S

applications beyond the popular scientific paradigm of our time. In this chapter, we examined M&S through the lens of a postmodern paradigm, developed ontological implications, and began developing a postmodern rhetoric of M&S. In so doing, we extended the philosophical foundations of M&S by questioning what is real, what is objective, and directly addressing the critical influence of the human condition on intelligent M&S applications.

References

Abbott, M.B., Vojinovic, Z.: Applications of numerical modelling in hydroinformatics. Journal of Hydroinformatics 11(3-4), 308–319 (2009)

Balci, O.: Verification, validation, and accreditation. In: Proceedings of the 1995 Winter Simulation Conference, pp. 41–48 (1995)

Banks, J., Carson, J.S., Nelson, B.L., Nicol, D.M.: Discrete-Event System Simulation, 3rd edn. Prentice-Hall, Upper Saddle Ridge (2001)

Baudrillard, J.: Simulacra and simulation (Glaser, S. F. (trans.)). The University of Michigan Press, Ann Arbor (1994)

Bera, P., Burton-Jones, A., Wand, Y.: Guidelines for designing visual ontologies to support knowledge identification. MIS Quarterly 35(4), 883–A11 (2011)

Berger, P.L., Luckmann, T.: The social construction of reality: A treatise in the sociology of knowledge. Anchor Books, New York (1966)

Bois, J.S.: The art of awareness, 2nd edn. William C. Brown Publisher, Dubuque (1975)

Brummett, B.: Some implications of "process" or "intersubjectivity": Postmodern rhetoric. Philosophy & Rhetoric 9(1), 21–51 (1976)

Davis, P.K.: Generalzing concepts and methods of verification, validation, and accreditation for military simulations. Technical Report. Santa Monica, CA: RAND (1992)

Debord, G.: Society of the spectacle. Black & Red, Detroit (2005)

Doyle, K.: The Reality of a Disappearance: Fredric Jameson and the Cultural Logic of Postmodernism. Critical Sociology (Brill Academic Publishers) 19(1), 113–127 (1992)

Feinstein, A.H., Cannon, H.M.: A hermeneutical approach to external validation of simulation models. Simulation and Gaming 34(2), 186–197 (2002)

Fishman, G.S., Kiviat, P.J.: The statistics of discrete-event simulation. Simulation 10, 185–195 (1968)

Ford, D.: The search for meaning. University of California Press, Berkeley (2008)

Foucault, M.: Power/Knowledge: Selected interviews & other writings 1972-1977 (Gordon, C., Marshall, L., Mepham, J., Soper, K. (trans.)). Pantheon Books, New York (1980)

Frankl, V.E.: Man's search for meaning: An introduction to logotherapy, 4th edn. Beacon Press, Boston (1992)

Gozzi Jr., R.: Has metaphor collapsed? ETC: A Review of General Semantics 50(3), 374–377 (1993)

Grandy, G., Mills, A.J.: Strategy as simulacra? A radical reflexive look at the discipline and practice of strategy. Journal of Management Studies 41(7), 1153–1170 (2004)

Hayakawa, S.I.: Language in thought and action (New Rev. ed.). Harcourt, Brace & World, Inc., New York (1964)

Heath, B.L., Hill, R.R.: Agent-Based Modeling: A Historical Perspective and a Review of Validation and Verification Efforts. In: Handbook of Research on Discrete Event Simulation Environments: Technologies and Applications. IGI Global, New York (2009)

Heath, B.L.: The History, Philosophy, and Practice of Agent-Based Modeling and the Development of the Conceptual Model for Simulation Diagram. Wright State University. OhioLink Electronic Theses and Dissertations Center (2010)

Heidegger, M.: Being and time (Macquarrie, J., Robinson, E. (trans.)). Harper Collins Publisher, New York (1962)

Jackson, R.A.: Contextualizing analysis in postmodernity. Presented at the 2011 Air Force Analyses, Assessments, and Lessons Learned Symposium, Dayton, OH (October 2011a)

Jackson, R.A.: Acquisition reform as performance art: Reforming acquisition reform. Defense AT&L 40(6), 25–27 (2011b)

Jackson, R.A.: A quantitative analysis comparing senior leadership tenures within the department of the air force from 1947 to 2005 with those of successful, private-industry corporations. Walden University. ProQuest Dissertations and Theses (2008)

Jackson, R.A., Dafler, J.R.: A slightly deconstructive treatment of the cost estimating project: Self-reflexive cost estimating and the potential rhetorical significance of cost estimates. In: Proceedings of ISPA/SCEA 2008 Joint International Conference (2008) ISBN 0-9720204-8-9

Kassel, G.: A formal ontology of artefacts. Applied Ontology 5(3/4), 223–246 (2010)

Kearney, R.: Strangers, gods and monsters. Routledge, New York (2003)

Korzybski, A.: Science and sanity: An introduction to non-Aristotelian systems and general semantics, 5th edn. Institute of General Semantics, New York (2000)

Kuhn, T.S.: The structure of scientific revolutions, 3rd edn. The University of Chicago Press, Chicago (1996)

Lakoff, G., Johnson, M.: Metaphors we live by. The University of Chicago Press, Chicago (2003)

Law, A.M.: Simulation, modeling, and analysis, 4th edn. McGraw-Hill, New York (2007)

Lyotard, J.F.: The postmodern condition: A report on knowledge (Gennington, G., Massumi, B. (trans.)). The University of Minnesota Press, Minneapolis (1984)

Nietzsche, F.: Will to power (Kaufmann, W., Hollingdale, R.J. (trans.)). Vintage Books, New York (1968)

Ostertag, S.F.: Processing Culture: Cognition, Ontology, and the News Media Processing Culture: Cognition, Ontology, and the News Media. Sociological Forum 25(4), 824–850 (2010)

Petković, K.: Michel Foucault and the Ontology of Politics: E Pluribus Unum? Politicka Misao: Croatian Political Science Review 47(3), 176–202 (2010)

Pidd, M.: Tools for thinking: Modelling in management science, 2nd edn., ch. 11, pp. 289–312. John Wiley and Sons, New York (2003)

Raadschelders, J.N.: The Future of the Study of Public Administration: Embedding Research Object and Methodology in Epistemology and Ontology. Public Administration Review 71(6), 916–924 (2011)

Radtchenko, D.: Simulating the past: Reenactment and the quest for truth in Russia. Rethinking History 10(1), 127–148 (2006)

Raubal, M., Kuhn, W.: Ontology-based task simulation. Spatial Cognition and Computation 4(1), 15–37 (2004)

Rzhetsky, A., Evans, J.A.: War of Ontology Worlds: Mathematics, Computer Code, or Esperanto? Plos Computational Biology 7(9), 1–4 (2011)

Sargent, R.G.: Verification and validation of simulation models. In: Proceedings of the 2005 Winter Simulation Conference, pp. 130–143 (2005)

Sartre, J.-P.: The transcendence of the ego: An existentialist theory of consciousness (Williams, F., Kirkpatrick, R. (trans.)). Hill and Wang, New York (1960)

Sartre, J.-P.: Existentialism is a humanism (Macomber, C. (trans.)). Yale University Press, New Haven (2007)

Schreiber, C., Carley, K.M.: Leadership style as an enabler of organizational complex functioning. E:CO 8(4), 61–76 (2006)

Turnitsa, C., Tolk, A.: Ontology applied – Techniques employing ontological representation for M&S. In: Proceedings of the 2006 Fall Simulation Interoperability Workshop (2006)

Turnitsa, C., Padilla, J.J., Tolk, A.: Ontology for modeling and simulation. In: Proceedings of the 2010 Winter Simulation Conference (2010)

Weick, K.E.: The social psychology of organizing, 2nd edn. Addison-Wesley Publishers, Reading (1979)

Weick, K.E.: Sensemaking in organizations. Sage Publications, Inc., Thousand Oaks (1995)

Wilcock, A.: Reflections on doing, being and becoming*. Australian Occupational Therapy Journal 46(1), 1–11 (1999)

About the Authors

Brian L. Heath is a Consultant in Operations Research and Analytics for Cardinal Health, a large distribution company in the USA. He obtained his PhD in Engineering with focus in Industrial and Human Systems and his M.S. in Industrial and Human Factors Engineering from Wright State University. His interests include simulation philosophy, validation, agent-based modeling, and work measurement. He is a member of the Institute of Industrial Engineers and the Institute of Operations Research and Management Science.

Ross Jackson (PhD Walden University, MA Ohio University) teaches economics and management in the School of Business at the University of Phoenix, and is a Visiting Assistant Professor at the Naval Postgraduate School. He graduated with distinction from the Air War College distance learning program, and holds professional cost estimating certifications from the Society of Cost Estimating and Analysis and from the International Society of Parametric Analysts. His current research applies Critical Management Studies approaches to strategy development and cost estimating. Deconstruction and postmodern analysis approaches are applied to better understand how strategic plans and cost estimates, as texts, are potentially misunderstood, co-opted, and distorted as these documents progress through organizational hierarchies.

Models as Partial Explanations

Paul Weirich

University of Missouri
Columbia, MO, United States

Models incorporate idealizations that do not hold in natural systems, and yet models elucidate natural systems. How can a model explain a natural system's operation despite a failure to represent the system accurately? This chapter presents an answer to this question. It argues that an explanatory model, despite idealizations, offers a partial explanation of a phenomenon by displaying the operation of some factors behind the phenomenon's production. The first section presents this view, the second compares it to rival views, and the third shows how it guides construction of models.

1 Explanatory Models

Suppose that an agent faces a decision problem such as selecting a meal from a menu, choosing a route to a destination, or deciding among offers of admission to graduate school. The agent must select an option from a set of options, and the agent has preferences over these options. The preferences are comprehensive, and so hold all things considered. A common model of rational behavior takes the agent's choice to be rational if and only if it comes from the top of the agent's preference ranking of options. For example, if a college graduate is choosing among graduate programs offering admission, the graduate's choice is rational if and only if it is a program such that all things considered the graduate prefers no other program.

The model is normative because it advances a standard of rationality; it prescribes rather than describes behavior. It has empirical interest, however, because people are rational by and large. Predicting that a person acts rationally is reliable in many cases. Economics uses such normative models for predictions.

The model incorporates idealizations. If an agent's preferences are irrational, then acting in accord with them does not ensure rational action. If a college graduate prefers a graduate program to all others because it has the shortest name among those offering admission, then in a typical case the preference is irrational and a choice following the preference is similarly irrational. To put aside such cases, the model assumes that the agent's preferences are rational. Also, if an agent has a good excuse for not following preferences, the agent may act rationally despite acting contrary to preferences. If the college graduate selects

A. Tolk (Ed.): Ontology, Epistemology, & Teleology for Model. & Simulation, ISRL 44, pp. 105–119.
springerlink.com

the most prestigious graduate program offering admission despite preferring another because of its superior placement record, the graduate may be excused if every advisor directs attention to a program's reputation so that the ranking of programs according to prestige dominates deliberations. To put aside such cases, the model assumes that the agent has no excuse for not attending to comprehensive preferences before choosing an option.

Because people often have irrational preferences and often have excuses, such as time pressure, for not attending to their comprehensive preferences, the model of rational action does not fit people well. It has assumptions that block its application to people. Its assumptions, taken to be about people, are false. Despite the model's false assumptions, decision theorists hold that the model helps explain rational behavior. How can a model with false assumptions have explanatory power?

Explanation has multiple senses. This chapter puts aside senses of explanation that make it a matter of human understanding. It allows for explanations that are beyond human understanding and for types of understanding, say, empathy, that do not rest on explanations. This chapter treats objective explanation. An explanation holds because of the relationship between the explanadum, the phenomenon to be explained, and the explanans, the part of an explanation that explains. Whether the relationship holds is an objective matter that may be beyond the ken of theorists or may be discovered only after long investigations. Explanation is a success term, so if a widely accepted account of a phenomenon meets a decisive objection, it never was a genuine explanation. A genuine explanation has true components. In particular, its explanans is true.

Hempel (1970) presents an account of scientific explanation according to which an explanation of a phenomenon, given the circumstances of the phenomenon's occurrence, either derives the phenomenon from universal laws, or else derives the phenomenon's high probability from statistical laws. An explanation shows that the phenomenon was to be expected. Subsequent accounts of objective explanation refine this account, retaining the requirement that an explanation's components be true.

According to a general characterization of objective explanation that covers normative fields such as decision theory as well as empirical fields such as behavioral economics, an explanation of a phenomenon states reasons for the phenomenon's occurrence. The explanation is complete if it states sufficient reasons for the phenomenon's occurrence and is partial if it states only some reasons for the phenomenon's occurrence. Every reason for a phenomenon's occurrence is actual: a law, principle, or circumstance that holds. An explanation, taken generally, presents real reasons for a phenomenon.

A partial explanation gives some components of a full explanation. Partial explanations give partial answers to why-questions, more precisely, components of full answers to these questions. Without knowing a full explanation of a phenomenon, one may be unable to verify that some factor is among its components. However, some cases warrant confidence that a full explanation will include a certain factor. This is so for the explanation of an act's rationality.

The agent's preferences are very likely to be part of a full explanation of the act's rationality even if they are not by themselves a full explanation.

That an option is at the top of an agent's preference ranking of options is a reason for the option's being rational. Various contravening reasons may make the option irrational all things considered. Nonetheless, the option's privileged position in the preference ranking of options counts in favor of the option's rationality. This section's model of rational behavior highlights this reason for an act's being rational and through idealizations eliminates all contravening reasons, so that the reason settles the option's rationality. The model incorporates the principle that an option is rational if and only if it is at the top of the agent's preference ranking of options. In fact, an explanation of an act's rationality in the model may use only the principle that being a top preference suffices for rationality. The explanation does not also need the principle that being a top preference is necessary for rationality. In the model, being a top preference ensures and so fully explains an act's rationality.

In the actual world, given the occurrence of irrational preferences, being a top preference does not suffice for an act's rationality, so it only partially explains an act's rationality. The foregoing model has explanatory power despite its idealizations because it offers a partial explanation of rational action. It identifies a reason for an act's rationality. This reason affects the balance of considerations that settles an act's evaluation for rationality even in cases where it does not operate in isolation from other reasons. It provides a partial explanation of an act's evaluation for rationality in realistic cases. A full explanation supplements the partial explanation with an account of the operation of all reasons bearing on the act's evaluation, including reasons that the model's idealizations put aside.

Models are diverse. They range from model airplanes and models of the San Francisco Bay to models of the atom. Some models, such as models of climate, have a predictive rather than an explanatory purpose. Some explanatory models may explain how possibly a phenomenon occurs rather than explain even partially the phenomenon's occurrence in the real world. This chapter focuses on models that help explain phenomena in the real world. It argues that these explanatory models are partial explanations.[1]

The chapter's argument needs a characterization of the models it treats. They specify individuals, their properties and circumstances, and principles that govern the individuals. They describe possible worlds. This chapter takes a model to be a possible world, and it takes a possible world to be a maximal consistent proposition. The possible world describes for every object in the world the object's properties and its relations to objects in the world. This characterization fits many models. The section's model for rational action specifies agents, their decision problems, their preferences among options in their decision problems, and classifies some options as rational according to its principles. If a model leaves some events indeterminate, it may be a class of worlds rather than a single possible world.[2]

---

[1] Grüne-Yanoff and Weirich (2010) review the types of model that social scientists construct.

[2] This view of models is similar to the view that Sugden (2002) expresses.

Events in the possible world a model describes depend on the principles that govern the world. The principles of the model explain what happens in the model—the events that occur in it. The model's initial conditions and dynamics settle its subsequent states. Analytic methods may derive these events using the model's governing principles. Computer simulations may also obtain the events from the model's governing principles and initial conditions. Simulations offer results in cases where analytic methods are intractable.

A model differs from a theory. A theory, but not a model, makes claims about the actual world. A model's presentation may claim that the model represents a natural system in the actual world, but the model itself does not make this claim. When a model is presented as a model of a phenomenon in a natural system, the model's presentation makes the claim that the model's production of the phenomenon resembles the natural system's production of the phenomenon.

A model, taken as a possible world, does not itself offer an explanation. Although the possible world is a maximal consistent proposition, the statement it makes does not offer an explanation; it does not specify an explanandum and an explanans. However, the model's presentation, if the model is explanatory, typically designates a phenomenon to be explained. It specifies the model's target phenomenon. The model's presentation indicates the target phenomenon by saying what the model treats, say, an act's rationality. An explanatory model's presentation also specifies the target phenomenon's explanation in the model, say, the act's being a top preference and the principle that a top preference is rational in the circumstances the model specifies. If the model is advanced as explanatory, its presentation also typically claims that the target phenomenon's explanation in the model assists the target phenomenon's explanation in the actual world. This chapter argues that the presentation's explanatory claim may be interpreted as a claim that the phenomenon's explanation in the model is part of the phenomenon's explanation in the actual world.

A model's explanatory ambitions impose constraints on its target phenomenon. The target phenomenon is a repeatable type with instantiations in the model and in the actual world. The target phenomenon may be an act's property of being rational. Acts may instantiate this property in the model and in the actual world. Elements of the model fully explain instances of the phenomenon in the model and partially explain its instances in the actual world.

An explanatory model specifies the events that would occur if its assumptions were met, granting that its assumptions are strong enough to generate the events. The hypothetical conditional is true if the model's principles settle the events in the model. The hypothetical conditional's truth does not explain the target phenomenon in a natural system, but the principles of the model that make the conditional true explain the phenomenon in the model. The model's presentation uses truths in the model to explain the target phenomenon in the model. Explaining the target phenomenon in a natural system requires extending the explanation to cases that do not meet the model's assumptions. It requires generalizing the model's explanatory principles. The step from the model to the natural system typically assumes that the explanatory factors the model treats operate the same way in the natural system as in the model, although in the natural system they interact with other explanatory factors. For example, a model of motion may describe the operation of some forces in the absence of friction, and then generalize its principles to accommodate friction, under the assumption that the forces originally treated operate the same way without or without friction's presence.

This section uses a normative model that treats rational behavior and that identifies some reasons for an act's being rational to illustrate its account of a model's explanatory power. However, its account of a model's explanatory power also fits descriptive models, taking an explanation to give reasons for the phenomenon to be explained.

Causes are a type of reason. Suppose that the puddle at the bottom of the driveway freezes overnight. A cause of its freezing is the air temperature's falling below 32 degrees Fahrenheit during the night. This cause is a reason for its freezing. So an empirical model that identifies a phenomenon's causes also identifies reasons for the phenomenon's occurrence. It may offer a partial explanation of a phenomenon's occurrence by identifying reasons behind the phenomenon's occurrence even if it does not identify sufficient reasons behind the phenomenon's occurrence as a full explanation would identify them.[3]

An empirical model for the freezing point of water, taken as a possible world, may specify samples of water, their temperatures, their purity, the atmospheric pressure to which they are subject, and laws governing the freezing of samples in the circumstances the model specifies. The possible worlds that the models describe may have assumptions that do not hold in the actual world. Making those assumptions may simplify the laws governing events in the possible world that the model describes. The simple laws of the model may fully explain events in the model. The simple laws of the model do not govern the actual world because the actual world does not meet their assumptions. Nonetheless, the simple laws may identify factors that affect a phenomenon in the actual world. These factors may partially explain the phenomenon in the actual world. A full explanation awaits identification of relevant factors that the model does not treat and laws that incorporate all the relevant factors. The laws in the model follow from laws governing the actual world when the latter are restricted to cases the model specifies. The explanatory work the laws do in the model they also do in the actual world if they are generalized to accommodate the circumstances that obtain in the actual world. The laws governing the model are just restrictions of the general laws governing the actual world. Because the laws in the model and the laws in the actual world have that relation, the features the laws use to explain the target phenomenon in the model also partially explain the target phenomenon in the actual world.

---

[3] This chapter claims that causes are reasons but does not claim that reasons are causes. Mathematical explanations give reasons for a theorem's truth without citing causes of the theorem's truth because the theorem's truth has no causes. An option's rationality is an interesting case. Perhaps an option's being a top preference causes it to be rational even though the option's being rational is not an empirical property. The chapter does not take a stand on the scope of causal relations.

A model that identifies some reasons behind a phenomenon's occurrence has a design that resembles a controlled experiment's design. The experiment controls for some factors influencing a phenomenon to examine the operation of other factors influencing the phenomenon. To investigate the freezing point of water, an experiment controls for atmospheric pressure and also the purity of the sample of water. It may measure pure water's freezing point at sea level. The temperature of an impure sample of water at a thousand meters above sea level is a factor affecting its freezing, although the controlled experiment's assumptions are not met. The results of the controlled experiment may offer a partial explanation of the sample's freezing although a full explanation includes air pressure and the effects of impurities.[4]

An explanatory model identifies some factors that generate a phenomenon. Its idealizations are controls that put aside other factors responsible for the phenomenon. It may have other assumptions that do not control for explanatory factors if the model is robust with respect to those assumptions so that the model's production of the phenomenon does not depend heavily on those assumptions. The model's principles propose an account of how the explanatory factors the model identifies contribute to the phenomenon's production. If the model's principles are accurate, it generates a partial explanation of the phenomenon. The factors it uses to explain the target phenomenon in the model are part of the phenomenon's explanation in natural systems.[5]

A partial causal explanation of an event has in the explanans some of the event's causes, say, some of the causes of the Chicago fire of 1871. A partial causal explanation of a variable has in the explanans some variables causally related to the target variable, say, the level of the sea at Venice. Although causal relations among events are basic, inferring those causal relations typically uses regularities that hold between event-types or regularities concerning variables. A model that offers a partial causal-explanation treats an event-type that may occur both in the model and in the actual world, or treats a variable that takes on values in the model and in the actual world. An indicator variable may represent an event-type's occurrence or nonoccurrence.

The distinction between partial and complete causal explanation of an event mirrors the distinction between a partial and a total cause of an event. A partial cause of an event is a set of events in the event's backward light cone, and the event's total cause is the event's entire backward light cone, or its backward light cone going back to a time in the past. Stopping the cone's backward progression at a time in the past halts addition of events that, given later events, are not necessary for the event's explanation. Because some events in the event's backward light cone may be causally irrelevant to the event, its total cause and its full causal explanation trim irrelevant events. Although the event's entire backward light cone fully explains the event, the event's full explanation, strictly speaking, trims causally irrelevant events. Full explanation presumes that all features of the explanans are relevant. An explanation does not include irrelevant

---

[4] Some theorists noting the similarity of models and controlled experiments are Orzack and Sober (2001) and Mäki (2002: 10–12).
[5] Weirich (2011) shows more fully how a model may offer a partial explanation of a phenomenon.

laws in addition to laws entailing the phenomenon to be explained and does not include events irrelevant to the phenomenon.

2   Alternative Views

Many theorists have wondered about the explanatory power of models that have false assumptions. What are the alternatives to taking explanatory models as partial explanations? They fall into three groups. The first denies the explanatory power of models with false assumptions. The second takes explanation to be subjective rather than objective. The third takes explanatory models to provide explanations by analogy. This section examines these three responses to the issue of a model's explanatory power.

The first alternative response to the issue simply denies that models have explanatory power if they incorporate false assumptions. A phenomenon's explanation does not have an explanans with false components. This response does not deny the value of models with idealizations, only the view that they are explanatory. A model with false assumptions may have predictive value. Also, a model with false assumptions may be part of a research program that continuously improves the model with the goal of eventually replacing all false assumptions with true ones so that the final model explains the target phenomenon. Successive generalizations of the model move closer to reality, as least close enough so that differences do not matter much in some applications of the model. A fully articulated model, presented as a representation of a natural system, yields a theory with minor idealizations that do not prevent the theory from being approximately right. The theory, not a restricted model leading to it, has explanatory power.

This chapter's account of a model's explanatory power acknowledges the force of this position. It acknowledges that a phenomenon's explanation requires a true explanans, but holds that a model with simplifying assumptions may present truths about its target phenomenon in the model and that these truths may yield parts of the phenomenon's explanation in a natural system. Explanatory progress need not await a theory's formulation.

The second alternative response to the issue replaces objective explanation with subjective explanation. It mentions the insight a model achieves despite having false assumptions. The explanatory power of a model with false assumptions arises from the model's power to generate understanding of the target phenomenon. Models with false assumptions may be useful heuristics leading to insights, just as novels may prompt insights into human nature despite being works of fiction. This account of a model's explanatory power takes manifestations of that power to be increased understanding of the target phenomenon. The understanding occurs in a person who grasps the model. It is a subjective matter. As Columbus's false belief that he could sail straight west from Europe to India led to the discovery of America, a model with false assumptions may serendipitously improve understanding of its target phenomenon.[6]

---

[6] Rice (2011) points out that models may be explanatory, in the sense of increasing understanding, without offering objective explanations.

This account of a model's explanatory power has two steps. The first step takes explanation to be subjective and so a matter of understanding. The second step takes an explanatory model to be a heuristic device that generates understanding. These steps go a long way from the view that a model represents truths about natural systems that explain a phenomenon occurring in the natural system. The chapter argues that these steps are not necessary to account for a model's explanatory power. Its explanatory power may be objective and rest on truths.

Given that models offer partial explanations, an account of their explanatory power need not shift from objective to subjective senses of explanation that focus on personal understanding rather than on nature's operation and need not treat a model as a heuristic device without any obligation to represent faithfully a natural system. It may attribute to a model an objective explanatory power that rests on truths. Instead of saying that a model offers understanding despite the false assumptions it incorporates, an account of a model's explanatory power may maintain that the power arises from the accuracy of the principles the model uses to describe the operation of selected explanatory factors. The model's presentation makes a true claim about the operation of these factors in the model and makes a true claim that the factors contribute to the explanation of the target phenomenon in natural systems.

According to the chapter's view, a model has explanatory power in an objective sense. A model's explanatory power comes from accurately describing some features of reality. Its explanatory power is not solely its power to produce understanding. A successful model, besides generating insight, yields a statement about the world. The statement, if true, may be part of a phenomenon's objective explanation. Models are more ambitious in their treatment of reality than are heuristic devices. They may be evaluated for accuracy and not just for their fruits. A model's presentation as an explanation claims that the model's features explain the target phenomenon in the model and that the model's features help explain the target phenomenon in the actual world. Whether the model's presentation is accurate depends on facts about the model and the actual world and not just the model's fruitfulness as a generator of insights and understanding.

The accuracy of the explanatory claims a model's presentation makes depends on whether the model helps explain the target phenomenon. To help explain it, the model must yield reasons for the target phenomenon. The reasons must contribute to the phenomenon's explanation in the actual world. Section 1, because it treats explanation in an objective sense, puts aside accounts of a model's explanatory power that treat only the model's ability to generate some type of personal understanding of the model's target phenomenon. It does not take an explanatory model merely as a heuristic device, but also takes the model's presentation to make claims about the actual world.

The third alternative response to the issue of a model's explanatory power stems from an observation about the function of models. The sciences use models to investigate features of natural systems. A model by design resembles a natural system. A model acquires explanatory power through its similarity to a natural system. In economics, a model of rational behavior introduces features of agents that resemble features of people. The model's agents have beliefs and desires, face decision problems, and deliberate about the problems' solutions, as people do. A model may explain people's choices because of an analogy between agents in the model and people in the actual world.

This alternative account of a model's explanatory power claims that a model explains because it is similar to a natural system. An inductive-statistical explanation has an explanans that makes probable, but not certain, the explanandum. An explanation by analogy may similarly have an explanans that makes probable, but not certain, its explanandum. Analogical reasoning may replace reliance on a statistical law. It may move from a model's presentation to conclusions about the real world because of the model's similarity to the natural system the model represents. Analogical reasoning uses a model to reach conclusions about a natural system in ways that explain the natural system. It may show that a phenomenon in a natural system was to be expected.[7]

Every model is similar to its target system in some ways and dissimilar in other ways. Explanation by analogy, to succeed, must identify the respects in which a model resembles a natural system. For example, a model of the flow of electricity in wires that compares it with the flow of water in pipes may use the width of the pipes to represent the width of the wires, but does not use the color of the pipes to represent the color of the wires' insulation. A model's creator identifies factors of the target system that the model represents, and principles of the target system that the model incorporates.

To be effective, an explanation by analogy must specify a relevant analogy. Merely asserting that some relevant analogy exists between a model and a natural system is insufficient for confirmation of an explanation. Showing the relevance of points of similarity requires specifying those points. Heated metal rods are similar to unheated metal rods, but only the heated rods expand. An analogy to be explanatory has to highlight similarities relevant to the phenomenon to be explained. If the target phenomenon is expansion, an explanatory model's presentation must identify similarities with a natural system that are relevant to expansion.

To specify relevant points of similarity, a model's presentation may state which natural system the model represents and which features of the natural system the model's features represent. A name may represent a person without explaining the person's features. A model explains a natural system because of the way the model represents the natural system. Accounting for its explanatory power requires specifying the representational features that give it explanatory power.

Section 1 shows that an explanatory model may have explanatory power because it offers a partial explanation of a phenomenon. Because models are very diverse, it does not claim that every explanatory model offers a partial explanation. Some explanatory models may possess explanatory power without offering a partial explanation. Section 1's view thus acknowledges the possibility

---

[7] Sowa and Majumdar (2003) argue for the importance of analogical reasoning. They argue that before any subject can be formalized to the stage where logic can be applied to it, analogies must be used to derive an abstract representation of the subject from a mass of irrelevant detail. Gilboa et al. (2012) take explanatory models to offer explanations by analogy.

of alternative accounts of a model's explanatory power. However, explanation by analogy, once made precise, is just a form of partial explanation. It is not a rival account of the explanatory power of models. Section 1's account of the explanatory power of models fleshes out an account of the explanatory power of models that appeals to explanation by analogy. It makes precise the features of the model that have explanatory power and specifies how they explain. It completes the claim that the model presents an analogical explanation by making the analogy precise.

An analogical explanation typically does not claim to make the explanandum probable because arguments by analogy do not proceed according to familiar statistical reasoning. Similarly, a partial explanation does not claim to make the target phenomenon probable. The target phenomenon's probability depends on its full inductive-statistical explanation. In typical cases the model does not provide a means of estimating the phenomenon's physical probability in the actual world and so does not supply an evidential probability for the target phenomenon in a natural system.

A version of the view that models explain through analogies holds that models explain because they approximate natural systems. This version of the view holds that the similarity between a model and a natural system is extensive so that the model nearly matches the natural system in relevant respects. A model of the San Francisco Bay may approximate the Bay in many features. If the approximation is close, one expects a phenomenon to occur in the natural system if its counterpart occurs in the model. In this case departures from the model's idealizations, because small, do not very much affect the target phenomenon.

Applications of this version of the view of a model's explanatory power, to be effective, must specify how the model approximates the natural system and must show that the model's imprecision does not affect production of the target phenomenon. It must show that small differences between the model and the natural system do not distort the target phenomenon. The model's approximation of a natural system does not account for the model's explanatory power unless its small differences from reality do not matter to the target phenomenon's production. Small differences sometimes matter and sap explanatory power. For example, in a model of rational behavior, replacing the false assumption that an agent has rational preferences with the true assumption that the agent has mostly rational preferences may affect the model's conclusion that an option is rational because the option is at the top of a preference ranking of options. An exceptional, irrational preference may incorrectly put the option at the top of the ranking so that the option's being there does not ensure its rationality.

Studies of a model's robustness show that some deviations from the model are not significant. Altering the model's assumptions does not affect production of the target phenomenon. For example, a study of robustness may show that introducing friction does not significantly change the motion of bodies predicted by a model that excludes friction.

The chapter's view that a model has explanatory power because it offers a partial explanation of a phenomenon fills out an account of a model's explanatory power that appeals to approximation. Specifying the model's forms of approximation and showing the model's robustness are ways of identifying the features of the model that affect the production of the target phenomenon in a natural system. The specification yields a partial explanation of the target phenomenon.

None of the alternative responses to questions about a model's explanatory power undercuts Section 1's view that a model may offer a partial explanation of a phenomenon in a natural system. Section 1's view accommodates the observations driving the alternative responses.

3   Design of Models

A good account of the explanatory power of models helps with the design and refinement of models. Knowing the origin of a model's explanatory power assists construction of models to enhance their explanatory power. This section shows how taking explanatory models as partial explanations improves model building. Its illustrations treat models in a variety of fields.

To begin, consider a normative model of rational behavior for agents who know the outcomes of their options. According to the model, an option is rational if and only if its utility is at least as great as any other option's utility. In the model, every option has a utility that measures the extent to which it meets the agent's goals, and the option's utility equals its outcome's utility. The agent's goals are rational, and the agent has no excuses for not using options' utilities to select an option.

The model's explanatory power arises from using idealizations to control for some factors that explain the rationality of behavior and specifying the operation of remaining factors—the utilities of options. Relaxing idealizations to generalize the model fills out the model's partial explanation of rational behavior. A more general model allows for an agent's uncertainty about an option's outcome. It handles cases in which an agent has only a probability assignment for an option's possible outcomes. Its normative principle states that in the model an option is rational if and only if the option's expected utility is at least as great as any other option's expected utility. The original principle of utility maximization follows from this more general principle's application to cases in which an agent is certain of each option's outcome. The more general model offers a more complete but still partial explanation of rational behavior. Its idealizations continue to control for explanatory factors. For example, they put aside cases in which an agent does not have a probability assignment for an option's possible outcomes.

As this illustration shows, Section 1's account of a model's explanatory power offers a way of comparing two models of the same phenomenon. Although the two models of rational behavior each include assumptions that real agents do not meet, both are explanatory because they use idealizations to control for explanatory factors and describe the operation of remaining factors. One model is better than the other because it is more general and offers a fuller explanation of rational behavior. This chapter's account of a model's explanatory power guides the design of models to make them explanatory and guides revision of models to enhance their explanatory power.

Next, consider models of signaling that Skyrms (2010) presents. The models depict the emergence of signaling systems in organisms. Each model's scientific goal is to elucidate signaling in the real world, such as, among Vervet monkeys, signaling the approach of predators so that colony members may take appropriate evasive action. A model makes progress toward this goal by offering a partial explanation of signaling that identifies factors appearing in a full explanation of signaling. The models are explanatory to the extent that their idealizations control for explanatory factors and depict the operation of other explanatory factors. A way of refining the models to enhance their explanatory power is to add to the models features that are part of the explanation of signaling in the actual world.

Skyrms's models consider signaling games in which a sender selects a signal to send, and a receiver selects a response to a signal. In a signaling game both the sender and the receiver benefit if the sender transmits information and the receiver processes it. The players have multiple strategies for sending and processing signals. A combination of strategies, exactly one for each agent, is an equilibrium if and only if each strategy is a best response to the other agents' strategies. A signaling system is an equilibrium of a signaling game. Organisms that establish a signaling system transmit and process information. They learn to signal effectively. Skyrms studies the resources necessary for learning to signal, the speed of learning, and the conditions under which organisms use the minimum number of signals needed for an efficient signaling system.

In repetitions of a signaling game, a signal may become effective through trial and error. Even organisms without cognitive capacities can signal effectively. Methods of learning differ according to the demands they place on organisms. Some methods of learning are differential reproduction, differential imitation, probing and adjusting, best-response reasoning, and reinforcement learning. An organism using reinforcement learning, for example, repeats behavior that succeeded earlier.

In a model, a method of learning, for the sake of realism, must fit an organism's abilities. Identifying a best response to another agent's strategy is taxing. A neighbor's successful response in a similar situation approximates a best response. Realism suggests models that simplify deliberations by having agents imitate successful strategies. It recommends low-cost methods of selecting acts. Simple reinforcement learning leads to signaling systems in many contexts. Reinforcement learning is a low-cost method of selecting an act. Unsophisticated organisms are capable of reinforcement learning.

Furthermore, realism encourages exploration of alternatives to a best-response equilibrium, such as a correlated equilibrium, which is a generalization of a best-response equilibrium. A correlated equilibrium may be efficient in a game with inefficient best-response equilibria.

Studies of the flow of information from organism to organism belong to interactive epistemology. Sophisticated agents know who knows what. They use their higher-order knowledge to make decisions. Representing higher-order knowledge adds realism to models of human signaling-systems. Also, signaling games have multiple representations. Strategies that tie according to some representations have different payoffs in other representations. A realistic representation attends to all consequences that matter to an agent.

Skyrms's models incorporate many idealizations. His research program progresses by relaxing idealizations to achieve greater realism. However, his models, even at early stages in their articulation, have explanatory power. They make progress with explanations of the formation of signaling systems. How do these models help explain real signaling systems? A model is an idealized world simpler to investigate than the real world. How should we understand the relation between signaling in a model and natural signaling? The model identifies factors, such as learning mechanisms, that influence signaling in the actual world and controls other factors through idealizations. An account of a model's explanatory power analyzes the function of idealizations that control explanatory factors and shows how to increase explanatory power by relaxing those idealizations.

Next, consider the causal models that Pearl (2000) constructs. They have the scientific goal of exhibiting causal relations among a set of variables. A model selects a few variables relevant to a target variable and uses specifications of these variables' values and the variables' causal relations, as inferred from statistical data, to explain the value of the target variable. The models idealize by supposing that they include relevant variables and by adopting causal assumptions on which empirical support of the variables' causal relations rests, such as the causal Markov condition, which states that in a causal model, given a variables' direct causes, the variable is independent of its non-descendants. Despite idealizations, a causal model provides a partial explanation of a phenomenon, if it is correct about the causal relations governing the variables in the model, by giving a partial account of the phenomenon's production. It controls for some explanatory factors and investigates the causal relations of other factors. Greater realism comes from introducing additional relevant variables.

A way of refining causal models to enhance their explanatory power is to add to the models features that are parts of the explanation of the target phenomenon in the actual world. For example, a model may represent the causal mechanism that makes a sidewalk slippery. The model may treat rain, taken as a value of a variable for weather, as a direct cause of the sidewalk's becoming slippery. In the actual world, rain is not a direct cause because air with high temperature and low humidity may dry the sidewalk as light rain falls. The model gains realism by introducing a variable for the sidewalk's being wet that stands between the variable for weather and the variable for the sidewalk's being slippery so that rain no longer is a direct cause of the sidewalk's slipperiness.

Models also play a major role in normative, as opposed to empirical, fields. For example, Horty (2001) presents models for the representation of utilitarian accounts of moral obligation. These models incorporate idealizations that put aside intentionality, probability, and temporally extended acts. How do these normative models, despite their idealizations, furnish insight into the utilitarian conception of moral obligation? They do this by using the idealizations to set aside some factors affecting utilitarian obligations and investigating the operation of remaining factors. The models of utilitarianism identify some factors that affect obligations in the real world and give an account of the operation of these factors while controlling for other factors. The explanation of obligations is partial because the model's idealizations ignore factors that affect obligations. A model enhances the explanation it provides by dispensing with idealizations controlling for some explanatory factors and generalizing to accommodate those explanatory factors. The model for utilitarian obligation may remove the idealization that assumes certainty about an option's outcome and may allow for probability assignments that represent an agent's uncertainty about the option's outcome.

Horty (2012) introduces a variable-priority default logic that includes mechanisms for excluding defaults. For simplicity, the logic makes exclusion of defaults complete and puts aside cases in which a reason reduces a default's priority but does not completely exclude the default. Despite the simplification, Horty's model of reasoning explains good reasoning by investigating some factors in the explanation of good reasoning. His model of default reasoning offers a partial explanation of good reasoning in the real world. The explanation is partial because it ignores nuances of good reasoning for the sake of simplicity. A generalization of the model may enhance the model's explanatory power by relaxing its idealizations, and so, for example, by allowing for partial exclusion of defaults. The model for default reasoning may generalize to cover defaults not completely excluded.

4   Conclusion

Using a model to explain a phenomenon's occurrence in a natural system requires a type of inductive inference from explanation in the model to explanation in the natural system. A similar inductive inference moves from explanation of a phenomenon in a laboratory setting to explanation of the same phenomenon in the field. Because assumptions that hold in the laboratory are absent in the field, and may be crucial, the inference may go wrong. The inductive inference is more reliable, however, the greater conditions in the field resemble conditions in the laboratory. Studies of robustness in the laboratory may demonstrate that allowing some factors to vary do not affect the phenomenon so that changes in those factors during moves from the laboratory to the field are unlikely to affect the phenomenon.

Theoretical conclusions resting on controlled experiments assume that laws work the same way in the laboratory and in the field. They assume the invariance of laws. The inductive inference is large, but at least the invariance it assumes concerns just the laws' operation in nature, either in the laboratory or in the field. The common setting supports the inductive inference. The invariance of laws' operation in a model and in a natural system has less support because the invariance stretches across two worlds. The laws do not operate in a common setting. Although induction typically supports extrapolation from a model to a natural system less strongly than extrapolation from the laboratory to the field, in some cases it adequately supports the nomic invariance that makes a model explanatory.

Although this chapter treats only explanations involving inferences from a model to a natural system, its points may extend to explanations involving inferences from laboratory conditions to field conditions, that is, inferences going from results of controlled experiments to observations in the field. Explanations in the laboratory may offer partial explanations of events in the field. The same explanatory factors may contribute to instances of a phenomenon in the laboratory and in the field.

The chapter concludes that taking explanatory models to offer partial explanations accurately depicts the ability of models to elucidate phenomena. Partial explanation fleshes out explanation by analogy, accommodates the diversity of descriptive and normative models, preserves the objectivity of explanation, and fruitfully directs design and refinement of models.

References

Gilboa, I., Postlewaite, A., Samuelson, L., Schmeidler, D.: Economic Models as Analogies (2012),
  http://itzhakgilboa.weebly.com/uploads/8/3/6/3/8363317/
  gpss_economic_models_as_analogies.pdf
Grüne-Yanoff, T., Weirich, P.: The Philosophy and Epistemology of Simulation: A Review. Simulation & Gaming 41, 20–50 (2010), doi:10.1177/1046878109353470
Hempel, C.: Aspects of Scientific Explanation. Free Press, New York (1970)
Horty, J.: Agency and Deontic Logic. Oxford University Press, New York (2001)
Horty, J.: Reasons as Defaults. Oxford University Press, New York (2012)
Mäki, U.: The Dismal Queen of the Social Sciences. In: Mäki, U. (ed.) Fact and Fiction in Economics, pp. 3–34. Cambridge University Press, Cambridge (2002)
Orzack, S., Sober, E.: Adaptationism and Optimality. Cambridge University Press, Cambridge (2001)
Pearl, J.: Causality. Cambridge University Press, Cambridge (2000)
Rice, C.: Optimality Explanations: A New Approach. Ph.D. dissertation, University of Missouri (2011)
Skyrms, B.: Signals. Oxford University Press, New York (2010)
Sowa, J., Majumdar, A.: Analogical Reasoning. In: Ganter, B., de Moor, A., Lex, W. (eds.) ICCS 2003. LNCS, vol. 2746, pp. 16–36. Springer, Heidelberg (2003), doi:10.1007/978-3-540-45091-7_2
Sugden, R.: Credible Worlds: The Status of Theoretical Models in Economics. In: Mäki, U. (ed.) Fact and Fiction in Economics, pp. 107–136. Cambridge University Press, Cambridge (2002)
Weirich, P.: The Explanatory Power of Models and Simulations: A Philosophical Exploration. In: Weirich, P., Grüne-Yanoff, T., Ruphy, S., Simpson, J. (eds.) Philosophical and Epistemological Issues in Simulation and Gaming, a special issue of Simulation & Gaming, vol. 42, pp. 149–170 (2011), doi:10.1177/1046878108319639

Theory Reconstruction of Several Versions of Modern Organisation Theories

Klaus G. Troitzsch

University Koblenz-Landau,
Koblenz, Germany

Abstract. This chapter compares the technique of reconstructing theory under the 'non-statement view' with the design and implementation of simulation models. For this purpose it uses several different versions of the famous 'garbage can' model, redefines this theoretical attempt in terms of the 'non-statement view' and compares it to simulation models of different authors who replicated the original 'garbage can' model and built on it to extend it.

1 Introduction

Some forty years ago, Cohen, March and Olsen invented the 'garbage can' model of organisational behaviour [4] which soon became a piece of discussion among scholars in organisation theory and has never lost the attraction of the field such that various reimplementations of the original model exist (Guido Fioretti offers a website with a collection of such reimplementations [8]), but several variants were published in recent years [9, 10, 11, 25], most of which used agent-based modelling and simulation for their implementations (whereas the original model was written in FORTRAN IV which is more or less forgotten by now, but other implementations used languages such as Basic, Objective C, Pascal and C). Several of these model variants will be used in this chapter to show the similarity and the differences between them, and this will be done using the technique of reconstructing theories developed (nearly at the time of the first occurence of the 'garbage can') by Sneed [19] and later on by Stegmüller [20], Balzer and Moulines [2]. At the same time the chapter will show the similarity between reconstructing a theory according to the 'non-statement view' and designing a simulation model which in a certain way represents this theory [23, 24, 1].

The chapter is organised as follows. The next section gives a short introduction of some of the main traits of the 'non-statement view' of theories, whereas the third section introduces the basic ideas of the 'garbage can' model first defined in [4]. What makes this model a good candidate for a 'new statement view' reconstruction

A. Tolk (Ed.): Ontology, Epistemology, & Teleology for Model. & Simulation, ISRL 44, pp. 121–140.
springerlink.com

is the fact that it has been extended several times by various authors and that its mostrecent variants come as agent-based models such that a discussion is possible whether this way of implementing the theoretical model can contribute to engineering a system which could optimise decision making and task allocation processes. The fourth section reconstructs the original model from its FORTRAN IV code and some of the follower versions written in different versions of NetLogo [27], while the concluding section summarises and gives an idea how in general simulation models and structuralist theory reconstructions can be transformed into each other.

2   The 'Non-statement View' of Theories

Unlike most other definitions of the term 'theory' the one developed in what was later on called 'non-statement view of theories' by Joseph D. Sneed [19] tried to overcome the dispute about language of theory and language of observation — or 'observable' and 'theoretical' terms or 'empirical laws' and 'theoretical laws' [3, ch. 23, p. 229] — by introducing a set-theoretic description of theories and by giving up the attempt to classify terms as either observable or theoretical and declaring the status of scientific terms only in the context of a specific theory, such that a term might be theoretical with respect to theory **A**, but non-theoretical with respect to theory **B**. This makes it possible to analyse the theoreticity in a more satisfactory sense than in [3] with the example of "an observable field [which] can be measured with a simple apparatus" or with the definition of 'observable' as "measurable by *relatively* simple techniques" (my italics). And for the example to be discussed here, the 'non-statement view' gives an opportunity to discuss the theoretical status of the terms of the 'garbage can' model and its successors.

In terms of the 'non-statement view' approach a 'theory element' is a structure consisting of a core and a set of intended empirical applications of the theory. The latter are easily conceived as empirical systems in the real world which seem to have something in common and which can be described with a number of (usually) natural-language terms. These intended applications are interpreted as a certain kind of *models* of the theory in question — note that the word *model* is used in another meaning than in everyday language, namely in the meaning assigned to this word by mathematical model theory. To make clear what is actually meant with *model* in this sense — without repeating the vast literature on the 'non-statement view' [19, 20, 2], see also the bibliography in [5, 6] — a *model* is a list of terms representing objects and functions, and these *models* come in three different forms, namely

- the *potential model* $\mathbf{M_p}$ lists all the terms which are needed to describe all objects and functions to which the theory refers whether they seem immediately observable (measurable) or not; but there is a distinction between two types of terms as in the positivist attempt to describe the languages of science:

    - some terms are already measurable before the theory is even thought of, because they are theoretical terms with respect to some other theory

– others only 'make sense' if the theory is already available and are undefined before this theory has been developed.

- the *potential partial model* $\mathbf{M_{pp}}$ lists all those terms in the former group; these terms, too, need some theory in order to become measurable, but not the theory in question; [25] used an example from mechanics in which one can see that lengths and angles can be measured with the help of geometry, and mechanics can take terms like these as givens, whereas terms such as force can only be made measurable when one has already accepted a theory of mechanics (many more examples are given in [19], all of them from physics, but as early as in 1981 a collection appeared which brought together applications from economics [21]);
- the (full) *model* $\mathbf{M}$, finally, complements the *potential model* by axioms or invariants, thus selecting among *potential (partial) model* candidates just those which behave as theoretically postulated.

The examples in the following sections will illustrate the role of *potential models* and of *models* proper, the difference between *potential partial models* and *potential models* plays a lesser role in this chapter, as this chapter is mainly about simulation models (in which everything is accessible to measurement), but the measurability of the concepts of the different versions of the garbage can model will also be discussed.

The 'non-statement view of theories' discusses also the relations between theories which form a 'theory-net' when some of the terms of one theory are re-used in another theory (such as length and angle both in mechanics and in geometry), but as this chapter concentrates on just one theory per section, the concept of links and constraints will also be only superficially touched.

3 The 'Garbage Can' Model

The 'garbage can' model is an early simulation model from organisation theory that from its first publication in 1972 has over and over again be quoted in different contexts: the so-called garbage-can model of organisational decision making behaviour [4]. The central term of this paper is organisation which the authors exemplify with "universities, a familiar form of organized anarchy", but the model can, of course, also be applied to other kinds of organisations even if these might be better structured and less anarchical than universities (perhaps it is due to the identification of universities with anarchical organisation that the model attracted university researchers over and over again, perhaps it is the somewhat weird name 'garbage can' of the model which derives from one of the possible outputs of the organisational decision making process, see below). The applicability to universities is stipulated on the first page of the paper when the authors say "Possible applications ... for more narrow predictions are illustrated by an examination of the model's predictions with respect to the effect of adversity on university decision making."

Other terms are listed in the first few lines of the paper, namely "decision makers", "problems" and "choices" (possible solutions to problems), and these are also used in their everyday meanings such that a more precise definition of these terms is certainly necessary, particularly if the theory[1] is intended to explain (or even predict [7, 26]) some empirical target system and the organisational behaviour in this system. The designata of these terms interact in an organisation in a manner which is theoretically described in the paper and formalised in a FORTRAN IV program.

According to [4] every organisation is characterised by a number of more or less complex attributes described and elucidated as follows:

decision structure: a matrix containing only 1's and 0's defining which decision maker may make which choice; depending on the contents of this matrix the decision structure of the organisation can be

specialised: in the extreme case there is only one 1 per row meaning that every choice has exactly one competent decision maker,

hierarchical: one decison maker is competent to make any choice, the employees on the lowest level of the hierarchy are in charge of only one choice each, while the intermediate level employees are in charge of the choices of their respective subordinates, and

generalised: all elements are 1. i.e. every employee may solve any problem.

access structure: a matrix containing only 1's and 0's defining which problem can be solved by which choice; here again several different contents of this matrix are possible and deserve further analysis.

problems: these can have three different states: passive (not yet recognised by any decision maker), active (currently being treated) or solved.

choices: these are the potential solutions to problems, they can be passive (not yet chosen), active (currently being applied to a problem) or made (having solved the problem). At any time there is a "most attractive choice for each decision maker", i.e. the kind of solution this decision maker knows to work with best, and a "most attractive choice for a problem". i.e. the most appropriate solution of this problem.

energy: It is not entirely clear what was originally meant with this term — certainly not the energy defined in physics! — but rather the energy we colloquially mean when we call a person "a very energetic one". The decision makers, according to Cohen, March and Olsen, devote only 60 per cent of their "energy" to their work for their organisation. An organisation is characterised by its energy distribution, which allocates each of its decision makers a certain amount of "energy". The "energy" necessary to solve a problem is a characteristic of the latter ("problem energy requested"), whereas the energy necessary to take this choice ("choice energy requested") is a characteristic of the choice. This means that the problem is not simply solved by applying the solution, but it will usually be the

---

[1] The 'garbage can' model is treated as a theory here, although the authors insist [18] that it is "'a' not 'the', and 'model' not 'theory'". But this does not exclude to reconstruct it with the help of the 'non-statement' view, one only has to take into account that the terminology of this approach mith respect to the words 'model' and 'theory' is slightly different from the mainstream terminology.

case that the decision maker has to take several attempts with the same choice to solve the entire problem. Consequently, the "choice energy expended" is the amount of "energy" already expended at a certain point of time to prepare the decision in favour of this choice.

"Energy" seems to be what we use this word for in everyday language to talk about people. In a scientific theory, this term might be useful if the "energy" necessary to solve a certain problem has the same numerical value in one university as well as in another university, or the "energy" of Dr Miller measured when he was provost of university A is the same as his "energy" measured when, several years later, he was provost of university B. Perhaps one could also find out the "energy" of decision makers as some function of their intelligence, then there would be a link to some psychological theory about intelligence guaranteeing that the "energy" of a certain decision maker is measured with the help of the garbage can theory and at the same time estimated from the intelligence measured with the help of the intelligence theory.

Perhaps all the decision makers in an organisation might agree on an ordinal measure of their "energy" in a way that all of them agree in assertions such as "A has more energy than B" or "C has the same amount of energy as D" for all A, B, C and D. In real-world applications, however, it will be extremely difficult to assign a certain person, a certain problem or a certain "choice" a numerical (real-valued) value as in the theory of Cohen, March and Olsen. But perhaps a redesign of this theory might enhance its applicability (see 4.3). And under certain circumstances it might even be possible to make this "energy" measurable with the help of a variant of the theory (as has been shown for simple theories in mechanics, see [25]).

To sum up the process of decision making in an organisation as it was understood by Cohen, March and Olsen, problems turn up and become visible for several decision makers one of which feels competent to solve it, choices (mainly the "most attractive choice for each decision maker") become possible, are taken into account and evaluated by decision makers and — if they are sufficiently attractive — will be used to solve problems if, from the point of view of the problems, they are "attractive". In more detail, Cohen, March and Olsen define three ways of decision-making:

by resolution: only then the problem is resolved — the choice fits the problem, and the choice is made: for instance, a vacant job position is filled again (for another, more detailed example see subsection 4.2),

by oversight: in this case a choice is made, but one which may perhaps solve other problems, but the problem in question remains unresolved: for instance, the vacant job position is left open, staff expenses are reduced, some budget restrictions are met, but the teaching situation in the university remains precarious, and

by flight: in this case, a problem had been associated with a choice which did not solve the problem, and a new choice became available which might solve the problem in question only later on, so again both problem and choice remain "active", for instance: the vacant job position is filled with a professor with a very different specialisation, but the new professor cannot attract any students as his courses are all selective courses — the attractive compulsory courses of the predecessor cannot be given any longer, and the selective courses of the successor do not attract students.

What seems to have been interesting for Cohen and his co-authors was the influence of decision structure, access structure and total energy on the decision making behaviour of the organisation as a whole. Without any doubt it is the merit of the 'garbage-can' model that it offers explanations that organisations with (e.g.) unsegmented decision structure solve their problems much faster than those with a specialised decision structure. "To understand processes within organizations, one can view a choice opportunity as a garbage can into which various kinds of problems and solutions are dumped by participants as they are generated. The mix of garbage in a single can depends on the mix of cans available. on the labels attached to the alternative cans, on what garbage is currently being produced, and on the speed with which garbage is collected and removed from the scene." [4, p. 2] Unfortunately the 'can' is not represented in the FORTRAN program such that the idea behind this sentence still remains somewhat unclear.

4 Structuralist Reconstructions of Some Versions of the 'Garbage Can' Model

This section formalises three versions of the 'garbage can' model, namely the original [4], an extension described and implemented by Fioretti, Lomi and Cacciaguerra [14, 9, 8] which is distinguished from the original by the feature that the 'energy' is a variable property of decision makers and (as 'energy requested') also of problems and choices, and finally one in which the one-dimensional 'energy' is replaced with a composition of 'skills'. All three formalisations use the manner applied to the definition of potential models of the 'non-statement' view and give references to the respective program code.

4.1 The Original

The first model to be formalised is the original which was formally described in a FORTRAN IV program by the authors. Reconstructing the defintion of a model of the theory postulated by Cohen, March and Olsen from the FORTRAN code is not so very straightforward, particularly as this is just one piece of code without any structuring (which would have been possible if the programmers had used FORTRAN functions or subroutines to make the code more easily understood).

The defintion of a theory or theory element usually begins with the definition of the theory core and the set of intended applications:

Definition 1. DefTE(GC) : TE(GC) := $\langle K(GC), I(GC) \rangle$, *where*

- $K(GC) := \langle Mp(GC), M(GC), Mpp(GC), Po[Mp(GC)], Mp(GC) \rangle$ *is the theory core, which contains all information which is needed to formalise what the theory will have to say, and*

- **I(GC)** *is the set of intended applications — i.e. a set of empirical data sets collected to describe the decision making in one or more universities.*

The next step is the definition what makes up a potential model of the theory, listing sets and functions — the terms used to speak about the theory and its intended applications. In the current case, there is no distinction between GC-theoretic and GC-non-theoretic terms, as for simulation purposes this distinction is not applicable (in a simulation all attributes of all object types are accessible, which is usually not the case in the intended applications of the real world where some properties of real-world entities can be measured without presupposing the theory, while others can be talked about but not measured, and the latter is certainly true for the term 'energy' of a decision maker).

Definition 2. Def M_p(GC):   *η is a potential model of* **GC***, i.e. $\eta \in M_p$(**GC***) iff $\mathcal{O}, \mathcal{P}, \mathcal{D}, \mathcal{C}, \mathcal{T}, \Delta, A, T_{ep}, T_{ec}, \Sigma_{ep}, \Sigma_{ec}, \Gamma_d, \Gamma_p, \Lambda, E_d, E_p, E_r, , E_e$ exist such that*

1. $\eta = \langle \mathcal{O}, \mathcal{P}, \mathcal{D}, \mathcal{C}, \mathcal{T}, \Delta, A, T_{ep}, T_{ec}, \Sigma_{ep}, \Sigma_{ec}, \Gamma_d, \Gamma_p, \Lambda, E_d, E_p, E_r, , E_e \rangle$.
2. $\mathcal{O} = \{ \langle P, D, C \rangle \mid P \subseteq \mathcal{P}, D \subseteq \mathcal{D}, C \subseteq \mathcal{C} \}$ *is a non-empty finite set [of organisations; note that different organisations have different subsets of problems, decision makers and choices].*
3. $\mathcal{P}$ *is a non-empty finite set [of problems].*
4. $\mathcal{D}$ *is a non-empty finite set [of decision makers].*
5. $\mathcal{C}$ *is a non-empty finite set [choices].*
6. $\mathcal{T}$ *is a set [of points in time].*
7. $\Delta : \mathcal{D} \times \mathcal{C} \rightarrow \{true, false\}$ *is a function [that assigns "true" to a pair decision maker / choice if this decision maker may make this choice, otherwise "false": decision structure].*
8. $A : \mathcal{P} \times \mathcal{C} \rightarrow \{true, false\}$ *is a function [that assigns "true" to a pair problem / choice if this problem may be solved by this choice, otherwise "false": access structure].*
9. $T_{ep} : \mathcal{P} \rightarrow \mathcal{T}$ *is a function [$\tau_{ep}(p)$ is the entry time of problem p].*
10. $T_{ec} : \mathcal{C} \rightarrow \mathcal{T}$ *is a function [$\tau_{ec}(c)$ is the entry time of choice c].*
11. $\Sigma_{ep} : \mathcal{P} \times \mathcal{T} \rightarrow \{passive, active, solved\}$ *is a function [that assigns the current state $\sigma_{ep}(p,t)$ to a problem $p \in \mathcal{P}$].*
12. $\Sigma_{ec} : \mathcal{C} \times \mathcal{T} \rightarrow \{passive, active, made\}$ *is a function [that assigns the current state $\sigma_{ec}(c,t)$ to a choice $c \in \mathcal{C}$].*
13. $\Gamma_d : \mathcal{D} \times \mathcal{T} \rightarrow \mathcal{C}$ *is a function [that assigns the currently preferred choice to every decision maker; $\gamma_d(d,t)$ is the "most attractive choice for decision maker" d].*
14. $\Gamma_p : \mathcal{P} \times \mathcal{T} \rightarrow \mathcal{C}$ *is a function [that assigns the currently optimal choice to every problem; $\gamma_p(p,t)$ is the "most attractive choice for problem" p].*
15. $\Lambda : \mathcal{T} \rightarrow \Re$ *is a function [that assigns solution coefficients to every point in time; this solution coefficient describes the proportion of "energy" that decision makers would devote to their work for the organisation; in the text of [4] this is a constant and equals 60 percent at all times — and for all decision makers! — whereas in a footnote and in the FORTRAN code different solution coefficients are given for the first half of the time].*

16. $E_d : \mathcal{D} \to \mathfrak{R}$ is a function [that assigns "energy" to every decision maker; this "energy" is constant over time; this function is called "energy distribution"].

17. $E_d : \mathcal{D} \to \mathfrak{R}$ is a function [that assigns "energy" to every problem; this is the "energy" necessary to solve the problem ("problem energy requested")].

18. $E_r : \mathcal{C} \times \mathcal{T} \to \mathfrak{R}$ is a function [that assigns "energy" to a choice $c \in \mathcal{C}$ at a particular point of time; this is the energy necessary to take this choice ("choice energy requested"); this total necessary "energy" can grow, particularly when the same choice is used to solve several problems in sequence].

19. $E_e : \mathcal{C} \times \mathcal{T} \to \mathfrak{R}$ is a function [that assigns "energy" to a choice $c \in \mathcal{C}$ at a particular point of time; this is the "energy" already expended to prepare the decision in favour of this choice ("choice energy expended"].

The following of these terms are most likely to be **GC**-non-theoretic terms:

- $\mathcal{O}$, as organisations like universities are defined as such by their charter which can be read and interpreted without any knowledge of their internal decision making structure,
- $\mathcal{P}$, as problems can be identified as such without any knowledge how and by whom they are to be treated,
- $\mathcal{D}$, as decision makers are identified by the organigram of the respective organisation,
- $\mathcal{C}$, as choices may be listed in handbooks of an organisation,
- $\mathcal{T}$, as for measuring time clocks and calendars are sufficient,
- $\Delta : \mathcal{D} \times \mathcal{C} \to \{\text{true}, \text{false}\}$, as the organigram, charter or handbook of the respective organisation defines which decision maker is competent (in the sense of 'in charge', not in the sense of 'skilled'!) for which choices,
- $T_{ep} : \mathcal{P} \to \mathcal{T}$, as the entry time of problem p can usually be determined by the time stamp of an incoming document describing the problem p,
- $\Sigma_{ep} : \mathcal{P} \times \mathcal{T} \to \{\text{passive}, \text{active}, \text{solved}\}$, as the current state $\sigma_{ep}(p,t)$ of problem $p \in \mathcal{P}$ will be documented in the information system of the respective organisation,
- $\Lambda : \mathcal{T} \to \mathfrak{R}$, as at least in principle it can be determined at any time whether decision makers devote their time or "energy" to the benefit of the respective organisation,
- $E_d : \mathcal{D} \to \mathfrak{R}$, as discussed above, if it can be measured with the help of some psychological theory, otherwise it seems to be a **GC**-theoretical term.

For the other terms not mentioned in the enumeration above it seems quite difficult to describe measuring devices or procedures without knowing more about this theory of decision making. For instance, it seems doubtful whether for all kinds of problems the choices to solve them are clearly defined in a function $A : \mathcal{P} \times \mathcal{C} \to \{\text{true}, \text{false}\}$ or in an access tructure matrix as applicable in all kinds of organisations, and when exactly a choice becomes visible (at entry time $\tau_{ec}(c)$) might sometimes be difficult to determine. On the other hand, $\gamma_d(d,t)$, the "most attractive choice for decision maker" d, might be easy to find out for some decision makers, but entirely random for others, and which is the currently optimal choice $\gamma_p(p,t)$ to a problem p may depend on the problems and choices currently being treated. Only if these two functions are thought time-independent they might count as **GC**-non-theoretical, as in this case they are predefined in the rules of the organisation, but if there is at least some administrative discretion in an organisation (which is typically the case in universities) it is difficult to measure the time-dependent 'attractivity" of a choice for a decision maker or for a problem — except that one admits that the choice a certain decision maker made to treat a certain problem was the most attractive one both for the decision maker and the problem, but this is often counterfactual, particularly in universities. For the remaining terms not discussed in this paragraph the **GC**-theoreticity just depends on the theoretical status of the "energy".

The 'non-statement view' reconstruction thus forces to give a detailed account of the measurability of the terms used in a theory such that one can evaluate the applicability of a theory to its intended applications.

The full model complements the potential model with invariants relating the more important terms:

Definition 3. Def M(GC) : ζ *is a model of* **GC**, *i.e.* $\zeta \in$ **M(GC)** *iff the following holds:*

1. $\zeta = \langle \mathscr{O}, \mathscr{P}, \mathscr{D}, \mathscr{C}, \ \mathscr{T}, \Delta, A, T_{ep}, T_{ec}, \Sigma_{ep}, \Sigma_{ec}, \Gamma_d, \Gamma_p, \Lambda, E_d, E_p, E_r, , E_e \rangle$, *i.e.* $\zeta \in$ **M$_p$(GC)**
2. *The invariants defined in the FORTRAN program published by the authors in the appendix [4, pp. 19–24] hold. These invariants are given as follows:*

   a. *A problem is passive before its entry time, it is solved when* $\exists c$ *mit* $\alpha(p,c) =$ *true* $\wedge \sigma_{ec}(c,t) =$ *made, otherwise it is active:*

   $$\sigma_{ep} = \begin{cases} passive & : \ t < \tau_{ep}(p) \\ solved & : \ t > \tau \wedge \sigma_{ec}(c,t) = made \wedge \exists c | \alpha(p,c) \\ active & : \ otherwise \end{cases}$$

   b. *A choice is passive before its entry time, it is made when* $\varepsilon_r(c,t) \leq \varepsilon_e(c,t)$, *otherwise it is active.*

   $$\sigma_{ec} = \begin{cases} passive & : \ t < \tau_{ec}(c) \\ made & : \ \forall t \in \{t \in T | \varepsilon_r(c,t) \leq \varepsilon_e(c,t)\} \\ active & : \ otherwise \end{cases}$$

   c. $\sigma_{ep}(p,t) = active \wedge \gamma_p(p,t) = c \rightarrow \varepsilon_r(c,t) = \varepsilon_r(c,t-1) + \varepsilon_p(p)$
   d. $\delta(d,c) \wedge \gamma_d(d,t) = c \rightarrow \varepsilon_e(c,t) = \varepsilon_e(c,t-1) + \lambda(t)\varepsilon_d(d)$
   e. *Which choice is most attractive for a decision maker and for a problem, respectively, follows from an algorithm which finds the choice* $c_{ma.d}$ *and* $c_{ma.p}$ *for which the difference* $\varepsilon_r(c,t) - \varepsilon_e(c,t)$ *is a minimum among those choices which are applicable according to the decision and access structure, respectively:*
   $c_{ma.p} = \gamma_p(p,t) iff \nexists c^* | \varepsilon_r(c^*,t) - \varepsilon_e(c^*,t) < (\varepsilon_r(c^*,t) - \varepsilon_e(c^*,t) \wedge \alpha(p,c)$
   $c_{ma.d} = \gamma_d(p,t) iff \nexists c^* | \varepsilon_r(c^*,t) - \varepsilon_e(c^*,t) < (\varepsilon_r(c^*,t) - \varepsilon_e(c^*,t) \wedge \delta(d,c)$

Looking at the FORTRAN code, one can easily identify the two-dimensional arrays
`IKA(20,20)`, `JIA(20,20)` and `XEA(20,20)` with the functions Δ, A and a
combination of E_d and Λ. The objects (decision makers, problems and choices) are
not represented at all in the FORTRAN code, except as integer numbers from 1 to
10 or 20, as the only organisation $o \in \mathcal{O}$ has 10 decision makers, 20 problems to
solve and 10 choices among which decision makers can choose.

λ is represented by the array `XSC(20)` which holds 20 real numbers for the 20
decision makers (the code contains the values in footnote 3 of [4], not the constant
0.6 mentioned in the main text). Most of the other functions are represented with arrays as well, e.g. `JET(20)` and `ICH(20)`, the problem entry time and the choice
entry time, are one-dimensional arrays whose values are input from the two punched
cards with integer values (it remains unclear why the array `ICH(20)` is of length
20, as only ten values are needed). This example shows that the simulation in [4]
does not proceed from time step to time step as one would have expected in nearly
all other simulation programs, but that the episode of 20 time steps considered in
the original 'garbage can' model is programmed in a global manner, and what happens in any future time step does not depend on what happened before, thus the
FORTRAN program is much like a script which tells decision makers, problems
and choices at what time they have to appear on the scene instead of having them
act dynamically. If one evaluates the structural validity (in the sense of [28, p. 5]
of this simulation implementation, one would be fairly reluctant, as — to use Zeigler's words — it does not "truly reflect the way in which the real system operates
to produce this behaviour".

In terms of the 'non-statement view' the usefulness of a theory depends on
whether one can find a relation between several of its non-theoretical terms which
can be tested against empirical measurements — where it is not necessary that these
measurements be quantitative. [4] does not give a specific intended application of
their model but refer only to what one could call a 'stylised fact' [13, 12] which
shows that "some university organizations" behave like the "organized anarchies"
[4, p. 16], so obviously these are potential partial models of **GC**.

FORTRAN IV was obviously not the most appropriate programming language
to describe a simulation of a theory about sets of different kinds of entities (a more,
although not entirely object-oriented programming paradigm is much more appropriate, as the other two versions will show); it would, of course, have been possible,
even in FORTRAN, to represent the decision structure and the access structure with
functions instead of matrices, but from the point of view of this implementation
where the functions only have to yield very few different values without any time
dependency or dynamic changes, this does not really matter.

4.2   The Fioretti- Lomi-Cacciaguerra Version

One of the more recent replications of the 'garbage can' model [14, p. 3] introduces
an additional term which couples the problems and their solutions. To understand
why this is new one has first to discuss how in this version problems arise and 'find'

their possible solutions. Unlike the original, this version defines the act of assigning a problem to a decision maker with the help of a topography in which decision makers and problems float, thus the entry times of the originals play a minor role (if at all). Instead decision makers find problems in their immediate neighbourhood (in a university environment one could imagine students moving over the campus trying to find staff helping them to solve their problems, problems are then personalised with the students who have these problems — although this interpretation of [14] is not entirely convincing). Elements of the fourth type — opportunities — come into being when problems and choices collide on the two-dimensional topography of the model, and problems are solved when opportunities and decision makers collide. Thus, in a way, problems can only be solved by decision makers when these problems bring their possible solutions with them. It is perhaps difficult to imagine a real-world system to which this version of the theory can be applied — one possible extension of the campus scenario above could be the following: identify 'problems' with secretaries carrying piles of papers to be copied, 'choices' with copying machines, 'opportunity' with piles of paper laid into the copier and 'decision maker' with the technician who re-starts the copying machine after a paper jam.

The ways of tackling problems are much the same in this version: If the technician in the scenario example is sufficiently 'energetic' or competent (this time in the sense of 'skilled'!) and thus succeeds in re-starting the machine, the secretary is happy with the copied papers, and the technician has gained what is again called 'energy' (here one would rather say 'competence' which is often highly correlated with the colloquial 'energy', as her knowledge in repairing copiers was reinforced) — the case of resolution. If the technician fails to re-start the copier because he is not competent or 'energetic' enough, the problem remains unsolved and becomes even more severe than it used to be (the opportunity gains 'energy requested'), he will never be called for help again (leaves the system) — the case of flight. And if secretary and technician both agree that the papers need not be copied on the same day but can wait until the next day when another copier is available which does not need repair, then the opportunity is split into problem and choice and neither the severity of the problem nor the competence of the technician is changed — the case of oversight.

The extension added to the original 'garbage can' model makes some extensions in the 'non-statement view' formalisation necessary. The changes refer to the new object types 'opportunity' as another finite set (which may even be empty) and 'space' as well as to the fact that all four object types now have co-ordinates in two dimensional space which they can change. This makes several additional functions necessary assigning co-ordinates and co-ordinate changes to the elements of the four object types. Thus definition 2 is converted to definition 4 (the definition of the respective theory element is practically tre same as definition 1 with **GC** replaced with **LC**).

Definition 4. Def $M_p(LC)$: β is a potential model of **LC**, i.e. $\beta \in M_p(LC)$ iff $\mathcal{O}, \mathcal{P}, \mathcal{D}, \mathcal{C}, \mathcal{U}, \mathcal{T}, \mathcal{S}, T_u, \Sigma_p, \Sigma_c, \Sigma_u, E_d, E_p, E_c, E_u, \Theta, \Phi, \Pi$ exist such that

1. $\beta = \langle \mathscr{O}, \mathscr{P}, \mathscr{D}, \mathscr{C}, \mathscr{U}, \mathscr{T}, \mathscr{S}, T_u, \Sigma_p, \Sigma_c, \Sigma_u, E_d, E_p, E_c, E_u, \Theta, \Phi, \Pi \rangle$.

2. $\mathscr{O} = \{\langle P, D, C \rangle \mid P \subseteq \mathscr{P}, D \subseteq \mathscr{D}, C \subseteq \mathscr{C}\}$ *is a non-empty finite set [of organisations].*

3. $\mathscr{P}$ *is a non-empty finite set [of problems].*

4. $\mathscr{D}$ *is a non-empty finite set [of decision makers].*

5. $\mathscr{C}$ *is a non-empty finite set [choices].*

6. $\mathscr{U} = \{\langle p, c \rangle \mid p \in P, c \in C, \theta(\pi_p(p,t), \pi_c(c,t)) < \theta_{max}\}$ *is a finite set (which might be empty at least sometimes) [of opportunities] where θ_{max} is the maximum distance between p and c [within which the problem and the choice form an opportunity].*

7. $\mathscr{T}$ *is a set [of points in time].*

8. $\mathscr{S}$ *is a set [of points in some real or virtual space].*

9. $T_u : \mathscr{C} \to \mathscr{T}$ *is a function [that assigns the length of the interval within which an opportunity is in the state 'not grabbed'].*

10. $\Sigma_p : \mathscr{P} \times \mathscr{T} \to \{not\ grabbed, grabbed, dead\}$ *is a function [that assigns the current state $\sigma_p(p,t)$ to a problem $p \in \mathscr{P}$].*

11. $\Sigma_c : \mathscr{C} \times \mathscr{T} \to \{live, dead\}$ *is a function [that assigns the current state $\sigma_c(c,t)$ to a choice (here called solution) $c \in \mathscr{C}$].*

12. $\Sigma_u : \mathscr{U} \times \mathscr{T} \to \{not\ grabbed, grabbed, dead\}$ *is a function [that assigns the current state $\sigma_u(u,t)$ to an opportunity $u \in \mathscr{U}$].*

13. $E_d : \mathscr{D} \times \mathscr{T} \to \mathfrak{R}$ *is a function [$\varepsilon_d(d,0)$ assigns initial "energy" to every decision maker (here called participant); this "energy" is no longer constant over time as it can be increased or decreased; this function is the "energy distribution"].*

14. $E_p : \mathscr{P} \times \mathscr{T} \to \mathfrak{R}$ *is a function [that assigns "energy" to every problem; this is the "energy" necessary to solve the problem ("problem energy requested")].*

15. $E_c : \mathscr{C} \times \mathscr{T} \to \mathfrak{R}$ *is a function [that assigns "energy" to a choice $c \in \mathscr{C}$ at a particular point of time; this is the energy necessary to take this choice ("choice energy requested"); this total necessary "energy" can grow, particularly when the same choice is used to solve several problems in sequence].*

16. $E_u : \mathscr{U} \times \mathscr{T} \to \mathfrak{R}$ *is a function [$\varepsilon_u(u,t) = \varepsilon_u(\langle p,c \rangle, t) = \frac{1}{2}(\varepsilon_p(p,t) + \varepsilon_c(c,t))$ is the energy requested to make use of the opportunity.*

17. $\Theta : \mathscr{S} \to \mathfrak{R}^+$ *is a distance function in the real or virtual space defined above.*

18. $\Phi : \mathscr{S} \to [0^o, 360^o)$ *is a function [which yields the direction in which an entity moves].*

19. $\Pi : (\mathscr{P} \cup \mathscr{D} \cup \mathscr{C} \cup \mathscr{U}) \times \mathscr{T} \to \mathscr{S}$ *is a function which assigns positions in two-dimensional space to elements of the four object types.*

The additional invariants — participants win additional energy when they successfully solve a problem, unsuccessful participants lose their energy to an opportunity and die in the case of flight — are defined in the `energy-drain` procedure in the NetLogo program which represents what has to happen in the case of definition 5 no. 5 .

The full model of the Fioretti- Lomi-Cacciaguerra variant of the 'garbage can' model can now be defined as follows:

Definition 5. Def M(LC) : υ *is a model of* **LC**, *i.e.* $\upsilon \in$ **M(LC)** *iff the following holds:*

1. $\upsilon = \langle \mathcal{O}, \mathcal{P}, \mathcal{D}, \mathcal{C}, \mathcal{U}, \mathcal{T}, \mathcal{S}, \ T_u, \Sigma_p, \Sigma_c, \Sigma_u, E_d, E_p, E_c, E_u, \ \Theta, \Phi, \Pi \rangle$, *i.e.* $\upsilon \in$ **M$_\mathbf{p}$(LC)**.
2. $\forall x \in (\mathcal{P} \cup \mathcal{D} \cup \mathcal{C} \cup \mathcal{U}) \theta(\pi_x(x,t), \pi_x(x,t-1)) = 1$
3. $\forall x \in (\mathcal{P} \cup \mathcal{D} \cup \mathcal{C} \cup \mathcal{U}) \phi(\pi_x(x,t)) \sim UD(0^o, 360^o)$
4. *[Collision of problem and solution, forming an opportunity:]*

$$\forall t \in \mathcal{T}$$
$$\theta(\pi_c(c,t), \pi_p(p,t)) < \theta_{max} \wedge \varepsilon_c(c,t) > 0 \wedge$$
$$\wedge \varepsilon_p(p,t) > 0 \wedge \sigma_p(p,t) = not\ grabbed \rightarrow$$

$$\rightarrow \begin{cases} \sigma_p(p,t) = dead \\ \sigma_c(c,t) = dead \\ u = \langle p, c \rangle \\ \sigma_u(u,t) = not\ grabbed \\ \varepsilon_u(u,t) = \frac{1}{2}(\varepsilon_p(p,t) + \varepsilon_c(c,t)) \\ \tau_u(u) = 0 \\ \sigma_u(u,t) = not\ grabbed \rightarrow \tau_u(u,t+1) = \tau_u(u.t) + 1 \end{cases}$$

5. *[Solution of the problem when participant and opportunity meet:]*

$$\forall t \in \mathcal{T}\ \theta(\pi_u(u,t), \pi_d(d,t)) < \theta_{max} \wedge \sigma_u(u,t) = not\ grabbed \wedge \varepsilon_d(d,t) > 0 \rightarrow$$

$$\rightarrow \begin{cases} \varepsilon_d(d,t) = \begin{cases} \varepsilon_d(d,t-1) + \varepsilon_u(u,t-1) &: \varepsilon_d(d,t-1) > \varepsilon_u(u,t-1) + \varepsilon\ \text{(resolution)} \\ 0 &: \varepsilon_d(d,t-1) > \varepsilon_u(u,t-1) - \varepsilon \quad \text{(flight)} \\ \varepsilon_d(d,t-1) &: otherwise \quad \text{(oversight)} \end{cases} \\[2em] \varepsilon_u(u,t) = \begin{cases} \varepsilon_d(d,t-1) + \varepsilon_u(u,t-1) &: \varepsilon_d(d,t-1) < \varepsilon_u(u,t-1) - \varepsilon \quad \text{(flight)} \\ 0 &: otherwise \quad \text{(resolution or oversight)} \end{cases} \\[2em] \varepsilon_p(p,t) = \begin{cases} \varepsilon_u(u,t-1) &: |\varepsilon_d(d,t-1) - \varepsilon_u(u,t-1)| < 2\varepsilon \quad \text{(oversight)} \\ 0 &: otherwise \quad \text{(resolution or flight)} \end{cases} \\[2em] \varepsilon_c(c,t) = \begin{cases} \varepsilon_u(u,t-1) &: |\varepsilon_d(d,t-1) - \varepsilon_u(u,t-1)| < 2\varepsilon \quad \text{(oversight)} \\ 0 &: otherwise \quad \text{(resolution or flight)} \end{cases} \end{cases}$$

Discussing the **LC**-theoreticity of the terms of this theory core leads to mainly the same results as in the case of the original 'garbage can' model, as the term 'energy' is still only loosely defined. The structural validity of the Fioretti-Lomi-Cacciaguerra version is certainly higher as in the case of the original, as this simulation implementation is dynamic, although the idea of a collision between a problem and a solution, both merging into an opportunity, raises the question how this could work in a real-world scenario apart from the quite artificial description above (with the secretary carrying papers to be copied, a copier with a paper jam and a technician removing the paper jam), as usually problems (and even more so, solutions) do not move around. One could represent problems with documents describing them and circulating through an organisation, but then solutions would be represented by something such as handbooks (which usually do not move around). But anyway, this version dispenses with the fixed decision and access structures and is open with respect to participants' (employees', decision makers') decisions to tackle problems when they show up.

4.3 Skills Instead of Energy, and the Introduction of Teams

As the one-dimensional 'energy' seems a little problematic, at least with respect to measurement issues, and it seems unrealistic to assume that any kind of problem can be tackled with the same kind of 'energy', a different concept characterising both decision makers (or more generally speaking: employees of organisations) and problems (more generally: tasks or assignments) might be more appropriate. To this end, [25] introduced the multidimensional concept of skills, and this concept makes it tempting to analyse the effect of co-operation in teams. With a one-dimensional 'energy' or 'ability' the effect of problem solving only depends on the time a decision maker can spend on a problem (if her 'energy' is not sufficient to solve the problem in one time step, she can continue working on it during the next time steps instead of giving up at the end of the day — which is certainly the case when the problem fails to be resolved in the 'flight' and 'oversight' outcomes). But if skills or abilities are conceived of as multidimensional, decision makers should try to tackle problems for which they are best skilled and leave other problems to colleagues who are fitter to solve them. This idea led (in [25]) to a three-dimensional description of the abilities of decision makers and of the challenges of a problem and to different strategies of finding the optimal match between employee and assignment (which turned out to be the minimum angle between the two vectors describing the abilities of the employee and the skill requirements of the problem — for more details about the alternatives see [25]). This match should minimise both the amount of wasted skills and the average processing time of tasks in an organisation.

But even if a simulated employee chose the most appropriate problem, it would waste part of its skills in many cases (when the fit was not exact). Thus it seemed interesting whether asking a colleague for help and processing the task in a two-person team could further minimise processing time and the amount of wasted skill. To add this element to the theory of organisational behaviour detailed in [25], it was again necessary to add more features to the definition of potentials models of this extended theory (and to remove some other features, as this version seems to be the shortest).

Definition 6. Def $M_p(TF)$ $: \tau$ *is a potential model of* **TF**, *i.e.* $\tau \in M_p(TF)$ *iff* $\mathcal{O}, \mathcal{P}$, $\mathcal{D}, \mathcal{I}, \mathcal{T}, \Delta, A, T_{ep}, T_{ec}, B_p, B_d, \Gamma$ *exist such that*

1. $\eta = \langle \mathcal{O}, \mathcal{P}, \mathcal{D}, \mathcal{I}, \mathcal{T}, \Delta, A, T_{ep}, T_{ec}, B_p, B_d, \Gamma \rangle$.
2. $\mathcal{O} = \{\langle P, D, I \rangle \mid P \subseteq \mathcal{P}, D \subseteq \mathcal{D}, I \subseteq \mathcal{I}\}$ *is a non-empty finite set [of organisations]*.
3. $\mathcal{P}$ *is a non-empty finite set [of tasks]*.
4. $\mathcal{D}$ *is a non-empty finite set [of employees]*.
5. $\mathcal{I}$ *is a non-empty finite set [of skills]*.

6. $\mathscr{T}$ is a set [of points in time].
7. $\Sigma_p : \mathscr{P} \times \mathscr{T} \to \{passive, active, solved\}$ is a function [that assigns the current state $\sigma_p(p,t)$ to a problem $p \in \mathscr{P}$].
8. $\Sigma_d : \mathscr{D} \times \mathscr{T} \to \{idle, busy\}$ is a function [that assigns the current state $\sigma_d(d,t)$ to an employee $d \in \mathscr{D}$].
9. $\Delta : \mathscr{D} \to [0,1]^{|\mathscr{I}|}$ is a function [$\delta(d)$ assigns a real valued vector with elements between 0 and 1 to employee d: competence structure].
10. $A : \mathscr{P} \times \mathscr{T} \to [0,1]^{|\mathscr{I}|}$ is a function [$\alpha(p,t)$ assigns a real valued vector with elements between 0 and 1 to task p: remaining task requirements at time t].
11. $T_{ep} : \mathscr{P} \times \mathscr{D} \to \mathscr{T}$ is a function [$t_{ep}(p,d)$ is the time when an employee d becomes aware of task p].
12. $T_{ec} : \mathscr{P} \times \mathscr{D} \to \mathscr{T}$ is a function [$t_{ec}(p,d)$ is the time employee d needs to perform task p].
13. $B_p : \mathscr{D} \times \mathscr{T} \to \mathscr{P} \cup \{\bot\}$ is a function [$\beta(d,t)$ tells which, if any, task d decides to perform at time t; it might be possible that at some time t no task is available at all].
14. $B_d : \mathscr{D} \times \mathscr{T} \to \mathscr{D} \cup \{\bot\}$ is a function [$\beta(d,t)$ tells which, if any, member d decides to ask for help in performing its current task at time t; it might be possible that at some time t no colleague is available at all].
15. $\Gamma : [0,1]^{|\mathscr{I}|} \times [0,1]^{|\mathscr{I}|} \to [-1,1]$ is a function [$\gamma(v_1, v_2)$ yields the cosine of the angle between the two vectors v_1 and v_2].

As usual $\mathbf{M_p}(\mathbf{TF})$ only describes the terms of this variant of the theory of organisational decision making behaviour, but not the invariants which distinguish potential models of the theory (which use the same terms but obey other axioms) from full models which are then defined in definition 7:

Definition 7. Def $\mathbf{M(TF)}$: φ is a model of $\mathbf{TF}$, i.e. $\varphi \in \mathbf{M(TF)}$ iff the following holds:

1. $\varphi = \langle \mathscr{O}, \mathscr{P}, \mathscr{D}, \mathscr{I}, \mathscr{T}, \Delta, A, T_{ep}, T_{ec}, B_p, B_d, \Gamma \rangle$, i.e. $\varphi \in \mathbf{M_p(TF)}$
2. $\forall t, d \in D^*$

$$\beta_p(d,t) = \begin{cases} \bot & : & P^* = \emptyset \\ p & : & otherwise \end{cases}$$

such that

$$\gamma(\alpha(p), \delta(d)) = \max_{\pi \in P^*} (\gamma(\alpha(\pi), \delta(d)))$$

and

$$\beta_p(d,t) = p \to \sigma_p(p,t) = active$$

where

$$D^* = \{d \in D \mid \sigma_d(d) = idle\}$$
$$P^* = \{p \in P \mid \sigma_p(p) = passive\}$$

3. $\forall t \in \mathscr{T}, d_1 \notin D^*$

$$\beta_d(d_1,t) = \begin{cases} \bot & : & D^* = \emptyset \\ d_2 & : & \textit{otherwise} \end{cases}$$

such that

$$\gamma(\delta(d_1), \delta(d_2)) = \max_{d \in D^*} (\gamma(\delta(d_1), \delta(d)))$$

and whenever $\beta_d(d_1,t) = d_2 \to \sigma_d(d_2) = \textit{busy with } D^* \textit{ and } P^* \textit{ as above.}$
4. $p \notin P^* \wedge \beta_p(d,t) = p \to \alpha(p,t+\Delta t) = \max(0, \alpha(p,i,t) - \Delta t (\delta(d) + \delta(\beta_d(d))))$

In the NetLogo implementation of this scenario there is a one-to-one correspondence between the description of the potential model of the theory and parcels of program code:

- Every run of the program is one element of the set $\mathscr{O}$. Organisations are characterised by their number of `workers` and by the speed at which `new-assignments-per-period` are generated.
- $\mathscr{P}$ and $\mathscr{D}$ can easily be identified with the turtle breeds `assignments` and `workers` both of which own an attribute called `skills-requested` and `skills`, respectively, which are lists (instead of vectors) of real numbers (currently three) and several other attributes needed for bookkeeping.
- $\mathscr{I}$ just defines the length and order of the elements of the lists called `skills` and `skills-requested`.
- $\mathscr{T}$ is represented by the internal `ticks` of NetLogo.
- Σ_p and Σ_d are represented by the respective attributes of `assignments` and `workers` (where the state 'solved' is not part of the implementation, as solved problems are immediately deleted.
- Δ yields the elements of the `skills` list during the initialisation `setup` procedure of the NetLogo implementation.
- A yields the elements of the `skills-requested` list; this function is called whenever a new `assignment` arrives.
- T_{ep} is not necessary in the implemented program as every assignment has its individual `processing-time` which is set to zero when it is generated, and incremented at every `tick`.
- T_{ec} is just the attribute `processing-time` of the `assignment` breed.
- B_p is represented by the function `to-report optimal-assignment` (with the caller as the implicit argument).
- B_d is represented by the function `to-report optimal-colleague` (again with the caller as the implicit argument).
- Γ is represented by the function `to-report similarity` (which can also deal with other measures of similarity beside the angle between vectors).

Although NetLogo's programming language is not a functional declarative but a procedural language, the core of the simulation program can be written in terms of functions.

As for the structural validity of this simulation, the problem with the one-dimensional 'energy' seems solved, but intended applications have never been analysed (as for all earlier versions). But the terms of **TF** can be applied to real-world organisations much more easily than in the two older versions. As in these, the entities $\mathcal{O}, \mathcal{P}, \mathcal{D}, \mathcal{S}$ represent the sets of organisations, assignments (tasks, problems, help desk requests), employees and skills where only the latter seem to be less straightforward, but an enumeration of skills expected can be found in most job advertisements, and their values $\delta(d,t)$ can even be found on university transcripts of records or other kinds of certificates. The empirical problem remains how to measure the skills requested by a task (function A). In general it seems difficult to answer this question, but organisational units such as helpdesks usually have a mechanism to implement the function Γ when they ask customers for a rough description of their problems and then allocate the problem (and the customer) to employees who can best help them. But perhaps A remains a **TF**-theoretical term whose values can only be measured after a successful application of **TF**, namely when problems have been classified and α values have been assigned to representative instances of these problem classes in a way that the overall sum of problem processing times $(\Sigma_{d,p} t_{ec}(p,d))$ has been minimised — if average processing time is the criterion against which the organisation is to be optimised. If the organisation prefers to be optimal with respect to wasted time (or wasted skills of its employees), then $W = \Sigma_{d,p,t,p=\beta(d,t)} w_{d,p,t}$ has to be minimised with

$$w_{d,p,t_e p} = (t_n \delta(d) - \alpha(p,t_{ep}))' \mathbf{1}$$
$$t_n = \max_i \frac{\alpha_i(p,t_{ep}(p,d))}{\delta_i(d)}$$

where t_n is the time worker d needs to solve problem p and $\mathbf{1}$ is a column vector whose elements are all 1 (and whose multiplication with the row vector of the difference between d's competence times t_n and the p's requested skills yields the sum of wasted skills).

5 Conclusion

This chapter presented 'non-statement view' reconstructions of three versions of the 'garbage can' theory of organisational decision making behaviour. The reconstructions were derived from two of the implementations, one in FORTRAN IV, one in NetLogo 1.1 and used to design the implementation of the third variant in NetLogo 5.0.[2] Program code written for these tree tools is not easily translated from one tool to another (not even in the case of the very old NetLogo version 1.1 and the most recent one). It seems that the 'translation' of each of these three programs into the language of the 'non-statement view' makes clear which are the differences and commonalities among them. One can, of course, discuss whether a translation of

---

[2] The program is available at http://ccl.northwestern.edu/netlogo/ models/community/index.cgi as TaskAllocation.nlogo

each of the two older programs into the current version of the NetLogo language would have been easier, but as a procedural language, NetLogo contains portions of code which would not have been necessary in a declarative functional language such as the obsolete MIMOSE (for a description of the correspondence between a 'non-statement view' theory reconstruction and the related MIMOSE program see [23, 24]).

As declarative languages are still rarely used in simulation (but see DRAMS [16], DREAMS [15], SDML [17], where the latter is meanwhile also obsolete), but there are new attempts at defining declarative rule-based environments for agent modelling systems which are expected to foster a very concise description of the invariants of a theory without mixing the details of procedural aspects into the program code (the latter is, of course, still necessary but can be encapsulated in the toolbox while only the important aspects are visible in the code).

The reconstruction of the three theory variants was first used to analyse the original 'garbage can' model and the extension implemented by Lomi, Fioretti and Cacciaguerra in more detail and find out under which conditions these theoretical approaches could be tested against the results of empirical research, as the reconstruction made clear which terms of the theory variants were "either directly observable by the senses or measurable by relatively simple techniques" [3, p. 226] or, more precisely, GC-non-theoretical, and thus can be measured with the help of other theories. For the third variant the reconstruction was used to design a simulation from whose invariants phenomena — relations between entities and their properties — could be derived which are observable, at least in principle. So a double use of the 'non-statement view' could be shown.

Finally, all versions can perhaps be used as templates for intelligent support (or rather: task allocation) systems, for instance in help desks and similar divisions of organisations where an automatic allocations of incoming orders or requests to the most appropriate among the available staff is dsired. And such systems are very likely to comply with most of the items of the definition of intelligent-based systems given in [22].

Acknowledgements. Parts of section 3 and of subsection 4.1 draw on [25] as for a comparison with other versions it was necessary to repeat some of its material.

References

1. Balzer, W., Manhart, K.: A social process in science and its content in a simulation program. Journal of Artificial Societies and Social Simulation 14(4) (2011), http://jasss.soc.surrey.ac.uk/14/4/11.html
2. Balzer, W., Moulines, C.U., Sneed, J.D.: An Architectonic for Science. The Structuralist Program. Synthese Library, vol. 186. Reidel, Dordrecht (1987)
3. Carnap, R.: Philosophical Foundations of Physics. Basic Books, New York (1966)
4. Cohen, M.D., March, J.G., Olsen, J.P.: A garbage can model of organizational choice. Administrative Sciences Quarterly 17, 1–25 (1972)

5. Diederich, W., Ibarra, A., Mormann, T.: Bibliography of structuralism. Erkenntnis 30, 387–407 (1989)
6. Diederich, W., Ibarra, A., Mormann, T.: Bibliography of structuralism II. Erkenntnis 41, 403–418 (1994)
7. Epstein, J.M.: Why model? Journal of Artificial Societies and Social Simulation 11(4) (2008), http://jasss.soc.surrey.ac.uk/11/4/12.html
8. Fioretti, G.: Computer code for the garbage can model (2012), http://www.cs.unibo.it/~fioretti/CODE/GC/index.html (last visit January 26, 2012)
9. Fioretti, G., Lomi, A.: An agent-based representation of the garbage can model of organizational choice. Journal of Artificial Societies and Social Simulation 11(1) (2008), http://jasss.soc.surrey.ac.uk/11/1/1.html
10. Fioretti, G., Lomi, A.: The garbage can model of organizational choice: An agent-based reconstruction. Simulation Modelling Practice and Theory 16(2), 192–217 (2008)
11. Fioretti, G., Lomi, A.: Passing the buck in the garbage can model of organizational choice. Computational and Mathematical Organization Theory 16(2), 113–143 (2010)
12. Heine, B.O., Meyer, M., Strangfeld, O.: Stylised facts and the contribution of simulation to the economic analysis of budgeting. Journal of Artificial Societies and Social Simulation 8(4) (2005), http://jasss.soc.surrey.ac.uk/8/4/4.html
13. Kaldor, N.: Capital accumulation and economic growth. In: Lutz, F.A., Douglas, C. (eds.) The Theory of Capital, pp. 177–222. Macmillan, London (1961) (reprint)
14. Lomi, A., Cacciaguerra, S.: Organizational decision chemistry on a lattice. In: The 2003 Swarmfest Conference, April 13-15 (2003), http://www.nd.edu/ swarm03/Program/Abstracts/LomiSwarm2003.pdf (April 21, 2012)
15. Lotzmann, U., Meyer, R.: A declarative rule-based environment for agent modelling systems. In: The Seventh Conference of the European Social Simulation Association, ESSA 2011, Montpellier (2011)
16. Lotzmann, U., Meyer, R.: Drams - a declarative rule-based agent modelling system. In: Burczynski, T., Kolodziej, J., Byrski, A., Carvalho, M. (eds.) 25th European Conference on Modelling and Simulation, ECMS 2011, pp. 77–83. SCS Europe, Krakow (2011)
17. Moss, S., Gaylard, H., Wallis, S., Edmonds, B.: SDML: A multi-agent language for organizational modelling. Comput. Math. Organ. Theory 4, 43–69 (1998), http://dl.acm.org/citation.cfm?id=593039.593059, doi:10.1023/A:1009600530279
18. Olsen, J.P.: Garbage cans, new institutionalism, and the study of politics. American Political Science Review 95(1), 191–198 (2001)
19. Sneed, J.D.: The Logical Structure of Mathematical Physics, 2nd edn. Synthese Library 35. Reidel, Dordrecht (1979)
20. Stegmüller, W.: The Structure and Dynamics of Theories. Springer, Berlin (1976)
21. Stegmüller, W., Balzer, W., Spohn, W. (eds.): Philosophy of Economics. Studies in Contemporary Economics, vol. 2. Springer, Heidelberg (1981)
22. Tolk, A., Adams, K.M., Keating, C.B.: Towards Intelligence-Based Systems Engineering and System of Systems Engineering. In: Tolk, A., Jain, L.C. (eds.) Intelligence-Based Systems Engineering. ISRL, vol. 10, pp. 1–22. Springer, Heidelberg (2011)
23. Troitzsch, K.G.: Structuralist theory reconstruction and specification of simulation models in the social sciences. In: Westmeyer, H. (ed.) The Structuralist Program in Psychology: Foundations and Applications, pp. 71–86. Hogrefe & Huber, Seattle (1992)

24. Troitzsch, K.G.: Modelling, simulation, and structuralism. In: Kuokkanen, M. (ed.) Structuralism and Idealization. Poznań Studies in the Philosophy of the Sciences and the Humanities, pp. 157–175. Editions Rodopi, Amsterdam (1994)
25. Troitzsch, K.G.: The garbage can model of organisational behaviour: A theoretical reconstruction of some of its variants. Simulation Modelling Practice and Theory 16, 218–230 (2008)
26. Troitzsch, K.G.: Not all explanations predict satisfactorily, and not all good predictions explain. Journal of Artificial Societies and Social Simulation 12(1), 10 (2009), http://jasss.soc.surrey.ac.uk/12/1/10.html
27. Wilensky, U.: NetLogo (1999), http://ccl.northwestern.edu/netlogo
28. Zeigler, B.P.: Theory of Modelling and Simulation. Krieger, Malabar (1985) (reprint, first published in 1976, Wiley, New York, NY)

Cutting Back Models and Simulations[*]

Andreas Pyka[1] and Simon Deichsel[2]

[1] University of Hohenheim
  Hohenheim, Germany
[2] University Bremen
  Bremen, Germany

1 Introduction

Agent-based models (ABMs) range from purely theoretical exercises focussing on the patterns in the dynamics of interaction processes to modelling frameworks which are oriented closely at the replication of empirical cases.[1] Advocates of the "Keep it descriptive, stupid!" (KIDS) approach openly recommend building models as empirically accurate as possible, they want to understand social processes from the bottom up.

This seems to be almost the direct opposite of Milton Friedman's famous and provocative methodological credo *"the more significant a theory, the more unrealistic the assumptions".*[2] Most methodologists and philosophers of science have harshly criticised Friedman's essay as inconsistent, wrong and misleading. By presenting arguments for a pragmatic reinterpretation of Friedman's essay, we will show why much of the philosophical criticism misses the point.

After that, we will use the developed arguments for contesting the claim that good simulations have to rely on descriptively accurate assumptions, which is, in a nutshell a plea for the "Keep it simple, stupid" (KISS) approach.

This plea is followed by a more general plea for dropping the philosophical idea of scientific realism. We give arguments challenging the idea that economic models should be "realistic" in the sense that they (more or less directly) represent mechanisms of the way the world works. We try to show that good economic modelling does not depend on seeing models as representing an external reality at all.

2 A Pragmatic Reinterpretation of Friedman's Methodology

Even more than 55 years after its original publication, Friedman's methodological essay still is *the* classic among all methodological texts for economists. As Daniel Hausman has stated, it is probably the only methodological work that a fair amount of economists has ever read.[3] More philosophically minded readers have

---

[*] Parts of this chapter have been published earlier in Deichsel and Pyka (2009) and Deichsel (2011).

[1] See Tolk, Adams and Keating (2012) for an overview of intelligence-based systems.

[2] Friedman (1953), p.14.

[3] See Hausman (1992), p.162.

A. Tolk (Ed.): Ontology, Epistemology, & Teleology for Model. & Simulation, ISRL 44, pp. 141–156.
springerlink.com   © Springer-Verlag Berlin Heidelberg 2013

usually rejected it as inconsistent, vague or false. A commonly held view reduces Friedman's essay to the point that the assumptions of a theory do not matter because all we should expect from economics is good predictions. This is a misleading interpretation, as we will show.[4]

Before constructively developing our pragmatic interpretation of Friedman's methodology, we will deal with some of the best-known criticisms of his approach.

Daniel Hausman's essay "Why look under the hood" is a paradigmatic example of critique of Friedman's essay.[5] According to Hausman, Friedman claims that the assumptions underlying a model are irrelevant and all that *is* relevant is predictive success.[6] Hausman tries to spot an error in this claim by providing an analogy: Suppose you want to buy a used car. Friedman would say that the only relevant test for assessing the quality of the car is checking whether the car drives safely, economically and comfortably. Looking under the hood and checking the status of the components is not necessary. Hausman claims it is obvious, that no one would buy a car without looking under the hood. In analogy to that, we should check the assumptions of theories as well and not merely rely on predictive success as the only criterion. Hausman takes this to be an argument against Friedman's position.

Such accusations, however, are attacking a straw-man, as Friedman does *not* at all hold the position that the assumptions of a theory are irrelevant. The error in Hausman's argumentation can be made clear by the following comparison: Hausman grants later in the text, that modelling always involves simplification, which is why the assumptions do not need to be perfectly true, but can be *"adequate approximations [...] for particular purposes".*[7] Ironically, in Friedman's essay there is a passage that states just that: *"the relevant question to ask about the "assumptions" of a theory is not whether they are descriptively "realistic," for they never are, but whether they are sufficiently good approximations for the purpose in hand."*[8] This should make it clear, that Friedman does *not* think the assumptions of a theory to be irrelevant. He aims rather at pointing to the deficits of the naïve demand for more realistic[9] assumptions in economics. Friedman takes the contrary position: Model building necessarily requires simplification, i.e. it is inevitably based on unrealistic assumptions. This point will be applied to our discussion of ABMs in section 3.a.

Additionally, Friedman expects good models to "explain much by little".[10] In section 3b we apply this point to the evaluation of ABMs.

---

4 See Schliesser (2005), Schröder (2004) and Hoover (2004) for some other interpretations of Friedman's classic that agree on this point.

5 See Hausman (1992), p.70-73.

6 See Hausman (1992), p.71.

7 Hausman (1992), p.72.

8 Friedman (1953), p.15.

9 Note, that Friedman equates „realistic" assumptions with descriptively accurate ones. Therefore, his thesis is not an ontological, but a methodological one.

10 Friedman (1953), p.14.

Both of these points give arguments, why interesting models must rely on assumptions, which are descriptively unrealistic and why making them more realistic does not lead automatically to better models. Modelling is different from mere abstraction, it necessarily involves construction, which means it is more than just extracting parts from reality. These are all well accepted arguments supporting the view that it can be all right to use unrealistic abstractions. What neither the arguments above nor Friedman imply, is the view that all unrealistic assumptions lead to good models.[11] His point is rather, that some unrealism is necessary and it is even an advantage if it is unrealism of the right kind.

Friedman's arguments about unrealistic assumptions has led to much contradiction and heavy confusion. It is indeed Friedman's fault that he did not formulate his thesis very carefully. He talks in a very general way of "unrealistic assumptions", which is problematic, because both terms "unrealistic" and "assumption" can be understood in many different ways.

Alan Musgrave criticised Friedman's position by distinguishing three kinds of assumptions and trying to show that in all three cases assumptions must be rather realistic than unrealistic.[12] Concerning negligibility and domain assumptions Musgrave's argument seems convincing at first sight: The colours of the traders' eyes *are* negligible, at least in the strict economic domain of analysing the stock-market. In this sense, a model that "assumes away" the influence of eye colour to stock-prices can be called realistic rather than unrealistic. However, Musgrave's argument is more a twist with words than a real refutation of Friedman's position: The negligibility of eyes' colours *can* be called realistic, because it *really* has no effect on the stock market, or unrealistic, because the traders *do* have coloured eyes in reality. Friedman tends to the latter view, but stresses the point, that it is not the "realisticness" of the assumptions that matters, but the implications they yield. From Friedman's perspective, it does not matter if a model that assumes away eyes' colours is called realistic or unrealistic, because, again, it is not the realisticness[13] of the assumption that counts, but whether they are the *adequate* assumptions for the modelling purpose. This is probably the main point of misunderstanding between Friedman and many critics (take (Hausman 1992) again as an example): Friedman recommends evaluating the assumptions only for specific purposes, whereas many of his critics aim at *broad* predictive success. The question is, however, whether broad predictive success is achievable at all. Friedman (implicitly) holds the view that it is not – there is no "theory of everything", so narrowing the domain of theories is always necessary.

---

[11] Sometimes it seems that this view is attributed to Friedman, even if it is obviously absurd. (See e.g. Samuelson (1963), p.233.) Such critics seem to forget that Friedman is accepting only those assumptions that lead to correct predictions. Besides that, it is a simply a logical error to conclude from Friedman's "the more significant a theory, the more unrealistic the assumptions" to the statement that unrealistic assumptions imply significant theories.

[12] See Musgrave (1981).

[13] "Realisticness" is a term introduced by Uskali Mäki in order to distinguish descriptively accurate assumptions from the philosophical position of realism. See e.g. Mäki (1998). This paper equates "realistic assumptions" with "empirically adequate assumptions" and hence avoids the philosophical discussion about realism.

The third class of assumptions Musgrave distinguishes, he calls "heuristic assumptions", which are employed when there is *no* domain where a factor that is assumed away is negligible to the outcome. The "heuristic assumptions" are rules for simplifications that guide researchers and tell them how to proceed, if a theory does not fit to the data. Musgrave states *"At any rate, his central thesis 'the more significant the theory, the more unrealistic the assumptions' is not true of 'heuristic assumptions' either."*[14] It is hard to see how Musgrave wants to judge the realisticness of a heuristic assumption, if he accepts that they are untestable. Heuristic assumptions are rules for simplification, they are *intentionally* unrealistic, they *define* the focus of a research programme. From this perspective, it seems difficult to decide whether they are realistic, *before* looking at the specific models (and their implications) that are based on them. With heuristic assumptions, the question is not whether they are realistic or unrealistic, but whether they are able to generate fruitful lines of research.[15] Musgrave's distinction between three classes of assumptions does not refute Friedman's position, as it fails to show why seemingly implausible heuristic assumptions such as the rationality assumption are always nothing but an error of a theory.[16] As the two other types of assumptions go, Musgrave seems to attack rather a straw-man than Friedman's position (at last in the pragmatic interpretation): Friedman states nowhere that making false negligibility or domain assumptions helps generating significant theories. He says that significant theories are mostly based on unrealistic assumptions, not that any unrealistic assumption creates a significant theory. When Musgrave stresses that wrong negligibility and domain assumptions usually lead to bad theories, this can be interpreted as stressing Friedman's point that the empirical correctness of the implications is relevant for judging assumptions: As soon as wrong negligibility and domain assumptions lead to wrong predictions (which they most probably do quite fast), they are immediately ruled out. If one is serious about the ability to predict, one cannot come up with wrong negligibility or domain assumptions.[17] The case is different with heuristic assumptions, as they are neither directly comparable to reality nor do they lead directly to empirical implications. Here it seems still correct allowing for assumptions that seem prima facie implausible or unrealistic.

Musgrave's paper clarifies what Friedman can*not* mean by unrealistic assumptions, but as we see it, he has failed to refute him. At no point in his essay,

---

[14] Musgrave (1981), p.385.

[15] See Mäki (2000), p.326.

[16] Of course, it is *sometimes* not fruitful to assume rational behaviour. E.g. modelling situations that involve decisions under uncertainty or trying to analyse innovation processes that require creativity are not likely to be fruitfully reconstructed by rational choice models. However, as there *are* situations that *can* be reconstructed by rational choice models, we cannot judge this methodological principle as such on grounds of its (im)plausibility, but only the aptness of a concrete application of the principle.

[17] Marcel Boumans adds that Friedman encourages empirically exploring the domain of a negligibility assumption. See Boumans (2003), p.320.

Musgrave delivers an argument why it should be false to claim that significant theories often rely on seemingly unrealistic heuristic assumptions.[18]

After having discussed the term "assumption" at some length, let us turn to the other term "realistic". As we have seen, Friedman claims that unrealistic assumptions are not a disadvantage of a theory per se. So there must be assumptions that are unrealistic in some sense, but still good ones for the purpose in hand. The following three cases show different interpretations of "unrealisticness" that meet this criterion:

1. In a trivial sense *all* assumptions are wrong, because they are necessarily incomplete. It is not possible to deliver an objective and complete description of the observable world.
2. Apart from the incompleteness, assumptions can be "unrealistic" in a different sense: As models propose hypotheses of causality, they must contain more than what is directly observable, as causal relations are not. This is why assumptions cannot be descriptively realistic in the sense of photographic depictions of the observable world.[19]
3. In a third sense assumptions could be called unrealistic, when they contradict common sense. This is the case e.g. with economic models that make heuristic assumptions such as constant preferences.

Friedman would not see disadvantages for economic modelling in all three cases of "unrealisticness". In the first case, unrealistic assumptions are unavoidable, in the second case going beyond observable reality is necessary for interesting models. The third case leaves open if the "unrealistic" heuristic assumptions are fruitful or not – the fact that an assumption seems implausible, however, is no good argument against it *before* its implications have been explored,[20] otherwise many scientific discoveries such as Galilei's laws of falling bodies would have never been made.

All this shows, that there is indeed support for Friedman's view, that the most important thing about assumptions is not their (seeming) realisticness, but the predictive success of the models that rely on them, because it is hard to judge the adequacy of assumptions before their implications have been checked. Realists like Uskali Mäki argue it is reasonable to assume that the heuristic assumptions isolate factors of mechanisms that are "out there" in the world. In this view, the

---

[18] Musgrave takes Newton's neglection of inter-planetary gravitational forces as an example for a heuristic assumption, but instead of refuting Friedman's claim this seems rather to confirm his view that significant theories are often based on unrealistic heuristic assumptions. See Musgrave (1981), p. 383.

[19] This statement does not touch the philosophical position of scientific realism, which is a *theory* about the truth-status of causal connections in scientific theories. The above argument is headed against a more naïve form of realism, which identifies realism with a one-to-one correspondence to observation.

[20] The rational choice assumption often leads to false implications and cannot be fruitfully adapted to some economic problem fields such as innovation economics, but as there are many problems that *can* be tackled (if only to a certain degree) by the (implausible) rationality assumption (see e.g. the works of Gary Becker), it should be clear that this assumption cannot be rejected beforehand by calling it unrealistic.

assumptions are strategic falsehoods that serve the purpose of isolating mechanisms from the rest of the world.[21] Friedman did not respond to this point, but a pragmatic interpretation would suggest, we should rather suspend judgement on ontological questions like this, because due to underdetermination of theory it is impossible to know whether the mechanisms of models are in fact true of the world or whether they lead to successful predictions without being literally true.[22]

Even if Friedman emphasises the relevance of predictive power, he is conscious that prediction with a high degree of precision is unachievable in economics. When he stresses predictive success as a quality criterion, he has rather pattern-prediction or conditioned predictions in mind than precisely forecasting stock prices.[23]

Friedman is aware that there are many competing aims in science and that predictive accuracy is not the only point of scientific enquiry: Depending on the problem even a less precise theory can be preferable, e.g. if it is easier to apply. Even theories that already have been (constructively) falsified such as Newtonian mechanics are still used today for this reason. This shows that Friedman's argumentation is essentially an *economic* one when it comes to theory evaluation: He asks what we gain by a new theory or more realistic assumptions compared to its costs relative to a given problem.[24] This economic argument will be discussed further and applied to ABMs in section 3c.

To sum up, Friedman pragmatically argues against judging the assumptions of a theory by their "realisticness", because, it is often hard to assess this independently from the rest of the theory and if it *is* possible to assess, realistic assumptions are neither a necessary nor a sufficient condition for the construction of good economic theories. They are not necessary because all theories rely on idealised rather than realistic assumptions and they are not sufficient, because more realisticness on the assumption-side does not automatically lead to better theorising.

3 Application: Descriptive Assumptions Yes or No?

How can we use this pragmatic interpretation of Friedman's methodology for the discussion of ABMs? In the following, we employ our interpretation of Friedman for delivering a critique of simulations that rely, in our opinion, too heavily on empirical data. Following a suggestion by Moss/Edmonds we distinguish between two antagonist simulation approaches called KISS (Keep it simple, stupid!) and KIDS (Keep it descriptive, stupid!).[25] For the sake of a more focused discussion we equate KISS with Friedman's view that the realism of the basic assumptions of a model is not a good criterion to judge it and KIDS with the opposite view,

---

[21] See e.g. Mäki (2008), p.14.

[22] This chapter largely ignores the philosophical discussion about realism and anti-realism because this can be separated completely from questions of theory-evaluation.

[23] See Friedman (1953), p.40.

[24] See Friedman (1953), p.17.

[25] See Edmonds and Moss (2004).

claiming that only models that are as descriptive as possible on the assumption side are likely to generate useful scientific insight.[26]

One of the first examples of simulation in the social sciences are the checkerboard segregation models by Thomas Schelling and they constitute a paradigmatic example for KISS.[27] A short summary is sufficient here to introduce the main idea: In Schelling's segregation models, black and white stones are distributed on a checkerboard, symbolising the black and white inhabitants of a (north-American) city. Now, a certain threshold-share of stones in the neighbourhood, that have a different colour from the stone under consideration, is defined (e.g. 70% have a different colour). If this threshold-ratio is reached for an individual stone, the stone is said "to feel uncomfortable" and as a result is moved away from its original position to the next free spot available. The result of this model was, that even if the threshold-ratio requires only 30% of stones in the neighbourhood having the same colour, complete segregation of the colours on the checkerboard results already after a few rounds of moving stones.[28]

Now, modelling in such a simplistic way is against the KIDS suggestion to include as much empirical data as possible in the assumptions. However, in accordance with our interpretation of Friedman, we hold that descriptively unrealistic assumptions (such as checkerboards as representations of cities) are not necessarily a disadvantage of a model. In the following, we will broaden Friedman's arguments by extending them to a defence of KISS modelling against the KIDS approach.

3.1 Simplification Is Necessary and Inevitable

Full realisticness is neither achievable nor is it desirable.[29] Stressing the need to include as much data as possible seems to suggest that we can come close to full realisticness by using empirical data where it is available and keeping the model general where it is not. However, it remains largely unclear how a model can be "left general" at all. ABMs are by definition models that assume a certain behaviour for agents in a simulation. Leaving an aspect of behaviour completely general would imply the inclusion of theory-free aspects of agent behaviour, which means nothing else than introducing random elements in the model (or recurring to ad-hoc assumptions without being aware of it).

A possibility to deal with this issue is to calibrate the aspects where a model is "left general" by running the simulation numerous times and thereby varying the

---

[26] The distinction between KISS and KIDS does not deny the fact, that there may be cases in which descriptively adequate assumptions are very simple. In these cases, there is no dissent between advocates of KISS and KIDS. In the vast majority however, there is a huge difference between building models that are as descriptively accurate as possible or as simple as possible.

[27] See Schelling (1971) for the locus classicus.

[28] This result is robust under various changes, see e.g. Flache and Hegselmann (2001).

[29] Note, that even the "toughest" sciences like physics make heavy use of idealisations or unrealistic assumptions. Just think of planets as "mass-points"or laws that apply only in vacuum. Without radical simplification, many of the basic laws of physics would have never been found.

parameters by means of a Monte-Carlo approach. This implies the belief that the right assumptions can be found in a quasi-automated manner. However, in our view it seems far too optimistic to believe that descriptively adequate models can be generated in this way. No matter whether we use empirical data for setting up the assumptions or if they emerge after calibration: theoretical considerations heavily influence the process of modelling:

First, every observation involves theoretical pre-assumptions, hence there is no pure observation. Even worse, transmitting the observations into programm-code as it is done in ABM modelling involves even more decisions, which sets limits to an accurate reproduction of the observable reality. If the model is fitted to observation by comparing the results of several runs with the observed patterns in reality, there is even a third layer of theory involved, that is the standard by which the model-results are compared with reality: Often the results of a model need heavy interpretation or statistical analysis so that a comparison with reality is far from being straight forward.

All this shows that it can be misleading to claim highly descriptive assumptions because the proposed ideal is never achievable, due to several inevitable restrictions of theory-ladeness. A high level of descriptive accuracy in ABMs would require a thorough understanding of all the processes involved, so it becomes hard to see what there is left to be learned from actually building the model. In practice, it seems more likely that we do *not* understand the processes under investigation to a high degree, which makes approximation and estimation inevitable. In a highly complex model this probably leads rather to potentiating errors than to generating accurate predictions.[30]

Therefore, when a high level of descriptive accuracy is suggested, this gives no guideline at all for judging exactly *how* high this level should be. The KISS modelling approach is clearer in this respect because it does not judge the assumptions in terms of descriptive accuracy at all. Besides, admitting to use "heroic" simplifications is surely more honest than making necessary simplifications while still claiming high descriptive accuracy.

3.2 Good Models Explain Much by Little

Even if we grant that it might be possible to build descriptively accurate models, there are other arguments why the usage of strategically unrealistic assumptions is sometimes preferable to highly accurate ones: Counterintuitive effects (such as complete segregation being caused by only mild preference for the own colour in the Schelling models) can get easily out of sight if a more realistic and less schematic model is used.

---

[30] Note however, that we do draw the conclusion that simple models are more realistic than complex ones. For advocates of the KISS modelling approach it is crucially important to keep in mind how incomplete such models are and how difficult it is to transfer their results into the real world.

The main argument of the KISS-advocates points to the core of the idea of modelling: we make models in order to *reduce* the complexity of the real world, not to mirror it. Of course, good models do not *neglect* the complexities of the systems they try to represent, but striving towards realisticness in every aspect is synonymous with the rejection of theorising, which can result in a mere collection of facts that may be descriptively highly accurate, but rarely helps explaining matters.[31] This is an argument why the call for more realisticness cannot be sustained as per-se-argument. It depends on the aim of the model, if the level of abstraction is rightly chosen. When the understanding of fundamental mechanisms is the aim, the KISS method still seems the approach of choice. Highly complex models may accurately *generate* output, but they do not enable scientists to understand how it comes about. This happens because complex models often develop their own life and produce artefacts, which can make them difficult to interpret and understand.

This point needs some substantiation, which is why we will illustrate it in some detail by explicitly discussing an agent-based model. For making our critique as strong as possible, we chose a model of aggregate water demand that Edmonds/Moss use to underline the *strengths* of their KIDS modelling approach.[32] Even if this model is far away from being descriptively accurate, which shows once again how difficult-to-achieve this "standard" is in practice, it produces dynamics where the observable regularities are caused by the external shocks that are programmed into the model, which is why it seems not very helpful for explaining the observed dynamics.

Here is a short summary of the model: Agents are distributed at random on a grid. Each agent represents a household and is allocated a set of water-consuming devices, in such a way that the distribution resembles empirically found data from the mid-thames region. The households are influenced in their usage of water-consuming devices by several sources: *"their neighbours and particularly the neighbour most similar to themselves (for publicly observable appliances); the policy agent[33]; what they themselves did in the past; and occasionally the new kinds appliances that are available (in this case power showers, or watersaving washing machines). The individual household's demands are summed to give the aggregate demand."*[34] There is no need to explain the model in full detail here; our main methodological point against this model can be made by looking at the outcome of many runs of the model starting with the same initial conditions:

---

[31] The term "explanation" is itself under philosophical discussion. The covering-law model of explanation is generally considered outdated due to several difficulties. We do not enter in this philosophical discussion here, but stick to a common-sense notion of explanation.

[32] See Edmonds and Moss (2004).

[33] The policy agent suggests a lower usage of water if there is less than a critical amount of rain during a month. This influences the agents to a certain degree.

[34] Edmonds and Moss (2004), p. 8.

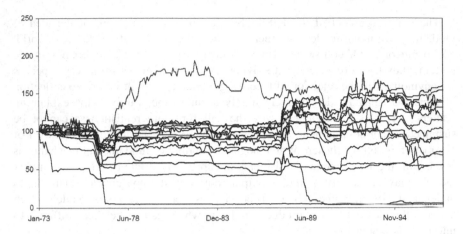

Fig. 1 Aggregate Water demand from model runs specified in (Edmonds and Moss 2004), p.10

Moss/Edmonds note, that "significant events include the droughts of 1976 and 1990, which often show up in a (temporarily) reduced water demand, due to agents taking the advice from the policy agent to use less water. Power showers become available in early 1988 and water-saving washing machines in late 1992 which can cause a sudden increase or decrease respectively."[35]

For us, it is hard to see, how this simulation helps for solving any problem it all. The recognisable effects result from the external shocks that are programmed into the model and even those are tough to identify for our eyes. Again, it is even doubtful wether the authors achieved their own goal of setting up descriptively adequate assumptions, as there are certainly more factors influencing people's demand for water than those integrated into the model. Additionally, the runs depicted above all start from the same specification and yet widely diverge in their outcomes. It seems hard to learn anything from the model as anything is possible from very low water demand to very high and from great changes in water demand to nearly constant consumption. Edmonds/Moss write that the model is made to capture the *range* of water demand responses.[36] Indeed they show that there is a very wide range, but as we see it, they *fail* to give *explanations* how the different results come about. By trying to create a descriptively adequate model Moss and Edmonds arrive at *"a complex model whose behaviour is not fully understood"*[37] and therefore, in our view, does neither explain matters nor does it succeed in being descriptively adequate on both the assumption- and the implication-side.

For us, it seems hard to learn anything from models whose dynamics are not fully understood. The advice given by KIDS-advocates to build models that are as

---

[35] Edmonds and Moss (2004), p.10.

[36] See Edmonds and Moss (2004), p.10.

[37] Edmonds and Moss (2004), p.8.

descriptively accurate as possible can lead to models that can neither predict nor explain in any meaningful sense.

The ability to explain much by little is therefore not only a pragmatic value, but has epistemic relevance as well. Therefore, the KISS approach has an advantage over KIDS in this respect.

3.3 Simplicity Is an Economic Value

Even from a much more down-to-earth point of view, there are advantages of keeping the assumptions simple instead of trying to make them descriptively adequate. Simpler models are not only easier to understand, but they are more tractable as well. In a highly complex model, instead, errors are more difficult to trace, the model is easier and cheaper to validate, it is probably easier to adapt to new situations and it leads more quickly to solutions.[38]

From an economic/pragmatic point of view, there is no such a thing as truth; models are tools for solving problems. Seen like this, simple models are clearly preferable to complex ones, if (and only if!) they achieve the same quality of solution for a given problem. Advocates of highly descriptive models are in charge for explaining the advantages of their models in terms of predictive and explanatory power when descriptive accuracy of the assumptions is rejected as valid criterion due to a pragmatic methodology like Friedman's.

It is highly important to note, that this does not mean that all models should be as simple as possible. As we stressed throughout the paper, models need the *right* level of complexity for the problems they tackle. While simplicity is a value for models, it is in our view secondary, compared to the model's ability to contribute to a better understanding of the phenomena under scrutiny. Nonetheless, taking the value of simplicity seriously means that starting with descriptive accuracy as first criterion is the wrong way for building helpful models.

4 Summary in between: Arguments in Favour of KISS

1. The quarrel about the realisticness of assumptions is often misleading. No theory and no model rests on realistic assumptions. Modelling is always centred on a specific problem. Whether the right level of abstraction was chosen can only be properly assessed with respect to the problem one wants to solve. For example, neoclassic equilibrium theorising should be criticised along these lines: Its abstractions are assuming away interesting aspects of many problems.

2. When models aim at predictive accuracy, more refined assumptions are probably needed compared to models, which aim at reproducing stylised facts. The advocates of a more descriptive modelling approach are right to point at the difficulty of comparing simple models to reality. However, this task does not necessarily become easier when models are based to a high degree on empirical data.

---

[38] See Chwif, Barretto and Paul (2000), p.452.

3. For *economic* reasons, it is more useful to start with building a simple model first and refine it by increasing its complexity if it is not successful.
4. Theories do not emerge out of empirical description. In our view, modeling is a creative process that involves construction and hypothesising. Therefore, non-observable elements must be included in ABMs as well, otherwise they are not likely to improve our understanding of the way the world works.

In accordance with authors who recommend a high empirical orientation, we hold that high generality is incompatible with models that make massive use of empirical data. We agree that abduction is the best way to characterise model building, but we contest the view that this requires the modeller to include as much data as possible when setting up the first version of the model.

We also believe that this is the essential message of Friedman's methodological essay: Assumptions should not be judged on their own, but by looking at their implications as early as possible, making model-building a process of continuous revision. So finally, our plea is neither in favor nor against highly descriptive models, but in favor of starting simple and reaching the right level of sophistication for the problem at hand. This raises the question, whether we think models and simulations can and should be realistic descriptions of the world after all. We will discuss this in the final section.

5 Philosophical Digression - What's Realism Got to Do with It?

It might be argued against the KISS methodology, that models which are built on simple assumptions cannot be realistic representations of the world. In the following, we deliver arguments supporting our conviction that good models do not have to be realistic in the above sense and that the project of scientific realism bears no normative weight whatsoever.

Every discussion of realist philosophy of science must necessarily begin by distinguishing the different forms of realism and by declaring what is exactly at issue. The following list provides an overview of different realist positions in philosophy of science, in ascending order, by the strength of claims being made:

1. Ontological realism: This is the most modest realist claim and merely entails the belief in the theory-independent *existence* of an external reality.
2. Weak epistemic realism: Scientific theories refer to an external reality and *may be* right in their claims about it, i.e. they are capable of being true or false. This includes the *semantic* thesis that theories are true if and only if they correctly refer to an external reality.
3. Scientific Realism/strong epistemic realism: Well-confirmed scientific theories refer to an external reality and *are* basically right in their claims about it.[39]

---

[39] This list is not meant to be exhaustive. The qualification that theories are only "basically" right allows for structural realism as well (see Worrall (1989) for the locus classicus).

Both weak and strong epistemic realism are deeply connected with a correspondence theory of truth because their central point is to make claims about the properties of an external reality. If those realists would rely on a coherence- or consensus-theory of truth, this would directly beg the question. In this paper, we take anti-realism as the thesis that we should suspend judgement on the truth and truth-worthiness of our theories or avoid talking about the truth of theories altogether in order to minimize the confusions that surround this concept.[40]

Let's start with the famous "no-miracle" argument for scientific realism. In its most basic version it simply states that realism is the "only philosophy that doesn't make the success of science a miracle"[41]. It states that the success of scientific theories can be *explained* by claiming that these theories capture elements of an external reality. It is true that anti-realism cannot offer such an explanation, but the crucial question is, whether the realist move is an explanation at all. Often, it seems that the realist's arguments are begging the question of the anti-realists, and vice versa. We believe that this is the case for the "no-miracle" argument as well. The anti-realist would claim that we are *not* justified in explaining the success of science by its truth[42] because theories could possibly be successful without being true, due to empirical underdetermination.[43] In short, scientists often accept those theories that work well and that is all there is to say. Accepting truth (in the sense of correspondence) as the best explanation for their success means to go beyond the borders of what we can legitimately infer. From this view, the suggestion that truth explains the success of theories is no explanation at all – it is rather an illegitimate *ad-hoc* statement. We could equally argue that the existence of God is the best explanation why our theories work, but anti-realists are convinced we should not do that on the same grounds why we should not "explain" success by a correspondence to an independent reality. In both cases, the explanation is based on uncertain ontological claims. However, what we *can* know, is whether a theory is helpful for solving our problems because that is a completely subjective judgement which does not involve an ontological claim.[44]

A stronger argument in favour of anti-realism is the fact that even inconsistent theories can "work" which shows that taking truth as an explanation for success is problematic because the truth can hardly be inconsistent.[45]

Once we talk about the acceptance of the "inference to the best explanation"[46] the quarrel between realists and anti-realists gets more complicated. In her daily work, an anti-realist may accept and use some theories because she holds them to

---

[40] Note that we do not claim that no theory can be possibly true - there may well be theories that are true (even if just by chance) but we should avoid talking about the truth of theories.

[41] See Putnam 1975, p. 73.

[42] Keep in mind that we assume that realism is committed to a correspondence theory of truth by definition.

[43] Underdetermination claims that two theories can both be empirically adequate while making different claims about reality.

[44] Laudan (1996)) provides details on the problem-solving approach to scientific progress.

[45] See Da Costa and French 2002, p. 105.

[46] The "inference to the best explanation" denotes the argument stating we are entitled to accept the current best explanation of a phenomenon as true.

154 A. Pyka and S. Deichsel

be the best explanation for a phenomenon under scrutiny. For example, the anti-realist may accept increased demand for a good as the best explanation for a rising price. Now the realist can ask why the anti-realist stops short of accepting realism as the best explanation for the success of theories and hence does not give up his anti-realist position. At this point, it becomes clear why the "no-miracle" argument is question-begging and cannot settle the argument between realists and anti-realists: both may be willing to *accept* best explanations, but the anti-realist never asserts the *truth* of the explanations she accepts and so will not accept truth as the best explanation for success.

Furthermore, the argument that scientific theories can fail does not refute anti-realism. It merely supports what we have dubbed "ontological realism", i.e. the view that there is an external reality, which can be *in*compatible with our theories. However, it does not show that those theories which *are* compatible with the external reality are such because they are "true" or "realistic".

The arguments given above show at minimum, that the traditional justifications for realism cannot settle the dispute. In order to show why realism is not needed, we will argue that "truth" is almost always replaceable by other terms that are ontologically more parsimonious (such as empirical adequacy[47] or fit with the totality of current knowledge[48]).

If scientists (in contrast to philosophers) want to assess theories, they almost always want to know how well they work, not why. The on-going battle between realism and anti-realism in traditional epistemology can be separated completely from issues pertaining to theory appraisal. Even if there was a conclusive proof in favour of scientific realism this would still allow for a purely instrumental way of assessing theories, i.e. deciding how well they are suited for solving given problems, since this question can be completely separated from their truth-status.

It should be clear that this does not imply that there is nothing acceptable in the realist's prescriptions, even if they may stem from the wrong reasons. For example, within the assumption debate, the realists carefully distinguish between assumptions that isolate real factors and others that merely serve the tractability of economic theory. A certain type of anti-realism may accept the message that it is important to filter out the crucial, the fundamental or the necessary assumptions of a theory even if it would hesitate to call them real. Such a procedure could be called "anti-realist ontology" as it is a venture into the status of the very fundamentals of economics and by this it would save the lessons from one of the realists' preferred projects, without committing to a version of ontological realism (as defined above).

Another possible form of anti-realism may even agree with Mäki's recommendation of developing useful surrogate models for analysing the real world instead of playing with substitutes[49], but in contrast to Mäki, the anti-realist would not ask whether a model is representing "the real world" but would focus on its ability to shed light on real problems. If the problem to be solved is one of

---

[47] (van Fraassen (1980)) is the locus classicus of a defence for this criterion.

[48] In the sense of (Quine and Ullian (1970)).

[49] See e.g. Mäki (2009), chapter 4.

policy-consulting, it should be clear to the anti-realist, too, that research on the formal aspects of some general equilibrium model can become a dangerous substitute for practically relevant economic research. However, if some formal aspects are indeed the problem a scientist wants to deal with, the anti-realist must accept this and cannot urge her to concentrate on surrogate models. A type of anti-realism could indeed accept a kind of "as-if-realism", which accepts many arguments and terminological points of realism, but rejects the interpretation that theories or parts of them are literally true.[50] With this in mind, the anti-realist could actually talk about more "realistic" assumptions when she uses a coherence theory of justification instead of a correspondence theory of truth.[51] The debate about realism against anti-realism would then be merely about semantics bearing no pragmatic implications whatsoever. In this case, the more realistic assumptions would be the ones that fit better to the totality of our current beliefs.[52] It is however another main point (that we argued for extensively above), that more realistic assumptions (even in this weak sense!) are not always the better ones, but that we should rather look for adequate idealisations for the problem at hand instead of mechanically heading towards more realisticness.

6 Conclusions

The arguments given above show on the one hand, why starting with simple models is in our view preferable to starting with descriptively rich models and on the other hand, why scientific realism concerning economic models is in our view ill-conceived. There are forms of anti-realism that can save much of the healthy critical impact of realist philosophy of science, but are far more epistemologically modest concerning the ontological status of theories. In short: Occam's razor cuts back realism as well.

References

Boumans, M.: How to Design Galilean Fall Experiments in Economics. Philosophy of Science 70, 308–329 (2003)

Carnap, R.: Empiricism, semantics and ontology. Revue Internationale de Philosophie 4, transcribed into a pdf by Andrew Chrucky 1997 (1950)

Chwif, L., Barretto, M.R.P., et al.: On simulation model complexity. In: Proceedings of the 32nd Conference on Winter Simulation, pp. 449–455 (2000)

da Costa, N.C.A., French, S.: Partial Truth and Partial Structures. Oxford University Press, Oxford (2002)

Deichsel, S.: Against the pragmatic justification for realism in economic methodology. Erasmus Journal for Philosophy and Economics, 23–41 (2011)

---

[50] A similar argument is made by (Carnap (1950)).

[51] The addition of a semantic correspondence theory of truth to a coherence theory of justification is in fact the only feature that clearly distinguishes Mäki's realism from the anti-realism presented here. See (Peter (2001)).

[52] This suggestion is inspired by (Quine and Ullian (1970)).

Deichsel, S., Pyka, A.: A pragmatic reading of Friedman's methodological essay and what it tells us for the discussion of ABMS. Journal of Artificial Societies and Social Simulation 12(4), 6 (2009)

Edmonds, B., Moss, S.: From KISS to KIDS – 'An Anti-simplistic' Modelling Approach. In: Davidsson, P., Logan, B., Takadama, K. (eds.) MABS 2004. LNCS (LNAI), vol. 3415, pp. 130–144. Springer, Heidelberg (2005)

Flache, A., Hegselmann, R.: Do Irregular Grids make a Difference? Relaxing the Spatial Regularity Assumption in Cellular Models of Social Dynamics. Journal of Artificial Societies and Social Simulation 4 (2001)

Friedman, M.: The Methodology of Positive Economics. In: Friedman, M. (ed.) Essays in Positive Economics, pp. 3–43. University of Chicago Press, Chicago (1953)

Hausman, D.M.: Essays on Philosophy and Economic Methodology. Cambridge University Press, Cambridge (1992)

Hausman, D.M.: The Inexact and Separate Science of Economics. Cambridge University Press, Cambridge (1992)

Hoover, K.D.: Milton Friedman's Stance: The Methodology of Causal Realism. Working Paper University of California, Davis 06–6 (2004)

Laudan, L.: Beyond Positivism and Relativism: Theory, Method, and Evidence. Westview Press, Boulder (1996)

Mäki, U.: Entry 'As If'. In: Davis, J.B. (ed.) The Handbook of Economic Methodology, pp. 25–27. Edward Elgar, Cheltenham-Northampton (1998)

Mäki, U.: Kinds of Assumptions and Their Truth: Shaking an Untwisted F-Twist. Kyklos 53/3, 317–335 (2000)

Mäki, U.: Realistic realism about unrealistic models. In: Kincaid, H., Ross, D. (eds.) A Handbook of the Philosophy of Economics, pp. 68–98. Oxford University Press, Oxford (2008)

Mäki, U.: Missing the world. Models as isolations and credible surrogate systems. Erkenntnis 70(1), 29–43 (2009)

Musgrave, A.: 'Unreal Assumptions' in Economic Theory: The F-Twist Untwisted. Kyklos 34/3, 377–387 (1981)

Peter, F.: Rhetoric vs realism in economic methodology: a critical assessment of recent contributions. Cambridge Journal of Economics 25(5), 571 (2001)

Putnam, H.: Mind, Language and Reality. Philosophical Papers, vol. 2. Cambridge University Press, Cambridge (1975)

Quine, W.V.O., Ullian, J.S.: The Web of Belief. McGraw Hill, New York (1970)

Samuelson, P.: Problems of Methodology - Discussion. American Economic Review 54, 232–236 (1963)

Schelling, T.C.: Dynamic Models of Segregation. Journal of Mathematical Sociology 1(2), 143–186 (1971)

Schliesser, E.: Galilean Reflections on Milton Friedman's "Methodology of Positive Economics", whith Toughts on Vernon Smiths's "Economics in the Laboratory". Philosophy of the Social Sciences 35, 50–74 (2005)

Schröder, G.: Zwischen Instrumentalismus und kritischem Rationalismus? – Milton Friedmans Methodologie als Basis einer Ökonomik der Wissenschaftstheorie. In: Pies, I., Leschke, M. (eds.) Milton Friedmans Liberalismus, pp. 169–201. Tübingen, Mohr Siebeck (2004)

Tolk, A., Adams, K.M., Keating, C.B.: Towards Intelligence-Based Systems Engineering and System of Systems Engineering. In: Tolk, A., Jain, L.C. (eds.) Intelligence-Based Systems Engineering. ISRL, vol. 10, pp. 1–22. Springer, Heidelberg (2011)

van Fraassen, B.C.: The Scientific Image. Oxford University Press, Oxford (1980)

Worrall, J.: Structural Realism - The Best of Both Worlds? Dialectica 43, 99–124 (1989)

Philosophical Aspects of Modeling and Simulation

Tuncer Ören[1] and Levent Yilmaz[2]

[1] University of Ottawa
  Ottawa, Ontario, Canada
[2] Auburn University
  Auburn, AL, United States

Abstract. To examine philosophical foundations of Modeling and Simulation, we present and clarify relations between reality, representations of reality, and simulation. The role experimentation and experience are delineated along with purposes of simulation, knowledge generation via simulated experimentation, and ethics. In relation to experimentation, the need for computational reproducibility and replicability are emphasized to improve credibility of simulation studies.

1 Introduction

A comprehensive background on simulation may be useful before one embarks to the study of the philosophical aspects of simulation; otherwise, claims about simulation may be similar to the fable of the descriptions of an elephant by blind people. Several articles were prepared about the big picture (Ören, 2007, 2009, 2010). Two recent articles cover several perceptions of simulation. One article (Ören 2011a) covers a collection of about 100 definitions where definitions are grouped under three classes and classified in nine types. Another article (Ören, 2011b) provides definitions of simulation from a contemporary comprehensive framework. A recent publication documents the richness of M&S and a list of about 500 types of simulation (Ören, 2012).

In the remaining sections, the following is done: In section 2, the close relationship of reality, representation of reality, and simulation is clarified and presented in twenty paradigms. Two important aspects of scientific, technological, engineering, and entertainment functions of simulation, namely experimentation and experience, are covered in sections 3 and 4. In sections 5 to 8, goals of simulation; knowledge generation and simulation; thinking, experience, and simulation; and ethics and simulation are elaborated on. Section 9 consists of our conclusions.

A. Tolk (Ed.): Ontology, Epistemology, & Teleology for Model. & Simulation, ISRL 44, pp. 157–172.
springerlink.com

2 Reality, Representation of Reality, and Simulation

In all types of simulation, references are made to reality and representations of reality. In this context "reality" is considered from a pragmatic point of view; since we realize that even at the beginning of the 21st century, we don't have a clear understanding of dark matter and dark energy which are claimed to make over 95 % of the universe (Nasa-science). Furthermore, cosmologist talk about multiverses that we don't know yet (Ellis, 2011). However, simulation is used even in cosmological studies.

In this section, we explore different types of relationships of reality and its representation. For this purpose, as outlined in Table 1, we consider the reality-model dichotomy from several perspectives such as historic paradigm, goal of knowledge processing, philosophy of science, psychological and artistic points of view, simulism, and modification of reality.

Table 1 Relationships of reality, representation of reality, and simulation: From different perspectives

History
        • imitation • pretence • fake
Goal of knowledge processing
        • experimentation in design problems, in decision support
        • experience in 3 types of training, in entertainment
        • augmented reality
Philosophy of science – simulation system is an:
        • executable hypothesis • executable theory
Psychological
        • simulator
Artistic
        • mimesis
Simulism
        • simulation is reality
Modification of reality
        • misperception
            simultanagnossia
        • misunderstanding – due to lack or limitations of:
            background knowledge, perception/conceptualization, evaluation ability
        • distortion – due to:
            psychological conditions (illusion), physiological conditions (hallucination), dissimilation

2.1 Reality-Model Dichotomy: Historic Points of Views

The term simulation existed in English since middle 14th century AD. Simulation has been used to mean imitation pretence, or fake. Table 2 outlines the historic paradigms of the evolution of the aspects of reality-model dichotomy as well as relationship with simulation. Most of these historic and original meanings of simulation are also used in some contemporary usage.

Table 2 Historic point of views

Reality Model	Para-digm	• Clarification of reality-model dichotomy - Definition of simulation
Reality → Imitation (*Imitated reality*)	2.1	• A model is an *imitation* of reality to be used instead of reality. Hence, model is imitated reality; e.g., imitated (simulated) leather, imitated (simulated) pearl. - *Simulation is* imitation of reality.
Reality → Pretence (*Pretended reality*)	2.2	• A model is a *pretence* to represent some aspect(s) of reality. Hence, model is pretended reality. - *Simulation is* pretension.
Reality → Fake (*Fake reality*)	2.3	• A model is a *fake* representation of reality; it is used to deceive. Hence, model is fake reality - *To simulate* is to fake.

From the *imitation* point of view, a model is an imitation of reality to be used instead of reality. Hence, a model is imitated reality; e.g., imitated (simulated) leather, imitated (simulated) pearl. Therefore, simulation is imitation of reality. The *pretence* aspect of simulation implies that a model is *pretence* to represent some aspect(s) of reality. Hence, model is pretended reality. Therefore, to simulate is to pretend; and simulation is pretension. When the pretention is done for the sake of *deceit*, a model is a *fake* representation of reality; it is used to deceive. Hence, model is fake reality. Therefore, to simulate is to fake.

2.2 Reality-Model Dichotomy: Goal of Knowledge Processing

From the point of view of goal of knowledge processing, three main categories of reality-model dichotomy can be identified (Tables 3 a, b). Five paradigms are particularly relevant with the modeling and simulation discipline.

Table 3a Goal of knowledge processing (experimentation, experience)

Reality Model	Para-digm	• Clarification of reality-model dichotomy - Definition of simulation
Reality to be generated → *Design (or a model)*	3.1	• In model-based engineering, a *design* (or a model) is the essence of a real system to be realized. - *Simulation is* goal-directed experimentation with dynamic models.
Existing, conceived or imagined reality → Representation	3.2	• Model is a *representation* of reality in *decision problems* in general as well as in *analysis problems* in model-based science, and in *control problems* in model-based engineering. - *Simulation is* goal-directed experimentation with dynamic models.
	3.3	• In *training,* a model is a *representation* of an existing or imagined reality. - *Simulation is* providing experience by using a model for developing and/or enhancing skills.
	3.4	• In *entertainment,* a model of an existing or non-existing reality is used to provide experience. - *Simulation is* providing or getting experience by using a model for amusement.

In paradigm 3.1, a design (or a model) precedes reality, as it is the case of most engineering problems. Simulation is used to test whether the design would satisfy the requirements. In this case, simulation is goal-directed experimentation with dynamic models. In paradigms 3.2, 3.3, and 3.4, a representation (or a model) of an existing, conceived, or imagined reality is used for different purposes:

Paradigm 3.2 corresponds to use of simulation in decision problems. Model is a representation of reality in decision problems in general as well as in analysis problems in model-based science, and in control problems in model-based engineering. Model is used for experimentation purposes. Hence, simulation is goal-directed experimentation with dynamic models.

In paradigm 3.3 and 3.4 a representation (or a model) of an existing, conceived, or imagined reality is used to provide experience for training (3.3) and for entertainment (3.4) purposes. Hence, in paradigm 3.3, simulation is providing experience by using a model (i.e., a representation) of an existing, conceived, or imagined reality for developing and/or enhancing anyone of the three types of skills: Motor skills by virtual simulation, decision making skills by constructive simulation, and operational skills by live simulation.

Paradigm 3.4 correspond to use of simulation for *entertainment* purposes; in this case, a model of an existing or non-existing, i.e., imagined reality is used to provide experience. Hence, *simulation is* providing or getting experience by using a model (i.e., a representation) of an existing or imagined reality for entertainment. Mutual benefits of simulation games (i.e., use of simulation to gain experience for entertainment purposes) and use of simulation to gain experience for training purposes is well known. Especially, the advanced visualization techniques and scenarios used in simulation games can indeed be very useful for simulation used to provide experience for training purposes. The American philosopher John Dewey expresses it as "There is no contrast between doing things for utility and for fun." (Dewey, 1910, p. 167.)

In the case of augmented reality (3.5), output of simulation can be superimposed to reality to enrich or to enhance it. Simulation is experimental (i.e., relating to, or based on experience or experiment) knowledge generation to enrich or augment reality. Mobil device-triggered simulation with e-lens or e-glass possibility for display opens a wealth of new application areas.

Tolk et al. (2011) posit that "for the simulation system, implemented model is reality." However, in an early brief article, Golomb (1970) had several recommendations about "don'ts of mathematical modeling" such as: "No model is ever a perfect fit to reality. Deductions based on the model must be regarded with appropriate suspicion. . . . Don't extrapolate beyond the region of fit. The world is flat, locally. . . . Don't apply any model until you understand the simplifying assumptions on which it is based, and can be tested. . . . Don't believe that the model is the reality. . . . and don't eat the menu." One can prepare food based on a recipe, but a recipe is not food. Similarly, based on a model, one can build a system, especially in engineering applications; however, the model (or the design) is still a model or design.

Table 3b Goal of knowledge processing (augmented reality)

Reality	Model	Para-digm	• Clarification of reality-model dichotomy - Definition of simulation
Augmented reality by simulation	Simulation with an existing or non-existing model of reality	3.5 augmented reality	• In augmented reality, output of simulation can be superimposed to reality to enrich or to enhance it. - *Simulation* is experimental knowledge generation to enrich or augment reality.

2.3 Reality-Model Dichotomy: From the Perspective of Philosophy of Science

Tolk et al. (2011, p. 5) make an important observation and posit that "following the philosophy of science, a simulation system is an executable hypothesis or –or once proven to be valid– an executable theory." Indeed, models can be used to generate model behavior; hence, model-bases have definite superiority over data-bases (Ören, 1984).

A fundamental importance of the philosophical implication on M&S is the positivist paradigm promoted by the French philosopher Auguste Comte (1844). In positivist paradigm, in philosophy, it is believed that "there is an objective reality" from the ontological point of view, and representational epistemology assumes that "people can know this reality and use symbols to accurately describe and explain this objective reality."

From a philosophical point of view, simulation model validation is influenced by the traditional view of model validity in operations research (Landry et al. 1983; Oral and Kettani 1993). To better understand the emergent validation approaches and their rationale, it is useful to be aware of the historical evolution of model validation philosophies. Here, we use the classification advocated in (Landry et al. 1983) and (Derry et al. 1993). In this classification the traditional reductionist/logical positivist school would see a valid model as an objective representation of the system under study. That is, the model is either correct or incorrect for its domain of application. On the other hand, pragmatist and holistic schools that promote systems thinking viewpoint would consider a model valid on the basis of qualitative and subjective evaluations of its contextual usefulness. In this school of thought, a model is not considered to be absolutely correct or incorrect, but rather subjective analysis of qualitative characteristics is considered essential for its acceptability and credibility. Since simulation model validity means "adequacy with respect to a purpose", validation needs to have qualitative and subjective evaluation components. The detailed discussion of the above philosophical perspectives is beyond the scope of this paper; therefore, we refer reader to (Naylor and Finger 1963; Derry et al. 1993), which provide a detailed overview of major schools of thought in philosophy of science that affected validation during the early years of simulation modeling.

2.4 Reality-Model Dichotomy: Psychological Paradigm

Use of terms evolves through time. For example, the word "computer" was first used to denote a person who computes. Similarly, the term "simulator" was originally used in early 19th century to denote a person who copies or feigns. Within the psychological paradigm, a simulator is a person who thinks as the object person to understand him/her; e.g., mindreading as it is the case in mental simulation. Hence, from the psychological paradigm, simulation is usually equated with role-taking, or imaginatively "putting oneself in the other's place." (Table 4).

Table 4 Psychological paradigm

Reality Model		Para-digm	• Clarification - Definition of simulation -- Relationship with simulation
A person	A simulator (another person) thinks as the object person	4.1	• A simulator (another person) thinks as the object person to understand him/her; e.g., mindreading. (mental simulation) -- Simulation is usually equated with role-taking, or imaginatively "putting oneself in the other's place."

2.5 Reality-Model Dichotomy: Artistic Paradigm

In art, an existing or conceived "reality" is a source of *inspiration* and is a "model" for the "artistic creation" which becomes the "reality", i.e., the "artwork (Table 5). In plastic art (such as sculpture), visual art (such as painting and photography), literature as well as its visualization (theater and movies), often existing or non-existing (i.e., imagined or conceived) reality is used as a *source of inspiration* to create an artwork (or work of art, or object d'art).Hence, simulation in art is imitation or re-enactment. The term mimesis –from Greek mimeisthai "to imitate"– denotes the imitative representation of nature and human behavior in art and literature (Rapp, 1984).

Table 5 Artistic paradigm

Reality Model	Para-digm	• Clarification of reality-model dichotomy - Definition of simulation -- Relationship with simulation
A person ← A simulator (another person) thinks as the object person	5.1	• A simulator (another person) thinks as the object person to understand him/her; e.g., mindreading. (mental simulation) -- Simulation is usually equated with role-taking, or imaginatively "putting oneself in the other's place."

2.6 Reality-Model Dichotomy: Simulism Paradigm

"The Simulation Hypothesis (simulation argument or simulism) proposes that reality is a simulation and those affected are generally unaware of this" (Wiki-simulism). In simulism (Table 6), a *created model* –a representation of something which may or not exist– becomes reality (Baudrillard, 1998). Simulism is "We live in a simulation" point of view. In simulism, "*Simulation is* reality".

Table 6 Simulism paradigm

Reality Model	Para-digm	• Clarification -- Relationship with simulation
A person ← A simulator (another person) thinks as the object person	6.1	• A simulator (another person) thinks as the object person to understand him/her; e.g., mind reading. (mental simulation) -- Simulation is usually equated with role-taking, or imaginatively "putting oneself in the other's place."

2.7 Reality-Model Dichotomy: Modification of Reality Paradigms

A representation of reality may be different than reality under several conditions such as: misperceived reality, misunderstood reality, distorted reality, deliberately distorted reality, apparent reality, and unknown reality (Table 7).

Table 7 Modification of reality paradigms

Reality	Model (or representation) of reality	Para-digm	• Clarification - Definition of simulation -- Relationship with simulation
Reality →	*Misperceived reality*	7.1	• *Misperceived reality* due to disorder of visual attention. Simultanagnosia, i.e., inability to experience perceptions as component of a whole.
Reality →	*Misunder-stood reality*	7.2	• *Misunderstood reality* due to: - Lack of or limitations of background knowledge, including theoretical knowledge; - Distorted perception or conception of reality due to limited discrimination abilities of (a) human senses and/or (b) sensing devices; - Lack of or limitations of evaluation ability of knowledge perceived/conceived with respect to the background knowledge.
Reality →	*Distorted reality*	7.3	• *Distorted reality* due to *psychological conditions*; such as illusion
		7.4	• *Distorted reality* due to *physiological conditions*; such as hallucination
Reality →	*Deliberately distorted reality*	7.5	• *Deliberately distorted reality* • Distorted reality is aimed to appear as the reality • Dissimulation
Reality	*Unknown*	7.6	• Lack of theoretical knowledge • Lack of instruments • Lack of will
Reality ←	*Apparent reality*	7.7	• In apparent reality, a representation (or model) of reality appears to be the reality

The case of *misperceived reality* (7.1) is a clinical condition due to disorder of visual attention. Simultanagnosia is inability in perceiving more than one object simultaneously and results in inability to experience perceptions as components of a whole. Two types exist and can also be simulated.

Misunderstood reality (7.2) is an important research area (Ören and Yilmaz, 2011) and is due to a combination of (1) lack of or limitations of background knowledge, including theoretical knowledge; (2) distorted perception or conception of reality due to limited discrimination abilities of human senses and/or sensing devices; and (3) lack of or limitations of evaluation ability of knowledge perceived/conceived with respect to the background knowledge. Misunderstanding is also affected by cultural background, personality, as well as emotional conditions.

Distorted reality can be due to *psychological conditions*, such as illusion (7.3) or due to *physiological conditions* such as hallucination (7.4). Simulation is applied in both of them.

Dissimulation is deliberate distortion of reality (7.5). Distorted reality is aimed to appear as the reality and includes "halo effect". Dissimulation is often done and the onus is on the target person (or the system) to detect it. Once was expressed as: "Computers can make mistakes; however, humans would lie, shamelessly" (Ören, 2005). However, under certain conditions, nations are recommended to benefit from dissimulation as the following quotation from The Art of War of Sun Tzu (c.500–320 B.C) illustrates it: "*All warfare is based on deception. Hence, when able to attack, we must seem unable; when using our forces, we must seem inactive; when we are near, we must make the enemy believe we are far away; when far away, we must make him believe we are near.*" Fanatics may also use dissimulation to impose their views.

Paradigm 7.6 covers the *unknown reality*, due to lack of theoretical knowledge, lack of instruments, or lack of will to recognize reality. In *apparent reality* (7.7), a representation (or model) of reality appears to be the reality; for example, on a clear sky, the view of the stars is an apparent view; since some of them may have ceased to exist even though their light may still travel. Belgian surrealist painter René Magritte (1898-1967) has a well known picture of a pipe. The caption on the picture is "this is not a pipe." Indeed, what we see is a picture of a pipe and not a pipe.

3 Experimentation and Simulation

As we have seen in section 2.2, experimentation is one of the pillars of simulation. A taxonomy of types of experiments are shown in Table 8 (Ören 2011b).

Experiments can be real experiments or virtual experiments. Real experiments, also called physical experiments, can be field experiments, lab experiments, or

Table 8 A Taxonomy of Types of Experiments

		In general	In life sciences
Physical experiment (real experiment)	Field experiment	In situ experiment	In vivo experiment
	Lab experiment	Ex situ experiment	In vitro experiment
	Computer-based experiment	(Computer-based) simulation (In silico experiment)	
Virtual experiment	Thought experiment		

computer-based experiments. Virtual experiments are known under the name thought experiments. Table 9 Provides definitions and explanations of different types of experiments in a systematic way (Ören, 2011b).

As outlined in Table 3, simulation is goal-directed experimentation with dynamic models. Simulation-based experimentation is used in model-based engineering to perform experiments on a design (or a model) of an object or system to be realized. In this case, reality is generated based on the conceived design. In decision problems in general as well as in analysis problems in model-based science, and in control problems in model-based engineering, model is a representation of reality and simulation provides a very powerful and flexible possibility for goal-directed experimentation with a model of reality. Furthermore, simulation-based experimentation can be done under conditions impossible, impractical, or inconvenient in real-world experimentation. Highlights of reasons of uses of simulation for decision support are listed in Table 10 (Ören, 2011c).

The role of simulation to test hypotheses (such as simulation of cosmological phenomena) is of particular interest. As clarified by Tolk et al. (2011) simulation study is an executable hypothesis and once validated simulation becomes a theory.

In relation to experimentation, reproducibility refers to the ability to independently replicate, reproduce, and, if needed, extend computational experiments. Emergence of reproducibility as a critical issue is predicated on the growing credibility gap due to wide spread presence of relax attitudes in communication of the context, experiments, and models used in computational research. Replicability, on the other hand, can be defined as the implementation of a conceptual model in a simulation study that is already implemented by a scientist or a group of scientists. Unlike the reproducibility of results using the original author's implementation, which is a strategy envisioned for reproducibility, replication refers to creation of a new implementation. However, the implementation of the replicated model differs in some way (e.g., platform, language) from the original model, but should be an executable representation to facilitate conducting the same experiments.

Table 9 Definitions/explanations of different types of experiment

> ▶ *Real experiments* can be field experiments, lab experiments, or computer-based experiments.
>   • *Field experiments* are called in situ experiments, in general, and in vivo experiments in the case of life sciences.
>   - *In situ experiment:* An experiment performed on a sample while it is still located in its native environment.
>   - *In vivo experiments* are performed in the living organism (of a plant or an animal).
>   • *Lab experiments* are called ex situ experiences, in general, and in vitro experiments, in the case of life sciences.
>   - *Ex situ experiment:* An experiment performed on a sample after it has been removed from the location wherein it was formed.
>   - *In vitro experiments* are performed in laboratories in life science applications.
>   • *In silico experiments* are computer-based experiments. (The term "in silico" means on a computer.)
> ▶ *Virtual experiments* are known –since antiquity– under the name "*thought experiments.*" Sometimes the German equivalent, i.e., "*Gedankenexperiment*" is also used. Even though the term "experiment" is used in "thought experiments" they are not experiments, but rather reasoning on given scenarios. A taxonomy of thought experiment is given at (Brown and Fehige, 2011).

While reproducing the results of a model distributed with the published document has benefits such as production of reports and visualizations for comparisons to versions listed in the document, provision of a strategy for replication of a study in a new context has its own merits. For instance, developers, who replicate models for cross validation, may have different implementation tools and infrastructure and hence may not be familiar with the platform specific constraints associated with the original model. Therefore, providing the ability to implement a conceptual model under specific experimental conditions and analysis constraints across multiple platforms is critical for practical and broader adoption of the practice of reproducibility. Also, by replicating a model and ignoring the biases imposed on the original model by its chosen toolkit, differences between the conceptual and implemented models may be easier to observe. To facilitate replicability, communication using an extensible and platform-neutral interchange language for the specification, distribution, and processing rules for model, simulator, and experimental frame elements is critical. A successful replication of a computational experiment advances scientific knowledge, because it demonstrates that the experiment's results can be repeatedly generated and thus the original results were not an exceptional case.

Table 10 Reasons of uses of simulation for decision support (adopted from Ören, 2009)

- **Prediction of behavior and/or performance** of the system of interest within the constraints inherent in the simulation model (e. g., its granularity) and the experimental conditions.
- **Evaluation of alternative** models, parameters, experimental and/or operating conditions on model behavior or performance
- **Test of hypotheses**
- **Sensitivity analysis** of behavior or performance of the system of interest based on granularities of different models, parameters, experimental and/or operating conditions
- **Evaluation** of behavior and/or performance of engineering designs
- **Virtual prototyping**
- **Testing**
- **Planning**
- **Acquisition** (or simulation-based acquisition)
- **Proof of concept**

4 Knowledge Generation and Simulation

From knowledge processing point of view, simulation is experimental knowledge generation and can be combined with other knowledge generation systems. For example combination of simulation and optimization techniques to lead simulation within optimization as well as optimization within simulation. Another possibility is to combine simulation with real system to provide augmented reality. Access to information and knowledge by mobile devices may lead to mobile simulation. An interesting development would be mobile-device-triggered simulation, or mobile simulation in short, using e-lenses or e-glasses to display outputs of mobile simulation to achieve ubiquitous simulation.

5 Thinking, Experience, and Simulation

Value of experience in thinking is elaborated by (Dewey, 1910, p. 12). Dewey posits that, "the origin of thinking is some perplexity, confusion, or doubt." He then adds, "Given a difficulty, the next step is suggestion of some way out – the formation of some tentative plan or project, the entertaining of some theory which will account for the peculiarities in question, the consideration of some solution for the problem. The data at hand cannot supply the solution; they can only suggest it. What, then are the sources of the suggestion? Clearly past *experience* and prior knowledge. ... But unless there has been *experience* in some degree analogous, which may now be represented in imagination, confusion remains mere confusion. ... Even when a child (or a grown-up) has a problem, to urge him to think when he has no prior *experiences* involving some of the same conditions, is wholly futile." (Italic of the experiences is from the authors).

Simulation by providing possibility to acquire experience under a variety of conditions –even under conditions, which shouldn't be attempted under non-simulation– is then an ideal way to provide experience to enhance human thinking. Several types of simulated experiences are outlined in section 2.2.

"Active, persistent, and careful consideration of any belief or supposed form of knowledge in the light of the grounds that support it, and the further conclusions to which it tends, constitutes reflective thought" (Dewey, 1910, p. 6). And, paraphrasing Dewey (p. 3), thinking to what goes beyond direct observation simulates reflective thinking. Here, the term "simulation" is used in the sense of "imitating" (category 1.1, in Table 1).

6 Ethics and Simulation

In a sustainable civilized society, respect to the rights of others is paramount. A rationale to have a code of professional ethics for simulationists is given by Ören (2002). A code of professional ethics for simulationists exists and is adopted by several important groups (SCS-Ethics). The code also known as SCS Code of Ethics consists of five sections as follows: (1) Professional development and the profession, (2) Professional competence, (3) Trustworthiness, (4) Property rights and due credit, and (5) Compliance with the code.

7 Conclusions

Perceiving simulation from a very narrow perspective may lead to its misappreciation; otherwise, claims about simulation may be similar to the fable of the descriptions of an elephant by blind people. For example, for a long time operations researchers mistakenly thought that simulation was a technique of operations research that could be used when all else fails. However, the onus of properly understanding something and avoiding false claims is on the experts who make this type of claims. Cerf and Navasky (1984) document this type of "Authoritative Misinformation." Along this line, an example of unappreciative view of the relationship of philosophy and M&S is expressed by Frigg and Reiss (2009).

References

Baudrillard, J.: Simulacra and Simulations. In: Poster, M. (ed.) Jean Baudrillard, Selected Writings, pp. 166–184. Stanford University Press (1998)

Brown, J.R., Fehige, Y.: Thought Experiments. In: Zalta, E.N. (ed.) The Stanford Encyclopedia of Philosophy (Fall 2011 Edition) (2011),
http://plato.stanford.edu/archives/fall2011/entries/thought-experiment/

Cerf, C., Navasky, V.: The Experts Speak – The Definite Compendium of Authoritative Misinformation. Pantheon Books, New York (1984)

Comte, A.: A general view of positivism [Discours sur l'Esprit positif 1844], London, 1856 (1844)

Derry, R., Landry, M., Banville, C.: Revisiting the Issue of Model Validation in OR: An Epistemological View. European Journal of Operations Research 66(2), 168–184 (1993)

Dewey, J.: How we think. Prometheus Books, Buffalo (1991) (originally published, 1910)

Ellis, G.F.R.: Does the Multiverse Exist? Scientific American (July 2011)

Frigg, R., Reiss, J.: The philosophy of simulation: hot new issues or same old stew. Synthese 169, 593–613 (2009), doi:10.1007/s11229-008-9438-z

Golomb, S.W.: "Don'ts" of mathematical modeling. Simulation 14(4), 198 (1970)

Landry, M., Malouin, J.L., Oral, M.: Model Validation in Operations Research. European Journal of Operations Research 14(3), 207–220 (1983)

Nasa-science. Astrophysics. Dark Energy, Dark Matter, http://science.nasa.gov/astrophysics/focus-areas/what-is-dark-energy/

Naylor, T.H., Finger, J.M.: Verification of Computer Simulation Models. Management Science 14(2) (1963), doi:10.1287/mnsc.14.2.B92

Oral, M., Kettani, O.: The Facets of the Modeling and Validation Process in Operations Research. European Journal of Operations Research 66(2), 216–234 (1993)

Ören, T.I.: Model-Based. Information Technology: Computer and System Theoretic Foundations. Behavioral Science 29(3), 179–185 (1984)

Ören, T.I.: Rationale for A Code of Professional Ethics for Simulationists. In: Proceedings of the 2002 Summer Computer Simulation Conference, pp. 428–433 (2002)

Ören, T.I.: Tuncer Ören. In: Pioneers of Computing 2005 –Honouring Those Who Influenced the History of Computing in Canada. IBM Centers for Advanced Studies. IBM Canada, Toronto, ON, Canada (2005)

Ören, T.I.: The Importance of a Comprehensive and Integrative View of Modeling and Simulation. In: Proceedings of the Summer Simulation Conference, San Diego, CA, July 15-18 (2007)

Ören, T.I.: Modeling and Simulation: A Comprehensive and Integrative View. In: Yilmaz, L., Ören, T.I. (eds.) Agent-Directed Simulation and Systems Engineering. Wiley Series in Systems Engineering and Management, pp. 3–36. Wiley, Berlin (2009)

Ören, T.I.: Simulation and Reality: The Big Picture (Invited paper for the inaugural issue). International Journal of Modeling, Simulation, and Scientific Computing (of the Chinese Association for System Simulation - CASS 1(1), 1–25 (2010), http://dx.doi.org/10.1142/S1793962310000079

Ören, T.I.: The Many Facets of Simulation through a Collection of about 100 Definitions. SCS M&S Magazine 2(2), 82–92 (2011a)

Ören, T.I.: A Critical Review of Definitions and About 400 Types of Modeling and Simulation. SCS M&S Magazine 2(3), 142–151 (2011b)

Ören, T.I.: A Basis for a Modeling and Simulation Body of Knowledge Index: Professionalism, Stakeholders, Big Picture, and other BoKs. SCS M&S Magazine 2(1), 40–48 (2011c)

Ören, T.I.: The Richness of Modeling and Simulation and its Body of Knowledge. In: Proceedings of SIMULTECH 2012, 2nd International Conference on Simulation and Modeling Methodologies, Technologies and Applications, Rome, Italy, July 28-31 (Invited Keynote Paper - in press, 2012)

Ören, T.I., Yilmaz, L.: Semantic Agents with Understanding Abilities and Factors Affecting Misunderstanding. In: Elci, A., Traore, M.T., Orgun, M.A. (eds.) Semantic Agent Systems: Foundations and Applications. Springer (2011)

Rapp, U.: Simulation and Imagination: Mimesis as Play. In: Spariosu, M. (ed.) Mimesis in Contemporary Theory. The Literary and Philosophical Debate, vol. 1, pp. 141–171. John Benjamins Publishing Company, Amsterdam (1984)

Tolk, A., Diallo, S.Y., Padilla, J.J., Turnitsa, C.D.: How is M&S interoperability different from other interoperability domains? In: 2011 Spring Simulation Interoperability Workshop, Bosto, MA, April 4-8, pp. 1–9 (2011)

Tzu, S.: The Art of War, Giles, L. (trans.) (1910),
    http://www.puppetpress.com/classics/ArtofWarbySunTzu.pdf

Wiki-simulism. Simulation hypothesis at,
    http://en.wikipedia.org/wiki/Simulation_hypothesis

Philosophical and Theoretic Underpinnings of Simulation Visualization Rhetoric and Their Practical Implications

Andrew Collins and D'An Knowles Ball

VMASC, Old Dominion University
Suffolk, VA, United States

1 Introduction

The considerations of the philosophical underpinnings of visualization have been left on the sidelines while researchers chase the latest technological applications, as with modeling and simulation (M&S). Visual rhetoric in M&S is also an effect of the latest technology that deserves closer observation into its uses. Our focus in this chapter is to develop the view, both now and in the future, of rhetoric's importance to simulation visualization. Visualization creates and uses images, diagrams, and/or animations to explain models, display simulations and their real-time results, and even, in some cases, for validation.

When we study the epistemology of visualization, the means by which we represent data and communicate it to others is not only a matter of how relevant data is displayed but why the producer of the simulation chooses particular means of visual communication. It is rhetorical methodological decisions that have the greatest impact on the end user and there is a necessity to examine closer the considerations that bring visual rhetoric to M&S. Visualization is a serious design activity that demands deeper conceptual investigation, trumping software and programming as the initial act of the visualization process. As foundations of M&S are addressed, the importance of how and why M&S is presented cannot be overlooked. A wealth of philosophical investigations exists in both arenas but joint consideration needs to be applied for a deeper understanding of current and future uses of visualization.

M&S has moved far beyond simple data representation into the world of visual communication over the past 15 years. The advancement of technology available for conducting visualization has been expanding at the same rate as the changes of methodologies available for conducting M&S. M&S practitioners are immersed in data, algorithms, and validation, and yet these are not the impressive gaming level high-definition realistic quality graphics that dazzle M&S customers; thus rhetoric emerges in the means of visual communication. Effective simulation visualization is a thing of true value, not simply eye candy or media fodder. The visual argument of data displays and statistics has been examined in the past, but the rhetorical appeals at play in M&S is uncharted territory, ripe for examination.

A. Tolk (Ed.): Ontology, Epistemology, & Teleology for Model. & Simulation, ISRL 44, pp. 173–191.
springerlink.com   © Springer-Verlag Berlin Heidelberg 2013

The need for a simulation to visually appeal and argue a directive clearly comes into play in the modern era, for example, a red square is no longer an adequate representation of a battle tank when compared to a highly rendered graphic, even though they both represent the same data. The function of simulation is of primary importance to its end result but it cannot be denied that the discipline of M&S now prizes fancy graphics to communicate. The philosophical underpinnings of visualization are deeply rooted in the power of an image to convey both quantitative and qualitative narrative structures as effectively as alphanumeric language systems. Is the visual clarity, organization, and visual understanding still secondary to the data? Are the form and the function so far removed from one another? It will be argued that the visualized form creates the narrative structure of the simulation, creating expectation and leading to bias. The case is presented for the algorithm, graph and model of M&S arguing in the same rhetorical fashion as the sign, signifier and signified at the heart of visual communication.

When focusing on visualized objects and on the visual nature of the rhetorical process, visualization presents part evidence and part story-telling that must both be considered and believed at once. Data visualizations and simulations, consciously or unconsciously, all reflect an agenda or an aimed conceptual focus, and these are rhetorical in nature because their inherent objective is communication. Visualization is the latest in a long line of media that exhibits the expectation of objectivity while being inherently rhetorical.

Like Frankenstein's monster, M&S visualization can have the ability to bring a simulation to life by using many pieces, parts, and considerations. In the midst of asking *how* the end result is brought to life, the modeling and simulation community should be asking *why*. Why should the producer and the customer care about the rhetoric of visualization? M&S is a relatively new subject that is still trying to find a foothold within the research community. Thus it is appropriate to understand the purpose and rhetoric of visualization, not just how to build bigger and better graphics. This chapter will argue that how we present our simulations is just as important as any composability or resolution/fidelity issues, because, ultimately, the acceptance of M&S within mainstream science and society will depend on the results that are produced *visually*. In order to expand this discourse, the goals of this chapter are to expose the M&S community to the existing research on the rhetoric of visualization and demonstrate the importance of contemplating the philosophy of visualization, to highlight and address existing problems with simulation visualization, and to bring the inherent rhetoric in visualization to the forefront of consideration and utilization.

2 An Introduction to Visualization

Computer simulations are a construction of mathematical algorithms and data; this statement is not meant to trivialize simulation, as many modern simulation are incredibly complex and involve tens of thousands of lines of code, it is meant to explicitly point out what they are. However, most people, with the exception of the simulation developers, will see a simulation though its visualization, the graphics used to represent the inner workings of the simulation. Simulation visualizations has been defined as "a process that generates visual representation such as imagery, graphs, and animations, of information that is otherwise more difficult to understand though other forms of representation, such as text and audio" (Sokolowski and Banks, 2010). Though this definition is not necessarily universally accepted, it acts as a working definition for the purposes of this chapter.

A computer simulation could simply be presented as a series of equations and tables but, even to those trained to read such things, this can be cumbersome and difficult to follow. By representing the different elements of simulation using visualization, we are able to gain a concise clarity of the simulation's purpose and function; this clarity can be organized in our minds to give an understanding of the simulation's purpose and results. Achieving this clarity for the viewer is no simple task and the art of visualization is discussed at length by Tufte (Tufte, 2001) and Cleveland (Cleveland, 1993).

The use of visualization can be a powerful one; as the cliché proverb says "a picture is worth a thousand words" but is the picture worth a thousand lines of computer code used in a simulation? As with most questions, the answer depends on many things. Most importantly, does the visualization actually represent the simulation?

Visualization is a requirement for many simulations due to a simple reason: the human mind is limited. The human mind can only process a small amount of information at any one time and the use of visualization aids in that process. This same limitation of humans is also the reason that rhetoric exists and we can be manipulated to see and understand things in a desired way. The billions of dollars spent on advertising each year is testimony to this fact. Thus visualization can be used to present views that are not actually real, or worse, that are false.

The use of visualization within a simulation is not all bad. Given the limited human mind, visualization and animation of a simulation greatly assist in a simulation's verification and its underlying model's validation (Sargent, 1992; Bell and O'Keefe, 1994). However, simulation visualization is secondary to the data and results associated with the simulation; just because an animation of model behavior is free of errors does not guarantee the correctness of the model results (Paul, 1989). In our modern era, the reverse of this statement has become of concern because simulation novices might consider the data and modeling secondary to fancy graphics: if the visualization does not look right, or advanced enough, then the simulation results are rejected.

The majority of the simulation visualization literature is concerned with the "hows" of simulation, that is "how do we produce more realistic 3-D graphics within our modeling;" The focus of this chapter is on the "whys", that is "why do we use visualization? Why is it effective?"

3 Current State of Affairs in Visualization

Over fifteen years ago, the film *Jurassic Park* and the computer game *Wolfenstein 3-D* were released. The graphics used in *Jurassic Park* were so revolutionary that people in cinemas all-round the country stood up and clapped at its finish. *Wolfenstein 3-D* impressed gamers with its fast-paced 3-D gaming, the likes of which had not been seen before, and it has become known as the grandfather of 3-D shooters (1UP Games, 2010). If *Jurassic Park* was released today, it would seem heavily dated and unlikely to receive positive reviews; similarly, *Wolfenstein 3-D* would be seen as retro gaming and would be placed in a genre of gaming where 3-D graphics are now the accepted, and expected, standard. Fifteen years has seen a dramatic change in our acceptance and expectations of visualization.

Our rapidly changing expectation of visualization has required that the related technology evolve at an alarming rate. As such, most of the focus of M&S visualizations has been on the "how to" and not the "why." There are also an incredible number of visualization packages on the market from Google Earth to Unity 3D, and, as such, the M&S professional education into visualization is dominated by the time required to gain a workable understanding of these packages. Even standard M&S tools like Matlab and Arena have more sophisticated integrated graphical tools that a professional must come to grips with.

Though there has been great advancements in the M&S visualization industry there are still several problems that have not been completely resolved, for example, perspective projection transformation and shading on reflective surfaces (Sokolowski & Banks, 2010). These problems have been keeping the M&S visualization academics occupied over the last decade and hence there has not been any focus on the "whys" but instead on the mimicry of real physical imaging systems.

As visualization becomes more realistic and easy to integrate within a simulation, its role within the simulation process is increased. Thus what started as a simple add-on to many simulations is now an integral part of it. This means that the influence of visualization on a simulation's design and output has grown over the years to a point where people are now starting to questions its role. Paul Roman highlighted the impact of visualization's influence in his paper: "Garbage In, Hollywood Out" (Roman, 2005). The title of the paper is a metamorphosis of the George Fuechsel's adage "Garbage In, Garbage Out" (Bulter et al, 2010), implying that bad data and design going into a simulation will result in unusable, and invalid, results being produced. Roman's play on the phase comes from the tendency of some commercial simulation vendors to mask the inadequate simulation designs behind advanced graphics. The use of visualization to express a simulation's output can be considered to be a rhetorical process.

The rhetoric of visualization can lead to problems with the simulation representation, especially with the high-resolution images that are displayed. A soldier might be represented using a graphic like given in the image below; however, within the simulation, the soldier does not have arms, clothing or even hair. The soldier in the simulation code will be nothing more than a blob in the environment that interacts with other blobs and the environment through a series of simple scripted rules and behaviors (see Figure 1).

Fig. 1 Simulation visualization of soldier

There is a tendency for simulation visualizations to include fancy details; Banks and Chwif highlight that high-fidelity combat models often contain smoke, dust and muzzle flashes when they are not even part of the simulation (Banks and Chwif, 2011). If the simulation actually represented muzzle flashes, then it would have to include various other factors within the simulation, for example, muzzle flare would mean that the firing entities visibility would be greatly increased and thus they would be more easily detected by the enemy. The effects of muzzle flashes could be included within a simulation by increasing the firing entities probability of detection by some small amount to reflect the visibility from the muzzle flare; however, this is likely to add a level of resolution unnecessary for the simulation's propose and thus should really be excluded.

It should not be assumed that more visual detail always makes a simulation better. A rocky terrain delivered as 8-bit scenery in no way conveys the actual nature of "rockiness" any better than an HD realistic rendering. Yet, as with visual art for generations, a main goal of modern visualization is to hone the craft to such a point wherein one can create a representation that is as close as possible to the object in real life. It is very easy for a visualization developer to add textures, or even foliage, to their simulation graphics when these factors are not taken into account within simulation. This could lead the viewer of the visualization to believe that certain factors are accounted for within the simulation when they are not. For example, adding a dessert terrain might imply that heat factors are taken into account within the simulation when in fact they are not.

The two previous examples, both muzzle flashes and terrain, lead to the questions: if you chose to excluded an aspect from the model, why should it be in its visualization? One answer would be that people include superfluous graphics in their visualization simply because they can. This leads to the phenomenon that Roman summarizes as "what you see may not be what you get."

4 Why Do We Model in This Fashion?

The visualizations being produced today are not just basic data representations. The high-end nature of the visualizations created with the use of top-of-the-line softwares allows for the process of production to be evaluated as a new medium. In the designers' eyes, because the ability to make glossy graphics exists, it is reason enough to bring them to life. But it begs the question: for every simulation visualization, is it always necessary to use all tools and tricks at hand to make it an intense experience?

There was once a concept used in the German design enclave Bauhaus at the turn of the Twentieth century that became pervasive in the designer's lexicon for years to come – the concept of form following function. In effect, the actual look and visual appeal of the final product was in no way as important as what the finished product actually did, its intended use and desired result. Of no consequence was the fanciness of an avant guard pitcher if it could hold no water. In many digital arenas, it now seems that the form has overtaken the function, and in some cases, replaced it altogether. The built-in planned obsolescence factor often leaves many commercially-viable final products reliant only on visual rhetoric to sell themselves as anything of value. Simple models now require flashy graphics not only for understanding but also desirability. What is making it work has suddenly become secondary – the analysis, the scenarios programmed in, the data is all but forgotten upon first viewing of an impressive constructed visual environment. These visualizations may look great but can they operate? Roman says, "Regardless of how good the outputs from a simulation look, high resolution in no way implies high validity for a particular purpose" (Roman, 2005).

All works are full of confusion and contradiction that the artist can't overcome because of a lack of meaning and truth inherent to the medium itself. There is no purely functional object for humans. Responsible designers taking on a project should evaluate the need for the clearest most-straightforward method(s) that allow for clarity of meaning. The fact that there are designers that simply use all tools in their arsenal clouds the nature of the original message, leaving room for rhetorical appeals to seep into a final product that may be sleek and visually beautiful without clearly communicating the facts and data. When this occurs, the message can only be understood in relation to the medium. The content matters less than evaluating the structures and practices of the medium itself – what does the medium amplify? What does it drive out of prominence? What does it do when pushed to its limits (McLuhan, 1964)?

The usefulness of a simulation's visualization is dependent on its purposes; this usefulness is not just a function of how much resolution a visualization has. To understand this point, consider the two pictures given in Figure 2. Both pictures are a representation of Mount Everest but depending on our purposes will determine which of the two pictures is most useful for us. If we intent to draw a picture of Mount Everest then the bottom-right picture will be of most use; if we are planning a route to climb the mountain then the top-left picture is likely to be of most use for our purpose.

Fig. 2 Different visual depictions of Mount Everest

When designing the visualization output from a simulation, there is most likely to be several choices that need to be made. For example, consider an operational level-military simulation that contains some tank units. Before the advent of modern advanced graphics, this scenario might be represented by a simplified 2-D bird's-eye-view of the battlefield, which only contains the area's key features, with the tanks being represented by rectangles or squares colored blue for your forces and red for the enemies. With modern Graphical Processor Units (GPU), this scenario could be represented in 3-D with both the battlefield and tanks given an almost life-like appearance. A high-level of detail may be necessary for analysis but designers must approach the initial phase of visualization development and decide whether using 3-D rendering is required when 2-D environments will work just as well. And yet, the end users/viewers of visualizations will *accept* a red square on a mock battlefield as representation of a tank, but a stronger rhetorical appeal to relatability is introduced by way of delivering a detailed model of the M1 Abrams tank that troops, for instance, will be using in combat so as to enhance familiarity, a frequent requirement of training in a simulated environment (see Figure 3). So which approach is better? That depends on purpose, but if we were forced to make an arbitrary choice between the two approaches, we might be tempted to go with 3-D version but is this choice made based on aesthetic considerations as opposed to practical considerations?

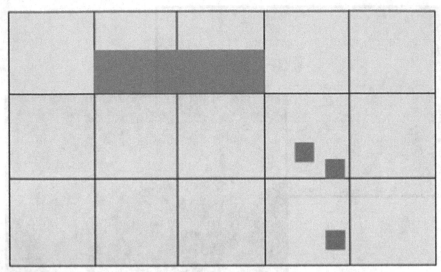

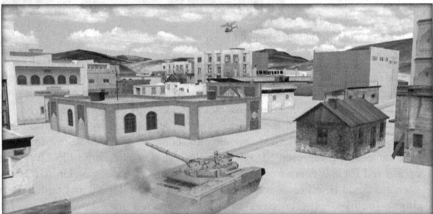

Fig. 3 Comparison of 2-D and 3-D tank visualizations

The 3-D simulation might give "information overload" to the commander or even mislead the commander. For example, the commander might be confused as to why a particular tank unit has its turret pointing in a particular direction when, in reality, a simulation at the operational-level is unlikely to even take into account a turret's direction in its underlying algorithmic processes and the direction of the turret is solely a consequence of the simulation's graphical engine.

The 3-D simulation, with its realistic details, might give the commander the impression that they are actually viewing the real battlefield when, in fact, they are only seeing a representation of it based on their current knowledge. If the simulation takes into account elements like "fog of war," where the enemy positions are not necessarily known, then a commander might be surprised when an enemy unit pops up in an area of the simulation wherein the visualization gave the impression they had a complete view and awareness.

The conclusion from this example is that there are limitations of using more advanced 3-D graphics over simple 2-D ones or, put another way: "just because you can, does not mean you should."

5 Visualization as It Relates to Form and Narrative

Form is of primary necessity to any design. By way of humanity's visual and cultural perception, any form can and does inherently create a narrative for the viewer/user. Whether it is a white line down a black background transforming into an abstract representation of a chasm or a road in the mind's eye of the viewer, or a simple photograph used in an ad conveying the concept of "boy loses girl" or "hero meets villain", the viewer/user creates and/or fills in the story line. The form itself can cause a connotative or denotative effect. Just as advertising campaigns encourage us to connotatively associate a product with other things that we value, visual media tell an entire sequence of events in one image, playing with our cognitive and social desire to form this narrative. This narrative appeal leaves room for social and aesthetic expectations of which designers are well aware and actively manipulate. The combination of narrative creation and end-user expectation leads to bias in the world of visual rhetoric. The same trope of biases can be applied to visualization, found both in the hand of the creators and in the perception of the customer.

Any modeler that claims that their simulation is without bias is either a liar or a genius. There will be biases in the assumptions made, the variables measured, data input, outputs measured, etc. The modern philosophical approach of deconstruction states that nothing can be viewed or said without bias; thus there cannot be any unbiased or objective visualizations either.

As with Kuhn's original genesis of the "paradigm shift" (Kuhn, 1962), instead of building the experiment at its base soley out of pieces and parts that would support a desired final outcome, discarding data that does not yield said desired result, gathering up the rejected bits can often lead to looking at a field of study in a whole new light. Between the rhetoric of the visualization and the evidence upon which it is built, one must ask what has been included and what has been left out that will impact the final model produced by a simulation running a breath-taking visualization.

6 Rhetoric and Evidence – The Bias of Both

Visual rhetoric is composed of social and cultural biases to visually negotiate, but evidence can also be skewed when certain bits are used for composability while others are discarded. So the visual appeal of the simulation could lead one to question faulty data far less often. The concept of presence and absence is a long-studied trope of visual rhetoric. For every item that is noted as present, its antecedent has to be considered as well for having been negated from the narrative frame. This revealed absence leaves open the question of biased data and end objectives.

A startling study by Jone Tiffany (Tiffany, 2011) showed that social biases were present towards avatars within the social virtual environment of Second Life®. In the study, the participants took on the roles of various minority groups, including those with obesity and disabilities, and interacted with other users of the Second Life® environment using an avatar that reflected their roles, e.g., one avatar was displayed using a wheel-chair, etc. The participants found that they were excluded from many conversations and group interactions within the virtual environment and these exclusions were attributed to the visual appearance of their avatars.

Helmers states, "Looking indicates the way things could be rather than proving the way things are." Whether consciously or subconsciously, in the end, data visualizations and simulations all reflect an agenda or a biased conceptual model. They are rhetorical in nature because their main objective is communication. As stated, visualization is a new medium, and like any visual media, it comes with the expectation of objectivity while by nature being inherently rhetorical (Helmers, 2004).

On the surface, visualization as a tool of visual communication is not "vague or ambiguous"; it is direct and seemingly based on solid data and statistics, but rhetorical "propositions can be expressed visually no less than verbally" or numerically, hence the productions of simulation visualizations in the first place (Blair, 2004). In visualization, data points are introduced into a visual narrative. For proper methodological analysis, both data and narrative must be considered both separately, together and as having an anchorage and relay effect on one another.

Visualization can and does display a propensity to be deceitful more so than ads on television or in magazines. The dependence on data for its basis is key. If the end viewer/user approaches the product with the expectation that solid sound research and analysis have gone into the visualization's creation, this naturally leads to the propensity to scrutinize it less. Therefore, the creators have even more of a responsibility to provide a sound foundation for the end product rather than something that will simply wow the crowd.

7 Visualization and Its Rhetorical Underpinnings

Many would be pleased to place rhetoric firmly in the realm of lawyers and politicians. Others may go as far as to posit it in the world of linguistics. In the past decade, rhetoric has branched out and taken a multidisciplinary approach to be applied to many aspects of the world around us. It now extends far beyond the path of verbal arguments and persuasion. As Zelizer states, "Visual *representation* gives way to visual *rhetoric* through subjectivity, voice, and contingency" (Zelizer, 2004). Meaning is visual and any visual representation is subject to having its meaning parsed for analysis and questioning.

Foss, Foss and Trapp define rhetoric quite broadly as "the unique human ability to use symbols to communicate with one another" (Foss, Foss, and Trapp, 1985). Blair opens the realm even further by stating, "Arguments in the traditional sense consists of supplying grounds for beliefs, attitudes or actions...pictures can equally be the medium for such communication" (Blair, 2004). Visual rhetoric used in visualizations does not force us to have certain interpretations as much as it creates the context for interpretive frameworks and, more importantly, shared expectations.

The basis of visual rhetoric can be found in the traditional methodologies of semiotics, or the study of signs. Semiotics is not only necessary for visual understanding, but it seeks to reveal the constructed character of meanings we use everyday. As a philosophy and method of critique, it questions and investigates the coded structure and meaning of *anything* that stands for something else – what is simulation visualization if not just that? Visual systems are signs existing in semantic space. The meaning is not on the surface but arises from collaboration between signs and interpreters. Semiotics allows for a more complex, subtle, and sophisticated mode of interpreting visual rhetoric present in simulation visualization.

As rhetoric relates to the arguments and appeals found within visualization's imagery, we turn once again to Blair who provides a modern definition of rhetoric as "the best means available to make the logic of the argument persuasive to the audience." We must be open to asking how rhetorical "constraints" and "opportunities" come into play in a particular visualization, because the developers are asking "what visual imagery will the audience understand and respond to" (Blair, 2004).

8   What Is Necessary for Visual Understanding?

Meaning in language, as in art, is derived from a culturally agreed upon structure of relationships, none of which occur naturally. So we have to visit the larger question of what exactly is required to visually make sense of things in the world around us before we can approach a simulated visualization of our world. If we take the visual semiotic work of Joseph Kosuth's *One and Three Chairs* for example in Figure 4, the definition of the chair being the signifier/concept, the chair being the signified/object and the picture of the chair becoming the sign whose meaning is produced via combined work done by both the signified, the signifier and the interpreter, then a correlation can be made to the model (signifier), the system under consideration (signified) and the visualization simulation (sign) as possessing and being subject to the same rhetorical analyses to which other tools of visual meaning-making are held (see Figure 5). The constructed meaning of the sign displaces the arbitrary meaning of the actual object. The rhetorical appeals found in the simulation hold more narrative weight than the real thing.

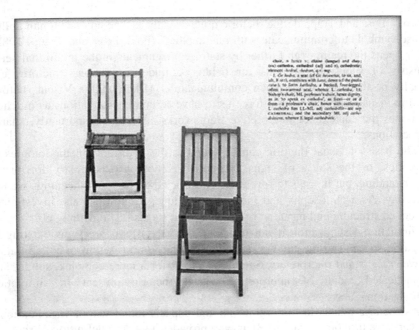

Fig. 4 Joseph Kosuth, *One and Three Chairs*, 1965

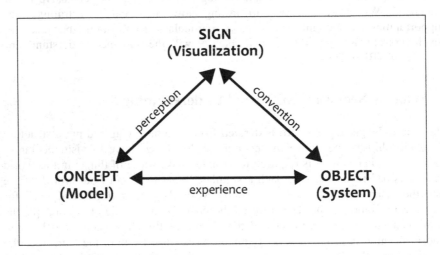

Fig. 5 Semiotic relationship and conveyances

As with signs and signifieds, in each case we are less interested in the relationship to the real object than to the significance of the relationship of image to assigned meaning and interpretation. Paradigmatic analysis must be used to compare and contrast each of the signifiers present in the visualization text with *absent* signifiers that, in similar circumstances, might have been chosen, and then we must consider the significance of the choices made. Nothing can, therefore, be

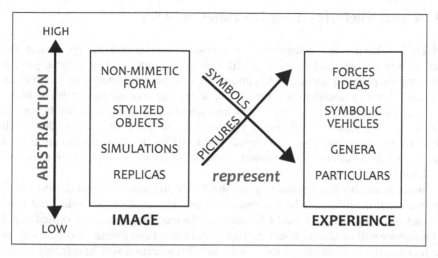

Fig. 6 Updated rhetorical levels of abstraction

exempted from meaning. Depicted in Figure 6, as simulations become more life-like and move further away from abstract forms, such as the tank example seen in Figure 1, the visual representations move closer to forcing ideas upon the viewer/user.

A visualization simulation has both a social rhetoric factor and an aesthetic component that must be deconstructed. The social aspect asks "what does it do?" (function) while the aesthetic requirements address "how does it look" (form). As for the appeals of social rhetoric, you find the end user falling prey to the concept that a visualization must be a good simulation simply because the graphics are so impressive. The aesthetic component falls on the designer in charge of massaging rhetorical implications visually. Both have equal roles to play in regards to rhetoric being introduced into a visualization. One must suspend the idea of function of all things to understand their meaning, how they function as signs and symbols to produce layered veiled meanings, why narratives are produced and how narrative alters the meaning of the images "informing" the interpretation.

For the French postmodern philosopher Jean Baudrillard, the very notion of reality has been complicated, particularly by the profusion of images of it - "the real no longer exists (Baudrillard, 1981)." Just as we look out at a beautiful tropical vista and say "wow...it looks just like a postcard", more often than not, we now reference the real world by comparing *it* to a simulation. We witness the erosion of meaning via its very excess...commodity fetishism. Baudrillard laments that the purveyors of visualization have destroyed reality in stages: "first they *reflected* it; then they *masked* and perverted it; next they had to *mask* its *absence*; and finally they produced instead the *simulacrum* of the real." Hutcheon responds by asking if we have ever "known the 'real' except through representations." Our senses may perceive it "but do we *know* it in the sense that we give meaning to it" (Hutcheon, 2002)?

9   Visual Rhetoric – How It Relates to M&S

M&S has been used extensively to support decision-making by giving the decision-maker new information, a different view-point or even a paradigm for framing the problem under consideration. It would be completely inappropriate to suggest that any simulation gives the "correct" answer to the problem under consideration but it does give insights into understanding some of the factors of problem. This idea is summarized by the famous quote of George Box: "All models are wrong but some are useful" (Box, 1979). Thus, as a simulation does not supply the absolute correct answer, its results are there to support the decision-maker in their decision.

Without an absolute answer to give, the M&S practitioners must decide which information, from the simulations results, to provide the decision-maker and what format this information should take: tables, dialogue or graphics. A skilled M&S practitioner will be able to select the right information and format in such a way as to increase the creditability of the simulation; this selection will be affected by the same biases that the practitioners had when developing the simulation in the first place. Thus the M&S practitioner has an opportunity to influence the decision-maker and the art of doing so is the rhetoric of M&S.

Visualization is just a small part of the M&S process; a generic overview of the whole process is given in Figure 7. Given that any visualization of the simulation results are the only thing most decision-makers will see of the simulation, there is a temptation to want to concentrate your efforts on developing the best visuals. This is not helped by the effect that visualization has on decision-makers, as highlighted by Banks and Chwif "[G]raphics can aid sales. Animated graphics seem to have a mesmerizing effect on the simulation novices" (Banks and Chwif, 2011).

This mesmerizing of simulation novices might initially seems innocent enough but it leads to a charlatan aspect of the M&S industry. Simulations, with fancy graphics, are being sold as tools for problems they are not equipped to solve. Analysis simulations with pretty front-ends but no substantial back-end are being peddled to unwary decision-maker. The results of such charlatanism might make a quick buck for some businesses but what is the affect on the industry as a whole? That decision-maker will most likely obtain bad results from the simulation and thus look unfavorably at the simulation and M&S as a whole. Is that decision-maker likely to recommend M&S to others? Quite the contrary. For a new and fledgling subject like M&S, the bad press could be devastating to its growth and, ultimately, survival.

The authors would like to say that, in most cases, the addition of extra graphics within a simulation is due to an innocent wish, by the simulation developers, to make the visualizations more life-like; however, from personal and anecdotal evidence, the authors believe that there are cases of commercial simulations whose visualization is purposely designed to mislead the potential user/buyer which we have defined as the charlatan aspect of our industry. No direct examples are given here to avoid any law-suit but by walking around any large industrial M&S conferences, any M&S expert should be able to spot these tricky practices.

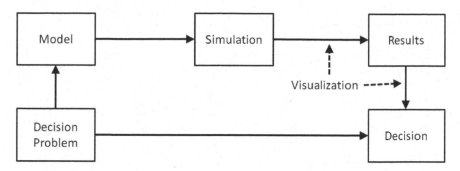

Fig. 7 Generic decision-making process involving M&S

Given that an M&S expert should be able to spot some of these misleading visualization practices, how can this be a problem? There are several reasons and two are highlighted here. Firstly, the people that control the purse-strings for M&S budgets are not necessarily M&S experts. To understand why this could be a problem, consider the analogy of purchasing M&S software to that of buying a second-hand car. The buyer might be well advised to take a mechanic (M&S expert) with them when going to the car showroom but due to the expense and availability of such an expert, this is not always possible. Secondly, M&S experts are not immune to the problems of rhetoric. To understand this point, consider the computer language in which a simulation is written. Many simple simulations can be constructed using Microsoft Visual Basic for Application (VBA) but snobs within the development community tend to give more kudos to those simulations constructed in C++. The same simulation could be constructed in either C++ or VBA, using the same internal algorithms, but the C++ version is likely to initially get more "respect" within the M&S community. Why does this happen? The C++ programming language is the more difficult to learn and thus it is human nature to assume that the C++ simulation will be more complex.

The use of rhetoric within M&S is not limited to visualization. There are many "Artificial-Intelligent-social-Bayes-net-agent-based-counter-insurgent-buzzword-piece-of-rubbish" simulations that sold on the market, too. It is very easy to include many techniques, i.e. Artificial Intelligence, within a simulation if your sole purpose for including them is to *include them* for sales purposes. Just because a simulation includes a technique does not mean that it has been implemented well or effectively.

10 Statistical Relationships with Visualization

Rhetorical issues present in information displays are not isolated to the M&S world. They have been wrestled with problems of visualizations rhetoric for years. The seminal work of Darrell Huff, entitled "How to lie with Statistics," highlighted many misleading practices that are used with the graphical representations of statistics (Huff, 1954). It was suggested within the book that the cause of these misrepresentations were rhetorical, e.g., the use of cut-off graphs to exacerbate gradient changes within the data.

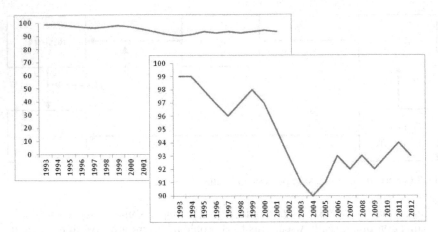

Fig. 8 Two graphs depicting the same data but with different y-axis scale

A cut-off graph is one where the y-axis does not include the origin, i.e. zero, of range of data under consideration. An example is shown on the right-hand side of Figure 8; a graph with the origin included is given in the left-hand side of the figure. The effect of a cut-off graph is that it makes changes in the data set seem more drastic then they might otherwise be viewed by exacerbating the gradients.

Statisticians recognize the need to counteract these problems with rhetoric because of the bad press their subject has received over the years. This can be summed up by the famous quote from 19th-century British Prime Minister Benjamin Disraeli: "There are three kinds of lies: lies, damned lies, and statistics." Most modern statistical books will discuss the proper use of statistics, and the statisticians have developed clear ways to express some of the issues with statistics, e.g., descriptions of type I errors versus type II errors. However, the battle has not been won and many of the problems that Huff presented over fifty years ago are present in our everyday life. To make matters worse, the rise of the internet has lead to even more ways to misuse statistics, e.g., dynamically changing graphs (Kostelnick, 2008). On a positive note, the abuse of visual rhetoric has lead to an active discussion on it within the academic community (Tufte, 2001).

As statisticians have discovered, the issues with visual rhetoric predate modern M&S. However, just because the M&S community has the benefit of being able to draw from the lessons learned by the statistical community does not mean they have learned anything. As a community we do not want Disraeli's quote to become "There are four kinds of lies: lies, damned lies, bad statistics, then simulation." What is important is that we are not alone in fighting against misleading rhetoric.

11 Rhetorical Visualization – Solutions

One must ask if anything can be done to counteract bad visualization. There are different schools of thought on how this might be achieved:

Verification and Validation (V&V): Paul Roman says that the rhetorical issues with visualizations can be overcome with good V&V in his statement that "[t]he primary defen[s]e against undue influence by impressive looking outputs is validation and verification" (Roman, 2005). However, V&V is a very subjective process and there is no agreed upon standard. The process of V&V is not an Instant one, and it might not be possible to apply it to a given simulation. This is especially true for simulation platform purchases. A simulation firm might release a limited version of the simulation platform for evaluations purposes but inadequate, or misleading, documentation of the simulations capabilities make it difficult for the simulation expert to evaluate the propriety components. As mentioned previously in the chapter, those that hold the purse-strings for purchasing simulation platforms are not necessarily M&S experts.

Transparency: The perceptual cognitive-based school of thought argues that all data displays should be as simple, thus transparent, as possible (Kostelnick, 2008). Given the complexity of the data outputs, this is just not always feasible. To follow this school of thought would require the analyst to present the results in graphs and diagrams as simply as possible; such a display would look dated and passé to the decision-maker and ultimately affect the simulation's creditability.

Neither of the solutions presented above really give an adequate solution to the problem of misleading visualization rhetoric and the rise of simulation charlatans so what about trying to educate the populous about rhetoric instead?

Design, itself, does in fact enable visualization rhetoric simply by nature of being a tool for communication, a generator of symbols, signs, codes, and narratives that become pervasive tropes within our culture. If one designer decides to exclude all appeals of a visual nature from his data visualizations, chances are his or her work will get passed over and go unnoticed, in a sea of flashy graphics – the kiss of death for most visual communicators. Therefore the flashiness will become the norm, being taught as a requirement in design schools.

The large majority of visualization M&S research is focused on areas of a technical nature. Other branches of research in visualization look closely at the foundations of M&S. But what about what is *actually shown?* The dialogue must be opened up for consideration and further investigation to take place regarding the rhetorical decisions that play out in the visual form that M&S ultimately takes and the implications of such decisions. With representation itself being on the order of illusion, the ability to understand the implications of *anything* is enhanced, or even made possible, by "re-contextualizing" the issues, moving them from one discipline into another to perform a comparative analysis.

Requirements may differ between the analyst and the customer, or metaphorically speaking, the car builders and the car buyers respectively. A first step towards preventing bad or unnecessary visualization would certainly be awareness of visual rhetoric's impact on visualization by both the builder and the buyer. Primary use and clarity of form must be brought to the forefront while agendas and subjectivity take a back seat once they become apparent. With newfound awareness, the analyst must place importance once again on creditability and acceptability of a simulation to move us closer to objective communication in visualizations for analysis and training.

12 Conclusions

Researchers and scholars are looking into the depths of M&S but what really matters is what those outside of the M&S community see, including customers & decision-makers that "use" M&S. The considerations of visualization's rhetorical underpinnings must be brought to the forefront of M&S study in order to effect change in the application of simulation visualization. Realizing that visual rhetoric is at play in many visualizations marketed today is a first step toward requiring greater verification, validation and transparency practices at the inception of the visualization process. Visual rhetoric in M&S as an effect of the latest technology deserves closer observation into its teleological uses.

Overcoming the problems relating to visualization's rhetoric is not trivial and could linger for a long time like the rhetorical problems of statistics. The problems cannot be ignored either, as there is a growth of charlatanism within our M&S industry which is especially due to the availability of fancy graphic for simulation purposes. Only time will tell if these problems are overcome but as the American engineer Charles F. Kettering stated "A problem well stated is a problem half solved."

Our focus in this chapter is to develop the view, both now and in the future, of rhetoric's importance to simulation visualization. The importance of these epistemological investigations lies not only in thinking about visualization in a new way but also in exposing audiences to the rhetorical nature of visualization via existing research as well as comparative analyses. These new connections therein expose existing problems in modeling and simulation by way of the application *and* the philosophy of visualization. Both facets must be examined at once for a true understanding of visual rhetoric's place in this field of study.

References

1UP Games. "Computer Gaming World's Hall of Fame." 1Up.com (2010), http://www.1up.com/do/feature?pager.offset=8&cId=3139081 (accessed on January 26, 2012)

Banks, J., Chwif, L.: Warnings about simulation. Journal of Simulation 5(4), 279–291 (2010)

Baudrillard, J.: Simulacra and Simulation (Glaser, S.F. (tr.)). University of Michigan, Ann Arbor (1994); Originally published in French by Editions Galilee (1981)

Bell, P.C., O'Keefe, R.M.: Visual interactive simulation: a methodological perspective. Annals of Operations Research 53, 321–342 (1994)

Box, G.E.P.: Robustness in the Strategy of Scientific Model Building. In: Launer, R.L., Wilkinson, G.N. (eds.) Robustness in Statistics: Proceedings of a Workshop. Academic Press, Waltham (1979)

Blair, J.A.: The Rhetoric of Visual Arguments. In: Hill, C.A., Helmers, M.H. (eds.) Defining Visual Rhetorics, pp. 41–61. Lawrence Erlbaum (2004)

Butler, J., Lidwell, W., Holden, K.: Universal Principles of Design, 2nd edn. Rockport Publishers, Gloucester (2010)

Cleveland, W.S.: Visualizing Data, 1st edn. Hobart Press, Summit (1993)

Foss, S.K., Foss, K.A., Trapp, R.: Contemporary Perspectives on Rhetoric, 2nd edn. Waveland P., Prospect Heights (1991)

Helmers, M.: Framing Fine Art Through Rhetoric. In: Hill, C.A., Helmers, M.H. (eds.) Defining Visual Rhetorics, pp. 63–85. Lawrence Erlbaum (2004)

Huff, D.: How to Lie with Statistics. W. W. Norton & Company, New York (1954)

Hutcheon, L.: The Politics of Postmodernism. Routledge, London & New York (1989)

Kostelnick, C.: The Visual Rhetoric of Data Displays: The Conundrum of Clarity. IEEE Transactions on Professional Communication 51(1), 116–130 (2008)

Kuhn, T.S.: The Structure of Scientific Revolutions, 3rd edn. University of Chicago Press, Chicago (1962)

McLuhan, M.: Understanding Media: The Extensions of Man. McGraw Hill, New York (1964)

Paul, R.J.: Visual simulation: seeing is believing? In: Sharda, R., Golden, B.L., Wasil, E., Balci, P., Stewart, W. (eds.) Impacts of Recent Computer Advance on Operations Research, pp. 422–432. Elsevier, New York (1989)

Roman, P.A.: Garbage in, Hollywood out! In: SimTecT 2005, Sydney, Australia (2005)

Sargent, R.G.: Validation and verification of simulation models. In: Swain, J.J., Goldsman, D., Crain, R.C., Wilson, J.R. (eds.) Proceedings of the 1992 Winter Simulation Conference, pp. 104–114. IEEE, Piscataway (1992)

Sokolowski, J.A., Banks, C.M.: Modeling and Simulation Fundamentals: Theoretical Underpinnings and Practical Domains. Wiley, Hoboken (2010)

Tiffany, J.: Second Life®: An Innovative Strategy for Teaching Inclusivity to Nursing Students. Clinical Simulation in Nursing 7(6), 265–266 (2011)

Tufte, E.R.: The Visual Display of Quantitative Information, 2nd edn. Graphics Press, Cheshire (2001)

Zelizer, B.: The As If of Visual Rhetoric. Paper delivered at Visual Rhetoric Conference, Bloomington, IN (September 6, 2001)

Modeling and Simulation as a Theory Building Paradigm

Saikou Y. Diallo[1], Jose J. Padilla[1], Ipek Bozkurt[2], and Andreas Tolk[3]

[1] VMASC, Old Dominion University
  Suffolk, VA, United Stated
[2] University of Houston Clear Lake
  Houston, TX, United States
[3] Old Dominion University
  Norfolk, VA, United States

Abstract. This chapter makes the case that theory can be captured as a model, which can be implemented as a simulation. This allows composing and recomposing theory components to process new theory out of existing theory. While current modeling and simulation applications focus on simulation as a computational activity that algorithmically produces output data based on valid input data, therefore providing information, the proposed approach utilizes the information and combines the application thereof, which provides knowledge. Relevant work is evaluated, but existing approaches neither us the conceptualization as the central component nor are they applied to ill-defined problems, thus the proposed approach is innovative and closes existing gaps. To show the feasibility and validity, theory is represented as axiomatic structures that can be executed under bounded conditions. As such, the chapter presents a methodological approach for building theory out of existing theory using modeling and simulation.

1 Introduction

Modeling and Simulation (M&S) is an emerging new discipline that is best known for its applications, in particular in the training domain. Most introductory texts focus on these aspects of applications, e.g. Sokolowski and Banks (2009, 2010). Alternatively, the introduction focuses on the computer science fundamentals of simulation development, as well covered in books like Banks et al. (2009) or Wainer (2009). One of the view approaches introducing M&S derived from its own theory, the Discrete Event System Specification (DEVS), has been developed by Zeigler et al. (2000) and represents a systems engineering approach to simulation specification. DEVS builds a significant part of the academic foundations of modeling and simulation. Nonetheless, the emphasis lies on the development and application of simulation systems to be applied as computational activities: a solution for a problem is solved by an algorithm that now can be

A. Tolk (Ed.): Ontology, Epistemology, & Teleology for Model. & Simulation, ISRL 44, pp. 193–206.
springerlink.com

applied to other data describing related problems. The difference between a simulation system and other information systems is that the algorithm implements a model of reality, a purposeful abstraction and simplification introducing assumptions and constraints. Therefore, model-based solutions are harder to compose into a new system than other information services, as the assumptions and constraints of the models need to be aligned in addition to other interoperability aspects. Recent research therefore emphasizes the need for computer interpretable conceptual models, like described in Tolk et al. (2008). But even in this new research, simulations are still perceived as applicable solutions to given problems. In science based disciplines, solutions are applied to solve problems as well, but the emphasis does not lie on the solution itself, but on the method on which the solution is based and the theory from which the method is derived. In general, theories are an important output of the knowledge creation process and finding new good and valid theories is among the most important goals of conducting research. As eloquently put by Popper (1968 p. 59):

> "theories are nets cast to catch what we call 'the world': to rationalize, to explain, and to master it. We endeavor to make the mesh ever finer and finer."

From the M&S standpoint, models and theories have a lot in common, so the question arises if theories can be represented by models. To make theories themselves easier to be accessible to such scientific evaluations, good definitions of theories, methods, and solutions as well as their connections are needed. Bacharach (1989) defines theory as a statement of relations among concepts within a set of boundary assumptions and constraints that are parsimoniously organized and clearly communicated. These concepts are studied in the form of directly observable variables that are related through hypotheses and in the form of constructs that are related through propositions. Hypotheses are concrete and operational statements derived from more abstract propositions; constructs are mental configurations of a given phenomenon that can be measured through variables; variables are observable entities that can take two or more values. Although an empirical perspective, Bacharach's account provide the basic elements one needs to consider when building a theory, namely constructs and propositions and if data is available then variables and hypotheses. It also shows the proximity of theory and model, as Bacharach's definition can be mapped to modeling principles.

However, Bacharach's perspective has different requirements that are not always fulfilled in problem domains: the phenomenon is directly accessible; objectively observable, directly or indirectly measurable, and more importantly the researchers studying the phenomenon have access to all these data. In particular when considering ill-defined problems the researcher copes with phenomena that are not directly accessible and with multiple and sometimes competing accounts on observations. This subjectivity leads to different constructs within different theories, which makes it difficult to determine what data to collect, if data is accessible at all. Recent research has shown that ill-defined problems are commonly found in different disciplines. In particular in new discipline that emerge from overlapping sub-domains of contributing related disciplines, like M&S with its roots in computer science, operations research, systems engineering, artificial intelligence, and more, such ill-defined problems have to be overcome when defining the body of knowledge representing a comprehensive and concise representation of concepts, terms, and activities is needed that make up a professional M&S domain.

The approach proposed and described in this chapter uses the idea to represent theory components as models. This allows implementing the components as simulation components that can be recombined under validity constraints. The main objective of the approach is to generate a theory, from existing theory, that can explain a phenomenon of interests by making explicit what the phenomenon is and how it works. The applicable phenomena, as mentioned, are those that have no forms a being measured, non-physical, no direct access to data, and due to these characteristics, multiple and often competing theories that attempt to provide an explanation. As a form to formalize the process and gain insight into these phenomena, M&S is presented as the conduit to develop the theory.

As a knowledge generation activity, Ören (2009, p.18) states that "from an epistemological point of view, simulation is a knowledge generation activity with dynamic models within dynamic environments." This suggests that M&S provides a way of exploration in areas of study that may not be accessible through empirical means while providing a formalism that rationalist means may not be able to achieve. The correspondence vs. coherence perspective provided by empiricism and rationalism is, therefore, also valid for simulation models. As Schmid (2005) states, a simulation model is accepted as true if these is correspondence to reality; the perspective of coherence also applies to simulation, in which a simulation model can be true only if it consists of a coherent system of believes.

When dealing with complex phenomena simulation becomes extremely useful given that allows the researcher to explore possibilities and test the boundaries of theories in development. Gilbert and Terna (2000) have stated that the reason why social sciences have not benefitted from computer simulation as a methodological approach enough may be that the main value of simulation in the social sciences is for theory development rather than for prediction. The proposed approach is a way of formalizing the use of modeling and simulation for the purpose of theory development. The flow of the chapter is as follows: In Section 2, three example approaches that are found in literature are discussed and critiqued. The proposed approach is described and explained in detail in Section 3. The validity of this proposed approach is discussed in Section 4, followed by the conclusions section.

2 State of the Art in Theory Building Using M&S

Within the body of knowledge of theory development, various approaches exist that propose methodologies and/or methods that use M&S. Davis, Eisenhardt, and Bingham (2007) propose a roadmap for developing theory using simulation methods. Simulation's primary value is in creative experimentation to produce novel theory. They suggest the following method:

- Research Question
- Identify simple theory (conceptual modeling)
- Chose simulation approach
- Create computational representation
- Verify computational representation
- Experiment to build novel theory
- Validate with empirical data (if available)

Davis et al. roadmap departs from an existing simple theory that can be simulated. The main purpose of the simulation is to generate data that can later be analyzed and if possible compared with empirical data. However, this approach's assumption of an existing simple theory may be more appropriate for theory testing than for theory building. If a theory already exists is usually suggested to proceed with its testing for which M&S can provide basis. In this sense, the testing of the simulation is equivalent to the testing of the theory. This approach does not elaborate on what a simple theory is or how to assess its level of simplicity to be able to be explored using the suggested roadmap. The simple theory for a researcher may be a complex theory for someone else. Lastly, this approach seems to focus its attention more on the simulation aspect than on the modeling aspect. Although simulation is key to establish a computer experimental environment and to generate the needed data to study, the modeling component may bear much of the biases of the researcher if this is not made transparent.

Bertrand and Fransoo (2002) present a methodology that builds objective models that partly explain the behavior of real-life operational processes or that can partly capture decision-making problems. They propose a methodology that follows axiomatic research using simulation:

- Conceptual Modeling
- Justification of research method
- Scientific model
- Justification of the heuristic or hypothesis
- Experimental design
- Analysis of results
- Interpretation of results

This approach focuses on the use of existing models or variant of models that have been studied before. This brings two assumptions: that there exist models that can be used and that they are correct for the problem at hand. This assumption is correct within the Operations Management community where new models are built on existing models that can be proven to be correct. However, this is not necessarily the case in areas where theories and models about those theories are scarce.

Sousa-Poza, Padilla, and Bozkurt (2008) present a methodology and a method. The rationalist/inductive methodology consists on generalizing from patterns found in the body of knowledge towards theory building, instead of generalizing from observations as it is the case of induction in empiricism. The method consists in building premises from these generalizations and put them together in a coherent system of premises where assumptions are made explicit and no contradictions are created. Coherence is then established via modeling and simulation and from the results of the simulation an interpretation is conducted. This approach is based on the traceability of the resulting theory to the set of premises and the set of premises to the body of knowledge as a form of validation of the theory. The Rationalist Inductive Methodology and Method is similar to the proposed approach in this study in terms of its focus on developing theory out of theory. However, it lacks the sufficient amount of detail needed for proper application. Figure 1 highlights the methodology and method.

It is noted that in all three cases modeling takes a supporting role to the simulation effort. In other words, modeling is important as long as the simulation is the one providing the insight. However, the approach proposed in this chapter

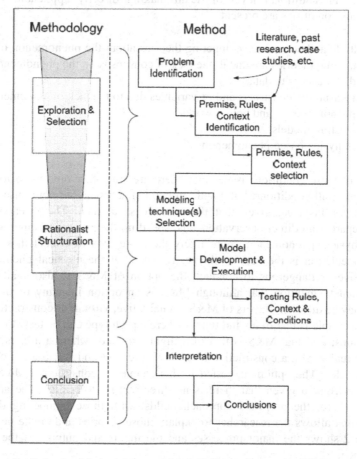

Fig. 1 Rationalist/Inductive Methodology and Method (Sousa-Poza et al., 2008)

in the following sections utilizes the modeling process to create theory that can be enriched with results from the simulation process. In addition, these approaches described in this section do not explicit provide mechanism for studying ill-defined problems. The proposed approach covers the development of theories for this kind of problems. As such, the explicit use of conceptualizations as dominant parts of theories and the application to ill-defined problems are innovative components of the proposed approach that have not been observed in related research.

3 M&S for Theory Building

The proposed approach has the advantage that does not assume an existing simple theory and does not depart from variants of existing models. In addition, it provides an additional level of detail making directly applicable where the following conditions are present:

1. Multiple and sometimes competing theories about the phenomenon of interest. Competing theories are due to the lack of consensus on the phenomenon
2. No direct access to data.
3. No measurable constructs and/or variables due to the lack of agreement of what the phenomenon is and how it works.
4. No existing models.
5. Non-physics-based phenomenon.

If one of these conditions is not met, there are still other options to follow. For instance, if all conditions, but condition two (2), a researcher could use grounded theory, for instance, given that there is access to data. The researcher could also depart from direct observations and speed up the process through M&S. If it is a physical phenomenon, more likely there are data to collect then empirical experimentation is the best candidate. However, if the physical phenomenon is expensive or dangerous to conduct, the option of M&S is also available. It is important to mention that although M&S is an option in many of these cases, they may be different flavors of M&S, namely, live, virtual, or constructive. A live M&S example is that of a wind tunnel where a prototype can be tested for real life conditions; a virtual M&S is that of a flight simulator where a user is immersed in a virtual world; a constructive M&S is where user and world are a creation of the modeler. The option suggested in this chapter is constructive M&S where a world is created given that there is no direct access to reality. The address the issues above, the proposed approach builds, with those competing theories, a world that allows the researcher to explain those theories and create new insight. Figure 2 shows the major processes and the inputs and outputs of the proposed approach.

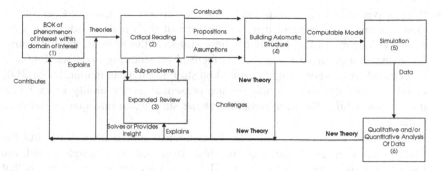

Fig. 2 Approach for Theory Building

Process one (1) refers to scouting the body of knowledge of the phenomenon of interest within a domain of interest. This means that the researcher first needs to identify the context within which the research is conducted and establish the boundaries of the domain of interest. This step allows the establishment of where the theory is intended to be applicable. This step is guided by the research problem for which a theory is needed and a research question that helps bound the scope of the research. The research problem establishes the need for a theory providing the significance of a possible solution. The problem is posed as a statement or series of related statements that must be supported by the BOK. The research question serves as a guide to address the problem within the domain of interest. For instance, a research problem related to interoperability would state: there is no theory of interoperability as it applies to M&S that explains what interoperability in an unambiguous and formal matter. This sentence is made up of concatenated statements that posit the problem at hand and for which a theory of interoperability is needed. As a follow up, a research question would read: what is interoperability and how it can be identified? Although it appears as two questions, the second one cannot be answered without the first and both need to be answered to explain interoperability. As the BOK is scouted, it has been identified that there is no accepted theory that provide a formal explanation of interoperability, but there have been attempts to an explanation. These attempts are mostly theories that need to be evaluated. These theories are the output of the study of the BOK that serves as the input of the process of Critical Reading (2).

Critical Reading is a key process within the approach. It is the process that evaluates explanations of how the phenomenon is defined and how it works within the BOK. From theses definitions and descriptions of the phenomenon, the researcher needs to evaluate their inadequacies and contributions by identifying within the selected theories:

- Lack of Precision: most theories instead of defining what something is they describe what something does. Further, they lack precision when presenting definitions. This is reflected by the use of undefined and ambiguous terms that may lead to subjective interpretations and/or circularities. Additionally, some attempts tend to classify or establish categories of the phenomenon when it is still undefined. This categorization adds to the lack of precision of the phenomenon.

- Perspective: identifying, if possible, the worldview of the proposer of the theory is important because it tells details about the mindset under which the theory was developed and its untold limitations. In the BOK of understanding, for instance, a group of researchers is focused on studying understanding as a process, whereas other group is focused on studying it as an output. In the BOK of interoperability, some definitions are presented as the ability to exchange information while others are presented as the state when information has been exchange.
- Assumptions: researchers postulate their theories and usually leave out the assumptions they use to build them. Most assumptions, although untold, are valid within the context of the theory. However, they are also weak points that may need to be challenged. Assumptions have different origins. One is the research method used to conduct the research. When the research is conducted via experiments, for instance, the main assumption is that the phenomenon can be directly observed and measured. This is regardless of the possible non-physical nature of it. Using the example of understanding. One of the observable processes used as a proxy for studying understanding is problem solving. However, the assumption that the identification of a solution is a reflection that a person understood is flawed given that a person can arrive to a solution by luck or by trial and error. Further, perhaps understanding is simply the identification that no solution is the solution to the problem in question.
- Preconceptions: during theory development, researchers are tempted to posit characteristics of the phenomenon that are neither the reflection of generalization from data, nor a logical deduction, nor a generalization from literature. These are ideas of how the phenomenon "should" work. In this case, this is no longer a theory building effort, but a theory testing effort where the how the phenomenon "should" work need to be tested first.
- Unique characteristics of the phenomenon: these are the components and processes that are part of the phenomenon. Common characteristics' selection is extremely important given that these are the main candidates for the constructs and propositions to be used to explain the phenomenon. For instance, when referring to understanding, one important construct that is commonly found in the literature is the concept of knowledge. Knowledge then becomes a construct used to explain understanding. The process of mapping is also commonly found in the literature when referring to understanding and its descriptions may become a proposition of how the process works. The combination of characteristics must identify the phenomenon in question uniquely and also isolate and bound the phenomenon from similar or concurrent phenomena. For instance, the phenomenon of understanding is usually defined as part of learning or as part of problem solving. However, its combination of characteristics must be different than the combination of characteristics of those processes; especially when components are shared, such as knowledge.

From critical reading there are four major outputs: generalized common components, generalized common processes, generalized assumptions, and generalized common sub-problems that were not addressed in the BOK. Common components are elements that can be turned into constructs of the phenomenon. As previously explained, constructs are not-directly measurable components of a phenomenon. Common processes are turned into propositions that bind constructs together. Propositions are statements that are believed to be true about the phenomenon. Assumptions allow theories to be formed, but they also limit their generalization. From the BOK, they are the main candidates subject to challenge given that some of the limitations need to be lifted for the new theory to take place. Sub-problems are issues about constructs and propositions that are not resolved within the BOK. They can be explained either on an expanded review of a broader BOK (3), through the construction of the axiomatic structure (4), or possibly through the simulation (5). A review of a broader BOK, and its corresponding critical review, means that the researcher needs to go beyond the boundaries of its disciplines and domain of interest to find an explanation for these sub-problems. In the understanding example, for a psychology researcher to investigate the concept of knowledge, it may need to seek supporting information in areas where knowledge has been studied such as epistemology and knowledge management among others. Sub-problems, if possible, must be addressed with the expanded review in order for constructs and propositions to be clearly defined and assumptions to be properly challenged within steps (4) and (5).

Building the axiomatic structure is a full modeling process and a major step in theory creation. Here, constructs and propositions brought over previous steps are formally defined in order to eliminate any ambiguity found within the BOK. This means that a construct and a proposition must be identified uniquely and mean only one thing. The axiomatic structure either solves some of the sub-problems that were carried over from (4) or it is the basis to become a computable model that later can be simulated. If a sub-problem is solved by means of the axiomatic structure, then a new theory has already been created. This is of extreme importance and of difference with traditional approaches that use M&S to build new theory. This means that theory is created "during" the modeling process. Traditional approaches generate theory based on the analysis of data from the simulation only (Davis et al., 2007; Bertrand & Fransoo, 2002; Sousa-Poza et al., 2008).

It is important to note that the axiomatic structure should explain existing theories from where constructs and propositions where derived. This means that the phenomenon is being explained within a general theory and not another instantiation of the theory. An explanation of existing theories with the axiomatic structure then becomes a test of the new theory. Theory and axiomatic structure are not the same. The explanation of the phenomenon through the axiomatic structure becomes the theory. The structure is just the conduit for that explanation.

It is suggested that this axiomatic structure be built in a manner that reflects a formal modeling process. Set theory or predicate logic are considered good candidates. Another candidate is modeling towards a computable implementation. In this case, the axiomatic structure is formal enough to be processed by a computer. This implies that the modeling can be done oriented towards a simulation using systems dynamics, discrete event, or using agents. Figure 3 roughly shows an algorithm for selection of the modeling paradigm.

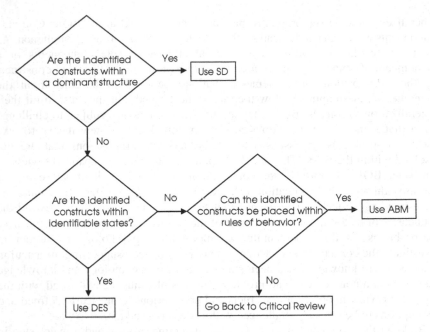

Fig. 3 Selection of the M&S Paradigm

The axiomatic structure then can be simulated. Simulation provides a glimpse into how the phenomenon works under bounded conditions contributing to the theoretical development. This is where this approach is the same as traditional approaches that use M&S. Simulation is mainly used to generate data that can be either assessed qualitatively and analyzed quantitatively used statistical analysis. Through this analysis, further theoretical insight about how the phenomenon works is derived.

Through generalizations from data, theory is created. This theory jointly with the theory created during the modeling process make the new theory. The new theory should address any existing sub-problem, be able to explain existing theory, and provide insight not foreseen before. This is particularly the case when emergence takes place during the simulation. Emergence, in this case is just a pattern that was not considered previous to the simulation, but that can be explained within the axiomatic structure.

Finally, either through the axiomatic structure or through the simulation means of how to measure the constructs of the phenomenon should be presented. However, given that these are still constructs, the accessibility to techniques and tools to measure them may not yet be available. Nonetheless, they provide the basis for future research and further empirical studies.

Padilla (2010) applied this approach systematically to evaluate the question of building a theory of understanding. The current literature identifies knowledge needed to solve a given problem, the world view allowing perceiving the problem, and the problem understanding and definition as such as the driving components for such a theory. Defining an axiomatic structure for knowledge, world view, and problem definition, the current interpretations of understanding were evaluated. Using a simple agent based approach, in which agents were used to represent the axiomatic structures for knowledge; theories for understanding were derived by re-composing the axiomatic structures of matching agents. Computational intensive experiments did not only produce know theories – such as understanding based on the knowledge needed to solve a problem, or understanding based on recognizing a problem to be similar to another problem for which a solution is known that can be applied to the current problem as well –, but new theories emerged that are not captured in detail in literature, such as the dominance of having the 'correct' world view in order to solve new problems (a problem known as 'cultural awareness' to many current defense related operations). Although the example used in Padilla (2010) is limited in its applicability and cannot be easily generalized to other domains, it presents a first application example proofing the feasibility of the approach.

4 Validity of the Proposed Approach

As most researchers in M&S would attest, validity is a contentious issue mainly because it mostly refers to its empirical roots. According to Moss (2008):

> Although model validation has been an ongoing issue in the social simulation literature, there has so far been no systematic consideration of whether different approaches to validation are appropriate to different approaches to modeling and whether some validation approaches, and their associated modeling approaches, are preferable to others.

Empirical validation of models is that in which "validation involves comparison of simulation results with empirical data. If the results of the simulation match the empirical evidence, then the simulation is validated for that empirical context" (Davis et al. 2007). Because of the reasons presented before, empirical validation by comparison with real world data is not possible. This is because until the constructs where postulated there were no objectively agreed constructs to measure. Further, given that the problem is ill-defined in the BOK, establishing an experimental case where it can be tested may not be possible. It is noted however, that the resulting theory is a generalization that considers the different instantiations of the theory. A testing of the theory will irrevocably result in falling into one of those instantiations. Empirical research departs from instantiations to establish generalizations. The theory in this case is already a generalization, an abstraction of the concept that reverts to one of the particular cases when tested. If this is the case, the researcher must identify which of the particular cases is being tested.

Schmid (2005) defines rational validation of a model if the model is true due to its membership of a coherent system of believes, in this sense a simulation must be consistent and non-contradictory in that system of beliefs. Schmid says that a model may be wrong in what regards to its correspondence with reality, but truth using coherence if it satisfies its subjective purpose. He presented two concepts in validation, *specific purpose* and *sufficient accuracy*; a model can be valid from one perspective (serves its purpose), but inaccurate on the other (lack of empirical data). In this case, the model is accurate and true (ergo valid) from the viewpoint of coherence while invalid from the viewpoint of correspondence. This position is consistent with Sousa-Poza et al., (2008). They suggest that the validity of this type of approach is based on the capability of the new theory to explain the theories from which it was derived and on the coherence of the new theory. Coherence, they suggest, is assessed by how well constructs and propositions fit together. From figure 2, the arrows that depart from New Theory that explain existing theories and explain the expanded review are forms of rationalist validation. In addition, the axiomatic structure and its formal structure is a form of rationalist validation as well.

One pragmatic form of validity built into this approach is the solution of the identified sub-problems. If the sub-problems are completely or partially solved, then the theory is found to be useful. Being useful is the base of pragmatism which is a common form of validation found in disciplines such as engineering; as long as it works it is valid. In figure 2, the arrow that solves or provides insight about sub-problems is a form of pragmatic validation.

5 Conclusions and Future Work

In summary, this chapter presents a methodological approach for building theory out of theory using modeling and simulation. The approach serves for conducting research in ill-defined domains of non-physical phenomena where: there are multiple and sometimes competing theories about the phenomenon of interest, no direct access to data, no measurable constructs and/or variables, and no existing models. The approach starts with scouting the Body of Knowledge, goes through a critical review thereof, formulates constructs, propositions, assumptions and sub-problems, incorporates them into a model, and conducts a simulation which result in a New Theory. This New Theory is a generalization of previous existing theories. This means that the theory developed using the proposed approach not only explains and encompasses all existing theories, but also provides explanations to sub-problems and bring new insight to the non-physical phenomenon at hand. The advantage of using M&S for theory building is highlighted. M&S provides formality and traceability making it a robust approach.

The proposed approach is more than a new application of M&S, it is a paradigm shift from applying M&S as a computational activity that applies algorithmic knowledge to solve problems by generating output data based on provided valid inputs towards real knowledge processing. In his well know chapter, Ackoff (1989) distinguishes between data, information, and knowledge: data are simply a set of symbols, facts or figures, while information is data that are processed in context to be useful and provides answers to questions such as "who," "what," "where," and "when;" and applying information useful results in knowledge. Computational simulation can only derive output data from input data, those providing information. By using the approach described in this chapter, the next quality leap from information to knowledge is supported, as the recombination of models representing theory is based on the application of useful information. Hence, M&S moves into the category of intelligent decision technologies with the potential not only to reproduce and conserve knowledge, but actually to produce new knowledge. Although predicted by visionaries like Yilmaz and Oren (2004), the chapter describes a real application of such ideas and proves the possibility thereof.

In order to take full advantage of this new application paradigm of M&S as an intelligent knowledge processing method, the conceptualizations of simulations need to be made explicit to allow their computer supported re-composition to derive new knowledge in a systematic way. First ideas are captured in Tolk et al. (2010) and the recent dissertations of Diallo (2010) and Padilla (2010), but more research in this direction is needed. However, the interpretation of models as representations of theory and knowledge allows perceiving model bases to extract knowledge as today we use data bases to extract information. As the structured query language (SQL) today allows to define what information is needed, based on metadata describing the formal specification of the conceptualization of simulations in the future the knowledge needed can be defined using something like a "model query language" (MQL). Such efforts will significantly impact domains of knowledge management, risk management, and many other fields. This will truly introduce a new intelligent decision technology envisioned in Tolk et al. (2009), namely decision support simulation systems. While the traditional view on decision support systems is still dominated by collecting and presenting data, simulation added a new feature by adding the model-based development of these data over time, focusing on the processes. Therefore, Tolk and colleagues define *"Decision Support Simulation Systems* as simulation systems supporting operational (business and organizational) decision-making activities of a human decision maker with means of modeling and simulation. They use decision support system means to obtain, display and evaluate operationally relevant data in agile contexts by executing models using operational data exploiting the full potential of modeling and simulation and producing numerical insight into the behavior of complex systems." (Tolk et al. 2009 p. 405). In particular in combination with the agent metaphor the approach described in this chapter will enable a new category of intelligent decision technologies.

References

Bacharach, S.: Organizational theories: Some criteria for evaluation. Academy of Management Review 14(4), 496–515 (1989)

Banks, J., Carson, J.S., Nelson, B.L., Nicol, D.M.: Discrete-Event System Simulation, 5th edn. Prentice Hall, Upper Saddle River (2009)

Bertrand, J.W.M., Fransoo, J.C.: Operations management research methodologies using quantitative modeling. International Journal of Operations & Production Management 22(2), 241–264 (2002)

Davis, J., Eisenhardt, K., Bingham, C.: Developing theory through simulation methods. Academy of Management Review 32(2), 480–499 (2007)

Diallo, S.Y.: Towards a Formal Theory of Interoperability. Doctoral thesis at Old Dominion University, Frank Batten College of Engineering and Technology, Norfolk, VA (2010)

Gilbert, N., Terna, P.: How to build and use agent-based models in social science. Mind and Society 1(1), 57–72 (2000)

Ören, T.: Modeling and simulation: A comprehensive and integrative view. In: Yilmaz, L., Ören, T. (eds.) Agent-Directed Simulation & Systems Engineering, pp. 9–45. John Wiley & Sons, New York (2009)

Mitroff, I., Betz, F., Pondy, L.R., Sagasti, F.: On managing science in the systems age: Two schemes for the study of science as a whole systems phenomenon. Interfaces 4(3), 46–58 (1974)

Moss, S.: Alternative approaches to the empirical validation of agent-based models. Journal of Artificial Societies and Social Simulation 11(1), 5 (2008), http://jasss.soc.surrey.ac.uk/11/1/5.html

Padilla, J.J.: Towards a Theory of Understanding Within Problem Situations. Doctoral thesis at Old Dominion University, Frank Batten College of Engineering and Technology, Norfolk, VA (2010)

Schmid, A.: What is the truth of simulation? Journal of Artificial Societies and Social Simulation 8(4), 5 (2005), http://jasss.soc.surrey.ac.uk/8/4/5.html

Sokolowski, J.A., Banks, C.M. (eds.): Principles of Modeling and Simulation: A Multidisciplinary Approach. John Wiley & Sons, New York (2009)

Sokolowski, J.A., Banks, C.M. (eds.): Modeling and Simulation Fundamentals: Theoretical Underpinnings And Practical Domains. John Wiley & Sons, New York (2010)

Sousa-Poza, A., Padilla, J.J., Bozkurt, I.: Implications of a rationalist inductive approach in system of systems engineering research. In: Proceedings of IEEE International Conference on System of Systems Engineering, Systems, Man, and Cybernetics (2008), doi:10.1109/SYSOSE.2008.4724186

Tolk, A., Diallo, S.Y., Turnitsa, C.D.: Mathematical Models towards Self-Organizing Formal Federation Languages based on Conceptual Models of Information Exchange Capabilities. In: Proceedings of the Winter Simulation Conference, Miami, FL, pp. 966–974 (December 2008)

Tolk, A., Madhavan, P., Jain, L., Tweedale, J.: Agents and Decision Support Systems. In: Yilmaz, L., Ören, T. (eds.) Agent-Directed Simulation and Systems Engineering, pp. 399–431. John Wiley & Sons, New York (2009)

Tolk, A., Diallo, S.Y., King, R.D., Turnitsa, C.D., Padilla, J.J.: Conceptual Modeling for Composition of Model-based Complex Systems. In: Robinson, S., Brooks, R., Kotiadis, K., van der Zee, D.-J. (eds.) Conceptual Modelling for Discrete-Event Simulation, pp. 355–381. CRC Press (2010)

Wainer, G.: Discrete-Event Modeling and Simulation: A Practitioner's Approach (Computational Analysis, Synthesis, and Design of Dynamic Systems). CRC Press, Taylor & Francis Group (2009)

Yilmaz, L., Oren, T.: Exploring Agent-Supported Simulation Brokering on the Semantic Web: Foundations for a Dynamic Composability Approach. In: Proceedings of the 2004 Winter Simulation Conference, pp. 766–773 (2004)

Zeigler, B.P., Praehofer, H., Kim, T.G.: Theory of Modeling and Simulation, 2nd edn. Academic Press, Sand Diego (2000)

Toward Replicability-Aware Modeling and Simulation: Changing the Conduct of M&S in the Information Age

Levent Yilmaz[1] and Tuncer Ören[2]

[1] Auburn University
Auburn, AL, United States
[2] University of Ottawa
Ottawa, Ontario, Canada

Abstract. The use of computational models in science end engineering is increasingly becoming pervasive. However, there is a growing credibility gap due to widespread, relaxed attitudes in communication of experiments, models, and validation of simulations used in computational research. Consequent disputes and article retractions due to unverified code and data suggest a pressing need for greater transparency. We introduce the e Portfolio concept, which is an ensemble documents that interweave the conceptual model, simulator design, experimental frames, and scientific workflow specifications. Strategies and potential mechanisms are delineated to enable authors, publishers, funding agencies, journals, and the broader scientific community to cooperate and establish a sustained model base, simulations, experiments, and documentation, so that scientists can build on each other's work and achievements.

1 Introduction

Modeling and Simulation (M&S) has emerged as a critical trans-disciplinary field that plays a critical role in advancing discovery and innovation in Science and Engineering (S&E) (WTEC, 2009). Numerous reports (Ören 2012; Cummings and Glotzer, 2010; NSF 2006) corroborate the significance and role of computer simulation not only in Science and Engineering (S&E), but also in societal decision-making and public policy formation. As the use of computer simulation is increasingly becoming central to scientific enterprise, lack of proper documentation, validation, and distribution of models and experiments may hamper reproducibility and hence cause a credibility crisis (Ören and Zeigler, 1979; Stodden, 2010).

Reproducibility, as a fundamental principle of the scientific method, refers to the ability to independently *replicate*, *reproduce*, and, if needed, *extend* computational artifacts associated with published work (Fomel and Hennenfent, 2009). Emergence of reproducibility as a critical issue is predicated on the growing credibility gap due to wide spread presence of relax attitudes in

A. Tolk (Ed.): Ontology, Epistemology, & Teleology for Model. & Simulation, ISRL 44, pp. 207–226.
springerlink.com   © Springer-Verlag Berlin Heidelberg 2013

communication of the context, experiments, and models used in computational research (Mesirov, 2010; Stodden, 2010; Donoho et al., 2009). Furthermore, as indicated in (Fomel and Claerbout, 2009), a published computational science article is not the scholarship itself; it is merely advertising of the scholarship. The actual scholarship is – in addition to the conceptual model – the complete software development environment and the complete set of instructions, which generates the article.

These observations, coupled with disputes such as Climate Gate (Economist, 2010), the microarray-based drug sensitivity clinical trials under investigation (Baggerly and Coombes, 2009), and article retractions due to unverified code and data (Alberts, 2010; Chang et al., 2006) suggest a pressing need for greater transparency in computational science. Besides, novel and beneficial progress in computational science demands generation of new knowledge in terms of elaboration and combination of both new and existing computational artefacts. Unless computational artifacts are designed and disseminated to be discovered, extended, or combined with other models, scientific progress can be hindered. Furthermore, the inability of others to independently reproduce and verify published results will slow down the adoption and the use of knowledge embedded within software and models (Peng, 2009). Therefore, as emphasized at a recent NSF panel (NSF, 2011), reproducibility should become the responsibility of the broader scientific community. Following the introduction of the term *reproducible research* (Schwab et al., 2000), increasing attention has been given to various dimensions of the problem. Recent emphasis on reproducible research culminated in recommendations and standards such as the *Reproducible Research Standard* (Stodden, 2010). While these guidelines are useful, implications on methodology and technical infrastructures need to be examined to engineer reproducibility in the first place (Yilmaz, 2011a; Yilmaz, 2011b).

Although there is a proliferation of computational system models used in application domains widely ranging from engineering to social sciences, most of these models have never been replicated by anyone but the original developer. However, replication is critical to scientific practice, and availability of a replicability-aware model development infrastructure is imperative to promote and enable the practice of reproducibility. In this chapter, one of our objectives is to promote replicability in the context of M&S; however, the proposed strategy and its computational infrastructure are generalizable for use by the broader computational science and engineering community. Considering that both science and engineering are becoming simulation-based, the importance of replicability awareness of M&S and associated infrastructures is of paramount significance.

The outlined strategy has two critical pillars. First, by extending the **Simulation Experiment Description Markup Language (SED-ML)** and applying increasingly sophisticated pilot projects, one can discern the minimal information needed to communicate specification of a conceptual model, its simulator, and the experimental frame, so that scientists can replicate the simulation experiment, possibly in a new context and platform. Specification of the simulation world view (e.g., discrete time, discrete event, continuous) and model behavior generation strategy should also be encapsulated with the specification. This requires provision of a schema that facilitates specification of the simulation model and experiment definition as an instantiation of the related ontologies. Developments in ontology-based model specification formalisms are promoted as potential mechanisms to further extend the SED-ML namespace and schema specification. Transformation of the models into RDF/EXtensible Markup Language (RDF/XML) distribution packages can facilitate extending the original SED-ML schema and namespace currently constrained to predefined Uniform Resource Names (URNs) such as SBML and CellML. The second phase of the proposed strategy involves development of a tool and graphical user interface to automate the generation of exported SED-ML documents from within the M&S development environments. Besides exporting such documents, transformation and verification utilities are needed to facilitate (1) modification of content and structure of model elements, (2) generation from raw data of output elements in a specified form, (3) transformation of the shared conceptual model into a selected target formalism, and (4) replicability quality assurance.

In this chapter, we discuss and promote a practical ontology-driven replicability for cross-validation of simulation models. Replicability is also expected to improve model longevity by enabling formal documentation, distribution, and exchange of conceptual models while providing tool support for their effective replicability in evolving platforms. Use of program generators from high-level specification languages would allow maintaining high-level specification of models, a practice which may have several advantages such as: (1) improved understandability by domain experts; (2) ease of model updating as well as model composition; and (3) elimination of coding errors, hence improved verification possibility. Furthermore, advances in model replicability are expected to increase societal trust and confidence in scientific knowledge.

The rest of the chapter is structured as follows. In section 2, we present a brief overview of the three major dimensions of reproducible M&S research that provides a framework and context to explore the issues and challenges in computational replicability, which are overviewed in section 3. Section 4 introduces a strategy based on the Simulation-Experiment Description Markup Language. Section 5 focuses on incentives and processes that scientific enterprise can adopt to promote reproducible computational research. In section 6, we conclude by discussing the significance of replicability and research challenges, in relation to the presented strategy.

2 Dimensions of Replicability in M&S Research

As data and models used in simulation research become more transparent and accessible, issues pertaining to functions of scholarly communication, organizational models of simulation research practice, and model development environments need to be revisited. As shown in Table 1, the major areas of focus in computational replicability research can be classified under three dimensions: Scholarly communication, methodology of scientific practice, and technical infrastructure.

Table 1 Methodological and Technical Dimensions

Dimensions of Reproducible Research		
Communication	Legitimization, Dissemination, Access	
Process	Roles	Scientist, Journal/Publisher, Funding Agency
	Ownership	Citation, Licensing, Open Source
Infrastructure	Content	Model, Simulation code, Data , Publication, Experiment
	Service	Search/Discovery, Collaboration, Sharing
	Tools	Authoring, Version control, View/Distribute Transformation, Quality control

Borgman (2007) examines data sharing for scholarship in the digital age and suggests a broad range of recommendations for quality assurance, dissemination, and preservation of data and code associated with publications. Among the recommendations are certification and continuous evaluation of data, use of metadata for accurate interpretation and dissemination, and open access with non-proprietary standards for preservation. With regard to methodology of scientific practice, reproducible research can be examined from the perspective of roles played by institutional elements of science and management of intellectual property (Altman and King, 2007; Stodden, 2009), as models, code, and data are transparently shared with the research community. In practice, scientists are encouraged to use third party archival systems such as Harvard's Dataverse network (http://thedata.org) as well as scientific collaboratories to share models, data, and code. However, preserving different components of reproducible research in separate domains without a unifying container that facilitates their co-evolution and synchronization impedes maintainability.

To encourage scientists to share components of their research, examination of licensing and copyright management schemes for reproducible research resulted in the Robust Research Standard (RSS) (Stodden, 2009). RSS recommends the use of CreativeCommons (2011) attribution license for media components, modified Berkeley Software Distribution (BSD) license for code, and Science Commons Database Protocol (http://commons.org/projects/publishing/open-access-data-protocol) to the data. Besides the characteristics of scientific practice and considerations of scholarly communication, new developments in infrastructure and model development environments should consider technical requirements associated with production, distribution, and access to components of reproducible research.

3 Computational Replicability

Replicability, which is the main focus of this chapter, has two implications. It is the essential characteristic of a conceptual model already developed by a scientist or a group of scientists (1) to be usable in a different simulation study by the same or different scientist or group of scientists and/or (2) to be validated and verified by an independent body. Unlike the reproducibility of results using the original author's implementation, which is the most common ambitious and long-term strategy envisioned for reproducibility (Fomel and Hennenfent, 2009; Fomel and Claerbout, 2009; Gentleman and Lang, 2005; Leisch, 2003), replication refers to creation of a new implementation. However, the implementation of the replicated model differs in some way (e.g., platform, language) from the original model, but should be an executable representation to facilitate conducting the same experiments. While reproducing the results of a model distributed with the published document has benefits such as production of reports and visualizations for comparisons to versions listed in the document, provision of a strategy for replication of a study in a new context has its own merits (Yilmaz et al., 2011b, 2011c). For instance, developers, who replicate models for cross validation, may have different implementation tools and infrastructure and hence may not be familiar with the platform-specific constraints associated with the original model.

Therefore, providing the ability to implement a conceptual model under specific experimental conditions and analysis constraints across multiple platforms is critical for practical and broader adoption of the practice of reproducibility. Also, by replicating a model and ignoring the biases imposed on the original model by its chosen toolkit, differences between the conceptual and implemented models may be easier to observe. To facilitate replicability, communication using an extensible and platform-neutral interchange language for the specification, distribution, and processing rules for model, simulator, and experimental frame elements is critical. A successful replication of a computational experiment advances scientific knowledge, because it demonstrates that the experiment's results can be repeatedly generated and thus the original results were not an exceptional case.

4 Toward a Computational Infrastructure for Computational Replicability in M&S

The main objective of the strategy envisioned and presented herein is to outline a practical approach and tool support using both open-source off-the-shelf technology and new utilities to demonstrate how replicability of computational models can start becoming mainstream. **e-Portfolios** are introduced and defined as containers that can help distribute, exchange, and deploy conceptual models of simulation models for both replication and cross-validation

4.1 An Integrated and Introspective Framework for Specification and Distribution Using e-Portfolios

Though exposure to the source code and the original model is important eventually, if done too early in the replication process, it may result in "groupthink" whereby the replicater unconsciously adopts some of the practices of the original model developer and does not maintain the independence necessary to replicate the original model, but instead essentially "copies" the original model. Therefore, the outlined strategy and its associated computational environment are intended to leverage conceptual models of simulations to avoid bias toward original implementation of the model.

Table 2 illustrates selected languages and modeling layers to develop practical language requirements for the communication of conceptual model, simulator, and experimental frames. At the core of the proposed strategy is the use of an extended version of SED-ML, which is an emerging mark-up language used for encoding procedures performed during computational simulation experiments and model development. SED-ML allows the definition of the model(s) to be used, the experimental task(s) to be run, and the result(s) to be produced.

The process-flow shown in Figure 1 presents selected high-level components of the envisioned information infrastructure that utilize SED-ML. By enabling the uploading of conceptual models of simulation experiments in the form of SED-ML encoded e-Portfolios new avenues are opened for tracing elements of technical documents to the conceptual models of computational research artifacts.

Table 2 Selected Markup Languages and Information Dimensions

Replicability-aware Modeling Layers	Model Specification	Simulator Specification	Experimental Frame Specification
Minimal Requirements	MIRIAM (Waltemath et al., 2011b)	MIASE (Taylor et al., 2008)	
Resource/Data Layer	XML, XSD, XMI, SBML, SED-ML	SED-ML	SBRML (Dada et al., 2010)
Ontology Layer	OWL, RDF, RDF Schema, SBO, ESG Ontology, NMM, Curator Metadata	DeMO (Silver et al., 2011), KISAO	TEDDY (Waltermath et al., 2011a)
Rule/Logic Layer	SWRL, MathML, RuleML	UML OCL, RuleML, SEDML	RuleML, SED-ML
Proof/Trust	Proof-carrying Models		Test-driven Development (Beck, 2003)

Such conceptual models can be interpreted and processed by simulators and analysis tools to reproduce results while also allowing reimplementation in new platforms.

4.1.1 The Case for Markup Languages and SED-ML

Many disciplines recognized the need for a common format for model description. This often manifests itself in the development of XML-based markup languages. Given the platform independence, extensibility, and flexibility of the XML formalism, SED-ML provides a viable strategy to describe components of both the conceptual model and the experimental frame of a simulation study. This view is consistent with the Common Information Model (CIM) perspective (Callahan and Murphy, 2011) that brings together experiment, model, and simulator aspects. Although SED-ML was originally devised for the Systems Biology domain, it

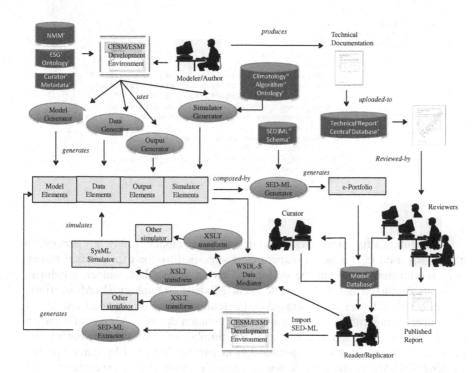

Fig. 1 Process Architecture

provides a sound basis for further extension to cover a broad class of models. As shown in Figure 1, a complex type in an SED-ML document has four major components, each of which has its own detailed schema.

e-Portfolio model elements identity the the location of the model to be simulated along with the native encoding format used in the specification of the model. The location can be given as a Uniform Resource Identifier (URI), which enables the software that interprets the SED-ML document to retrieve the conceptual model (e.g., XMI – XML representation of the UML-based conceptual model). Currently, SED-ML encodings are restricted to XML-based languages; however, to improve interoperability, the particular language of a model needs to be specified using one of the predefined SED-ML language Uniform Resource Names (URN). Since new URNs can be registered with SED-ML, one can specify a language precisely by extending the current SED-ML baseline.

Furthermore, inclusion of tags in the XML document can facilitate listing of changes that can be applied to a model specification before simulation, so that values of attributes, parameter values, and model structure can be altered. This feature can be achieved by adding or removing XML elements. XPath (Clark and DeRose, 1999) expressions can identify the target XML to update designated model entities required for manipulation in SED-ML documents.

```
<xs:element name="sedML">
  <xs:complexType>
    <xs:complexContent>
        <xs:extension base="SEDBase">
            <xs:sequence>
                <xs:element ref="listOfSimulations" minOccurs="0" />
                <xs:element ref="listOfModels" minOccurs="0" />
                <xs:element ref="listOfTasks" minOccurs="0" />
                <xs:element ref="listOfDataGenerators" minOccurs="0" />
                 <xs:element ref="listOfOutputs" minOccurs="0" />
            </xs:sequence>
             <xs:attribute name="level" type="xs:decimal"
                    fixed="1" />
                <xs:attribute name="version" type="xs:decimal"
                       fixed="1" />
        </xs:extension>
    </xs:complexContent>
  </xs:complexType>
</xs:element>
```

e-Portfolio simulator elements define the model behavior generation algorithms used in conducting experiments specified by the experimental frame. An experimental frame can be defined in terms of the data, output, batch-run elements discussed below. Discrete-event Modeling Ontology (DeMO) efforts (Silver et al., 2011) can be leveraged and extend metadata to construct ontologies of simulation algorithms for the selected application domain. Also, configuration details such as start and end times, or the number of time steps to output and parameter sweeps for batch experimentation can be described by extending the existing SED-ML schema. An important aspect of a simulation experiment is the simulation algorithm (i.e., behavior generator). But the sole reference of a simulation algorithm through its name in the form of a string is error prone and ambiguous. Firstly, typing mistakes or language differences may make the identification of the intended algorithm difficult. Secondly, many algorithms exist with more than one name, having synonyms or various abbreviations that are commonly used.

These problems can be solved by using a controlled vocabulary to refer to a particular simulation algorithm. As shown in the example below, one attempt to provide such a vocabulary is the Kinetic Simulation Algorithm Ontology (KiSAO) (Courtot et al., 2011). KiSAO is a community-driven approach for classifying and structuring simulation approaches by model characteristics and numerical characteristics. A similar foundation can be explored and documented for the selected application domain. Relevant characteristics include, for instance, the type of variables used for the simulation (such as discrete or continuous variables) and the spatial resolution (spatial or non-spatial descriptions). Numerical characteristics specify whether the system's behavior can be described as deterministic or stochastic, and whether the algorithms use fixed or adaptive time steps. Related algorithms can be grouped together, producing classes of algorithms.

e-Portfolio experimental frame defines the input/output data elements and experiment design configurations that specify the context and relate to intended purpose of the simulation study being replicated.

- Data elements and transducers help define transformations of raw data into numerical forms relevant to objectives promoted by the simulation study. The data generators (transducers) prepare the raw simulation results for later output. They also encode the postprocessing to be applied to the simulation data. The postprocessing steps could be simple normalizations of data as well as mathematical calculations. Since replicability may involve using the implementation of the model for a new purpose, data elements of the SED-ML document should be amenable to update to specify new transducers. Transducers may simply be references to a model variable, but also can define complex expressions specified in terms of languages such as MathML (Ausbrooks et al., 2003), which can be embedded within the SED-ML based e-Portfolio package. The following is a hypothetical example for demonstration purposes.

```
<listOfDatGenerators>
    <dataGenerator id="d1" name="time">
        <listOfVariables>
            <variable id="time" taskReference="task1"
                symbol="urn:sedml:symbol:time" />
        </listOfVariables >
        <math xmlns="http://www.w3.org/1998/Math/MathML">
            <ci> time </ci>
        </math>
        ...
</listOfDataGenerators>
```

- **Output elements** describe the strategy to group and structure the numerical data produced by the transducers. For instance, one can use two or more data streams to generate 2D or 3D plots. Integration with GNUPlot and R programming environments via mediator components that are able to map XML-formatted scripts into native GNUPlot or R interpreters facilitate extending SED-ML for integration with sophisticated output processing capabilities.
- **Replicability Standard (RS)** should be defined to signify the level of replication in a simulation study. A successful replication is one in which the replicaters are able to establish that the replicated model creates outputs sufficiently similar to the outputs of either the original model or available empirical data. Typically, a scientist working on an experiment will not attempt to exactly reproduce numerical results produced by another scientist. Instead, the RS aims to reproduce the level of precision necessary to establish the hypothesized regularity. This means that the RS itself changes depending upon the question being asked. Proper notation is needed to specify in SED-ML at least three general categories of replication standards: *numerical identity*, *distributional equivalence*, and *relational alignment*. The goal of *distributional equivalence* is to show that the implemented models generate data that are sufficiently statistically similar to each other. To meet this RS, researchers often show statistical indistinguishability, i.e. that given the current data there is no proof that the models are not distributionally equivalent. *Relational*

alignment exists if the results of the two implemented models show qualitatively similar relationships between input and output variables, e.g., if you increase input variable x in both models then if output variable y increases in the first model it should also increase in the second model.

4.1.2   The Case for Domain and Simulation Ontologies

If the knowledge domain is well-defined (here this is equivalent to existence of a domain ontology), the domain concepts can be mapped onto simulator concepts. Consider for example a domain with relatively well-defined concepts and relations. A small ontology can be used to represent the knowledge about this domain. These concepts may be directly linked to model components in the simulation ontology. For example "GridCell"-"Entity", and/or "TriPolar Projection"-"Process", "Arrival"-"Event", etc. These mappings may give a modeler a direct choice of one or more modeling formalisms. Extending SED-ML to facilitate mappings between simulator and model elements through domain and simulation ontologies (e.g., DeMO can provide proper semantics for shared models prior to their accurate replication and interpretation.

4.1.3   The Case for Rule/Logic Layer – Semantic Rule Language

Often model structure and behavior have assumptions (King and Turnitsa, 2008) and constraints imposed by the original author/developer, and such constraints are implicit in the software that implements the conceptual model. Without explicit communication of such constraints, replicaters will experience the risk of producing an incorrect implementation that is unlikely to produce accurate results during cross-validation against the original model. There are recent proposals for languages at the Rule/Logic Layer including the Semantic Web Rule Language (SWRL), with the ability to specify rules using a subset of RuleML (http://ruleml.org/). It also permits Horn logic rules to be added to ontology descriptions. This allows more complex predicates to be created and used for more precise definitions of concepts. To incorporate constraints into shared conceptual models, SED-ML can be extended with XML language constructs to implement a rule-logic layer.

4.2   Minimum Information Requirements for Communicating e-Portfolio Model, Simulator, and Experimental Frame Elements

By leveraging coherent minimum reporting guidelines used in biomedical investigations (Taylor et al., 2008), it is feasible to identify minimal conceptual information needed to facilitate replicability of simulation experiments. This information does not only serve as guidelines and checklist for authors/model developers, but also helps further refine the SED-ML schema that enable encoding conceptual models. Next, for the sake of brevity, we present a selected list of high-level requirements. Specifically, we start with considering requirements for models, simulators, and experimental frames from the perspective of *reference correspondence, attribution annotation*, and *external resource annotation*. In regard to information about models, specifications used in the experiment must be identified, be accessible, and fully described. This requires having the description of the simulation experiment together with the models.

The models required for the simulations must be provided with all governing equations (e.g., MathML), parameter values, and necessary conditions (e.g., RuleML for initial state and/or boundary conditions). Moreover, any modification of a model (pre-processing) required before the execution of a step of the simulation experiment must be described. In relation to simulator definition, a precise description of the simulation steps and other procedures used by the experiment must be provided. Specifically, simulation steps must be clearly described, including the simulation algorithms to be used, the models on which to apply each simulation, the order of the simulation steps, and the data processing to be performed between the simulation steps. Finally, in relation to experimental frame, information necessary to obtain the desired numerical results must be provided. That is, postprocessing steps applied on the raw numerical results of simulation steps have to be described. That includes the identification of data to process, the order in which changes were applied, and also the nature of changes. If the expected insights depend on the relation between different results, such as a plot of one against another, the results to be compared have to be specified. XML-based workflow specifications, such as those provided with the KEPLER workflow system (Anand et al., 2009) can augment SED-ML with explicit experiment workflow definitions that can be reproduced by replicators.

For *reference correspondence*, the following criteria are relevant:

1. The model must be specified in a public, standardized, machine-readable format (e.g., XML,XHTML).
2. The model must comply with the standard in which it is encoded (e.g., ESG, NMM component schema, SED-ML schema)
3. The model must be related to a single reference description. If a model is comprised of different parts, there should be a descriptive specification of the derived/combined model.
4. The encoded model structure must reflect the activities described by the reference description.
5. The model must be in a form that can be instantiated in a simulation: all quantitative attributes must be specified, including their initial conditions.
6. When instantiated, the model must be able to reproduce results given in the reference description in accordance with the RS criteria.

Given the significance of intellectual property rights associated with digital research artifacts, proper attribution annotations need to be included in the distributed SED-ML packages. Along with the license, the attribution meta-model encapsulated within the conceptual model should be structured in a way to satisfy at least the following requirements:

1. The model has to be named.
2. A citation to the reference description must be provided (complete citation, unique identifier, unambiguous URL). The citation should identify the authors of the model.
3. The name and contact information for model creators must be provided.
4. The date and time of model creation and last modification should be specified.
5. The model should be linked to a precise statement about the terms of its distribution.

Cross-referencing between the conceptual model and the technical documentation (or source code) facilitates tracing model elements to concepts presented in the reference so that data and constraints listed in the reference can be traced to elements in the conceptual model. Among the requirements for external resource annotation are the following:

• The annotation must unambiguously relate external knowledge to a model element.
• The referenced information should be described using a triplet collection, identifier, qualifier (e.g., RDF triples of the Annotea tool (Annotea, 2011)):

    o The annotation should be written as a Uniform Resource Identifier (URI).
    o The identifier should be considered within the context of the framework of the collection.
    o Collection namespace and identifier should be combined into a single URI.
    o Qualifiers (optional) should refine the link between the model constituent and the externally defined knowledge: has a, is version of, is homolog to, etc.

4.3 Processing Tools for SED-ML Documents

SED-ML transformation and processing methods and tools are necessary for the implementation of the integrated e-Portfolio support environment. Interoperability and accessibility from within model development environments is paramount. To create SED-ML documents, it is critical to combine model elements, simulator elements, and the experimental frame comprised of data and output elements in such a way that SED-ML processing tools can be applied. To demonstrate the capability of SED-ML to facilitate exchange of simulation experiment descriptions, freely available independent applications that support SED-ML can be used: SED-MLWeb tools (http://sysbioapps.dyndns.org/SED-ML-Web-Tools/), lib- SedMLScript, which is available at (http://libsedml.sourceforge.net/), and SBSIVisual that can be accessed at (http://www.sbsi.ed.ac.uk/).

The software environment for working with SED-ML documents requires relatively few tools: (1) *SED-ML composition software:* A tool to enable authors to integrate components of the conceptual model to create SED-ML documents. (2) *External software:* A mechanism for organizing references to auxiliary software such as GNUPLot, R runtime system, Matlab S functions, documentation, and datasets so that they can be interfaced with the data analysis system. (3) *Reconstruction software:* A tool with multiple filters for processing the conceptual model to yield different outputs, typically involving transformations of the components of SED-ML documents. (4) *Replicability verification and review software:* A tool for testing and validating a SED-ML document, for both the author and the reader. (5) *Packaging and distribution software:* A tool for distributing SED-ML documents and for managing them on both the client (i.e., reader) and the server side; on the server side this includes versioning; and on the client side, it includes tools to access and view the model, simulator, and experimental frame elements.

4.3.1 Interpreting SED-ML Documents

Processing a SED-ML document consists of two sets of computations. One set pertains to processing the structure of the conceptual model by identifying and manipulating the different model, simulator, and experiment elements. The second set of computations involves evaluating the scripts within the SED-ML document structure. The evaluation of each code script will take place in the appropriate language (e.g., ESMF, Java, GNUPlot, R) for that code and is delegated to that programming system. So, this set of computations for the scripts may involve one or more different programming languages such as ESMF, C, Java, C++, R, Perl, or Matlab. The first set of computations on the structure of the SED-ML document can be developed in any general programming language. What is imperative is that some form of markup language is needed to identify different components of an active document.

4.3.2 Conceptual Model Composition Software

For SED-ML documents to become an accepted conceptual model distribution mechanism, it is necessary to have easy to use tools for creating and authoring them. Graphical front-end interface tools that use Java libraries such as, **jlibsedml**, for creating, manipulating, validating and working (http://sourceforge.net/projects/jlibsedml/) with SED-ML documents is practical. This library provides support for retrieval and preprocessing of models, by application of XPath expressions, and also postprocessing of raw simulation results as specified by SED-ML data generator elements.

The *jlibsedml* application programming interface (API) follows a similar organization to that of *libSBML* (Bornstein et al., 2008), a successful and popular library for manipulation of SBML documents. An alternative tool that can be considered with the integrated model development environment is SProS (the SED-ML Processing Service), which is an Application Programming Interface (API) described in Interface Definition Language (IDL) for creating, reading and manipulating SED-ML documents, and so can be used by multiple software packages.

4.3.3 Reconstruction Software

SED-ML documents can be used as e-Portfolio containers for elements from which one can generate inputs for tools that can be plugged to the model development environment. These tools may include simulators, analytical programs, visualization tools etc. Once the model is mapped onto the a ontological schema, as well as identified elements of to the Discrete-Event Simulation Ontologies using tools such as *Annotea*, the replicater can define the instance of the model, and then automatically translate it into the Ontology Working Language (OWL) instances of the selected ontologies. These instances, in turn, can be translated to XML formats suitable for particular simulators (interpreters). It is feasible to use XSLT to transform the XML document containing elements of the conceptual model into another document that is recognized by a browser, like HTML and XHTML. Normally XSLT does this by transforming each XML element into an (X)HTML element. In the reconstruction process, XSLT will use XPath to define parts of the source document that should match one or more predefined tags. When a match is found, XSLT will transform the matching part of the source document into the format of the selected output type.

4.3.4 Replicability Verification and Review Software

In addition to conventions and tools for creating SED-ML documents, a tool for replicability verification is needed. Specifically, the replicaters or reviewers need a tool to verify if the requirements discussed in previous section are realized. The provision of author-provided test scenarios that demonstrate accurate replicability with respect to selected RS criteria (e.g., numerical identity, distributional equivalence, or relational alignment) could emulate the *Test-Driven Software Development* methodology (Beck, 2003) and constitute part of the Replicability Review documentation encapsulated in the SED-ML document. While the testing process is generally open-ended and context-specific, there are some relatively simple and achievable benchmarks. For example, one can compare output from components of the conceptual model with a master copy. While we are not aware of any general, widely-adopted strategies for doing this, most software packages have some self-verification mechanism that can be run at installation.

4.3.5 Packaging and Distribution Software

Model developers will generally prefer to provide others with access to their SED-ML documents and hence require mechanisms for distribution. Readers will need tools to help search for and locate interesting conceptual models, and then to download and process the document for reading or other purposes. Many languages support transparent distribution and installation of modules. A Metadata Registry system is a viable option, but tools for deploying a complex package (e-Portfolio) of conceptual elements of the model, simulator and experiment frame is needed. For example, CRAN for R and CPAN for Perl are software archives that provide the search and distribution facilities, albeit in a single centralized location. Tools provided with these languages (e.g distutils for Python and install.packages in R) provide the client-side installation mechanism that might be extended to support SED-ML documents.

However, using these packaging methods requires a transport mechanism. *XML Serialization* can be used as the process for converting an object into a form that can be readily transported. An advantage of XML serialization is that it does not constrain the type of application as long as the generated XML stream conforms to the SED-ML schema. A mechanism for attaching version numbers to SED-ML documents will be very useful. Such a system will allow users to identify newer versions (perhaps with new data or with errors fixed) and differentiate between different versions. Version numbers can be used in general distribution systems that allow users to automatically obtain updates, as the SED-ML author makes them available.

5 M&S Research Practice

Availability of an infrastructure to support and promote reproducibility is necessary, but not sufficient. In the short term, without proper incentives and processes involving scientists, publishers, journals, institutions, and funding agencies, the infrastructure may be underutilized. Furthermore, issues pertaining to citation, licensing, and ownership to digital objects beyond scholarly manuscripts require careful consideration. Individuals and institutes need to play a critical role to foster an environment that promotes reproducibility of simulation-based science and engineering projects. To sustain such an environment proper licensing, citation, and open source frameworks should be available to provide incentives for broad sharing and dissemination of an e-Portfolio. We consider how authors and institutes can follow a set of systematic guidelines in regard to legitimization, dissemination, and access dimensions discussed above to improve reproducibility and hence credibility of the results drawn from simulation projects. The significance of credibility in M&S studies is also emphasized in the Code of Professional Ethics for Simulationists (SCS, 2012).

5.1 Authors

Authors, as producers of reproducible research products, play a critical role. The following are basic suggestions for the M&S research practice. These suggestions aim to improve reproducibility in the short term using available technology.

- Authors can provide hyperlinks to simulation models and code underlying the figures and data presented in the paper. If possible, the provision of online access to data and visual analysis software could help regenerate results. Authors can post research artifacts on an institutional or university web page or openly accessible third-party archived open-source software management sites such as Source-Forge.net or BitBucket.org. As a side effect, a scientific community that forms around a Source-Forge project could sustain and help maintain the code, model, and the data for a longer term. Similarly, for publishing, citing, and discovering research data, authors can use infrastructures such as the DataVerse

Network Project (http://thedata.org). The use of nonproprietary formats consistent with the research practice in code and data representation can encourage reuse and facilitate broader dissemination.

- Authors can publish generic and customizable versions of their models and their simulators as services so that readers can interact with the authors' computational calculations. The use of version control systems such as The Concurrent Versions System (CVS) could assure citing the correct version of a model and its simulation code.
- Authors can provide scenarios, test suites, and associated test drivers as part of the experimental framework to help users instill confidence in the functional correctness of the simulation code. Adoption of a test-driven model development (Beck, 2003) can link code functionality to specific attributes of the phenomena espoused in the manuscript.
- Authors can leverage recent developments in cloud computing and service-oriented systems technology to provide access to code, data, and visualizations in terms of digital objects that are amenable to customization.
- Authors can use standard protocols for documenting their models within the text of the article in a way that maps design concepts and implementation elements to formal constructs in the manuscript. This could help engineer traceability into the process to facilitate linking assumptions and conceptual arguments presented in the article to their counterparts in simulation code.
- The use of open licensing schemes adopted by the OpenAccess and Open-Data projects under the CreativeCommons (http://creativecommons.org) framework can secure intellectual property rights, while maximizing access and citation to various components of the work.

5.2 Institutional Environment

The use of research practices that support reproducibility would benefit from an institutional environment that incentivizes their further development and adoption. Hence, funding agencies could encourage reproducibility research, while publication outlets such as peer-reviewed archival journals promote guidelines and author instructions that facilitate reproducibility of published work. Specifically,

- Funding agencies can support research groups that fully implement reproducible research. Such implementations can provide experience and insight into requirements for model development environments and methodologies that enable reproducible research. Funds can be made available to support research groups that specialize in toolkit and M&S environment development to generalize and leverage such experience to streamline reproducible computational research.
- Cyber-infrastructure research can serve as a testbed for implementing scientific workflow systems such that transparent and explicit workflow specifications are augmented with necessary constructs to facilitate reproducible research. For instance, automated orchestration and deployment of services can serve as a basis toward attainment of reproducibility.

- Funding agencies can support formation of research communities or communities of practice that sustain and ensure maintainable and ongoing reproducible research.
- Funding agencies can support development of tools that better link various artifacts of reproducible research. Engineering traceability into e-Portfolios to better link simulation software, data, and formal requirements and assertions presented in the manuscript is key to understandability required for reproducing and building over the results of prior work.
- Journals may require provision of stable and accessible repositories for code and research data associated with published articles. Such data need to be encoded using standard ontologies to improve understandability and reuse. In addition, instructions for reproducibility or its demonstration may be required for reproducibility review.
- Journals can establish new categories for articles to label their level of reproducibility. A higher level of reproducibility lends credence to a published manuscript.
- Journals can utilize systems analogous to automated program grading environments (Edwards, 2003) used in computer science education to assess completeness and validity of the scenarios used in testing simulation code. Such environments can help assess authors' performance at testing their own simulation code and hence determine the extent to which reproducibility is considered.

6 Conclusions

While parts of the computational infrastructure described in the preceding sections are off-the-shelf, other aspects of the design and implementation of the integration platform require further research. The first challenge is to handle traceability. Since the primary data-flow path will be from concrete low-level of abstraction to higher levels, derivation of links requires effective use of annotations and specification of a general-purpose resource dependency schema. Furthermore, since outputs in a workflow may be due to processing of multiple input sources and synthetic analysis flows coordinated by the workflow system, the knowledge embedded in a workflow needs to be extracted for inclusion in traceability matrix.

Second challenge pertains to tailoring the general-purpose process and software infrastructures with the selected application domains. This requires engineering a set of packaged configurations and workflows that capture basic protocols that arise in a selected domain of application. These configurations need to be customizable in a way that does not require detailed understanding of the underlying tools.

The third challenge will be in evaluating how the process, infrastructure, and tools impact the research practice and attainment of the reproducibility criteria. Science is a collective phenomenon. Progress in simulation-based science and engineering requires the ability of scientists to create new knowledge, elaborate and combine prior computational artefacts, and establish analogy and metaphor across models. Unless models are designed, disseminated to be discovered, extended, or combined with other models, scientific progress can be hindered. The inability of others to independently reproduce and verify published results will slow down the adoption and the use of knowledge embedded within models. Therefore, reproducibility should become responsibility of the broader scientific community.

References

Alberts, B.: Editorial Expression of Concern. Science 327(5962), 144 (2010), http://www.sciencemag.org/content/327/5962/144.1.full

Altman, M., King, G.: A Proposed Standard for the Scholarly Citation of Quantitative Data. D-Lib Magazine (March/April 2007), http://www.dlib.org/dlib/march07/altman/03altman.html

Anand, M.K., Bowers, S., McPhillips, T., Ludäscher, B.: Exploring Scientific Workflow Provenance Using Hybrid Queries over Nested Data and Lineage Graphs. In: Winslett, M. (ed.) SSDBM 2009. LNCS, vol. 5566, pp. 237–254. Springer, Heidelberg (2009)

Annotea. The Annotea Project (2011), http://www.w3.org/2001/Annotea/

Baggerly, K., Coombes, K.: Deriving Chemosensitivity from Cell Lines: Forensic Bioinformatics and Reproducible Research in High-Throughput Biology. Annals Applied Statistics 3(4), 1309–1334 (2009)

Beck, K.: Test-Driven Development by Example. Addison Wesley (2003)

Borgman, C.L.: Scholarship in the Digital Age: Information, Infrastructure, and the Internet. The MIT Press, Cambridge (2007)

Chang, C., et al.: Retraction. Science 314(5807), 1875 (2006), http://www.sciencemag.org/cgi/content/full/314/5807/1875b

CreativeCommons (2011), http://www.creativecommons.org

Cummings, T.P., Glotzer, S.C.: Inventing a New America through Discovery and Innovation in Science, Engineering, and Medicine. World Technology Evaluation Center (2010)

David, A.P., Spence, M.: Toward Institutional Infrastructures for e-Science: The Scope of Challenges. Technical Report: Oxford Technical Institute (May 2003)

Data Cite (2011), http://thedata.org/citation/tech

De Roure, D., Goble, C., Stevens, R.: The Design and Realisation of the myExperiment Virtual Research Environment for Social Sharing of Workflows. Future Generation Computer Systems 25, 561–567 (2009)

Donoho, D., Maleki, A., Rahman, I., Shahram, M., Stodden, V.: Reproducible Research in Computational Harmonic Analysis. Computing in Science & Engineering 11(1), 818 (2009)

Economist, The Clouds of Unknowing. The Economist (March 18, 2010), http://www.economist.com/node/15719298

Edwards, H.S.: Improving Student Performance by Evaluating How Well Students Test Their Own Programs. Journal of Educational Resources in Computing 3(3), 1–24 (2003)

Fenton, E.N., Bieman, J.: Software Metrics: A Rigorous and Practical Approach. Chapman & Hall (2010)

Fomel, S., Hennenfent, G.: Reproducible Computational Experiments using Scons. In: IEEE International Conference on Acoustics, Speech, and Signal Processing, vol. 4, pp. 1257–1260 (2009)

Fomel, S., Claerbout, J.: Guest Editors' Introduction: Reproducible Research. Computing in Science and Engineering 11(1), 5–7 (2009)

Foster, I.: Service-oriented Science. Science 308, 814–817 (2005)

Gentleman, R.: Reproducible Research: A Bioinformatics Case Study. Statistical Applications in Genetics and Molecular Biology 4(1), 123–137 (2005)

Kiczales, G., Des Rivieres, J., Bobrow, D.G.: The Art of the Meta-object Protocol. MIT Press (1992)

King, D.R., Turnitsa, C.D.: The Landscape of Assumptions. In: Proceedings of the Spring Simulation Multiconference, pp. 81–88 (2008)

Leisch, F.: Sweave and Beyond: Computations on Text Documents. In: Proceedings of the 3rd Intl Workshop on Distributed Statistical Computing, pp. 87–96 (2003)

Mesirov, J.: Accessible Reproducible Research. Science 327(5964), 415–416 (2010)

NSF. Simulation-Based Engineering Science, Revolutionizing Engineering Science through Simulation. National Science Foundation (2006)

NSF. Changing the Conduct of Science in the Information Age. Summary Report of Workshop Held on November 12, 2010. National Science Foundation (June 28, 2011), http://www.nsf.gov/pubs/2011/oise11003/

Oinn, T., Addis, M., Ferris, J., Marvin, D., Senger, M., Greenwood, M., Carver, T., Glover, K.: Taverna: A Tool for the Composition and Enactment of Bioinformatics Workflows. Bioinformatics 20(17), 3045–3054 (2004)

Overstreet, M.C., Nance, R.E., Balci, O.: Issues in Enhancing Model Reuse. In: Proceedings of the First International Conference on Grand Challenges for Modeling and Simulation (2002)

Ören, T.I., Zeigler, B.P.: Concepts for Advanced Simulation Methodologies. Simulation 32(3), 69–82 (1979)

Ören, T.I.: The Richness of Modeling and Simulation and its Body of Knowledge. In: Proceedings of SIMULTECH 2012, 2nd International Conference on Simulation and Modeling Methodologies, Technologies and Applications, Rome, Italy, July 28-31 (invited keynote paper - in press, 2012)

Peng, R.: Reproducible Research and Biostatistics. Biostatistics 10(3), 405–408 (2009)

reST. reStructuredText: Markup Syntax and Parser Component of Docutils (2011), http://docutils.sourceforge.net/rst.html

Saeki, M., Hiroi, T., Ugai, T.: Reflective Specification: Applying a Reflective Language to Formal Specification. In: Proceedings of the 7th International Workshop on Software Specification and Design, pp. 204–213. IEEE CS Press (1993)

Schwab, M., Karrenbach, M., Claerbout, J.: Making Scientific Computations Reproducible. Computing in Science and Engineering 2, 61–67 (2000)

SCS. The Code of Professional Ethics for Simulationists (2012), http://www.scs.org/upload/03-Code_0.pdf

Stodden, V.: The Legal Framework for Reproducible Scientific Research: Licensing and Copyright. Computing in Science & Engineering 11(1), 35–40 (2009)

Stodden, V.: The Scientific Method in Practice: Reproducibility in the Computational Sciences, Technical report, MIT Sloan Research (February 2010)

Sphinx. Python Document Generator (2011), http://sphinx.pocoo.org

Wilensky, U., Rand, W.: Making Models Match: Replicating an Agent-Based Model. Journal of Artificial Societies and Social Simulation 10(42) (2007), http://jasss.soc.surrey.ac.uk/10/4/2.html

WTEC. Panel Report on International Assessment of Research and Development in Simulation-Based Engineering and Science. World Technology Evaluation Center (2009)

Yilmaz, L.: On the Need for Contextualized Introspective Simulation Models to Improve Reuse and Composability of Defense Simulations. Journal of Defense Modeling and Simulation 1(3), 135–145 (2005)

Yilmaz, L.: Reproducibility in Modeling & Simulation Research. Simulation 87(1), 3–6 (2011a)

Yilmaz, L.: Simulation Reproducibility: Implications for Theory, Methodology, and Model Development Environments. In: Epistemology of Modeling & Simulation Conference, Pittsburgh, April 1-3 (2011b), http://www.modelingepistemology.pitt.edu/program/

Immersed in Immersion: Simulation as Technology and Theory of Mind

John Z. Elias

University of Central Florida
Orlando, FL, United Sates

1 Introduction

Cognitive theories involving the notion of simulation have developed hand in hand with the advancement and pervasiveness of simulation technologies. This intimate interrelation suggests the promise of implementing simulation technology in cognitive research, as well as in the facilitation and manipulation of cognitive and affective mechanisms for learning and training. In this chapter I describe the general interdependence of forms of technology and theories of mind, the former often furnishing metaphors for the latter, and offer a brief historical sketch leading up to the recent emergence of the centrality of simulation. I then follow with a critical evaluation of the role of simulation in current cognitive theories, and relate these critiques to philosophical concerns about the ontological, epistemological, and methodological status of modeling and simulation as a research tool. I end with some illustrative examples from cognitive research and therapy, and point towards potential future applications.

2 Mind and Technology, Technology and Mind

Technology is our predominant means of predicting and controlling our environment, and as such is an immensely potent and important force in shaping our lives and our world. Yet we tend to take for granted the informing and shaping power of technology, often quickly becoming accustomed to new advancements and fluently incorporating them into our ways of life. Indeed we need to strain to remind ourselves, for instance, that there was once a time when the only way to hear a piece of music was to attend an actual performance of it. Photography, in its contemporary impact and its influence on our sense of history, is another example of the deeply determining power of technology. There exists a stark dividing line between pre- and post- photographical history: imagine how different our sense of pre-photographical historical figures would be if we had access to actual photos of them, a photograph of Shakespeare, say, or of George Washington. These are just a few among endless examples of the thoroughgoing and all-encompassing influence of technology upon our lives.

A. Tolk (Ed.): Ontology, Epistemology, & Teleology for Model. & Simulation, ISRL 44, pp. 227–240.
springerlink.com   © Springer-Verlag Berlin Heidelberg 2013

Technology, in shaping our world, hence shapes our sense of ourselves. Conceptions of the mind through history often reflect the current state of technology at the time (Searle 2004) (Draaisma 2000). Freud's psychoanalytic theory of the mind, for example, may be viewed as informed by industrial age technology, with its emphasis on mechanisms and drives, on pressures built and pent up, seeking release by whatever valves available. More pertinently, the development of computers and the development of cognitive science have paralleled each other intimately. Alan Turing and others ushered in an age of computers, both technologically and theoretically. Turing's notions of the *Turing Machine* and of the *Turing Test* established a thoroughly computational conception of cognition, in which the critical criterion for intelligence was the capacity for purely symbolic processing. Computers also appeared to provide a potential "solution" to the mind-body problem: simply put, the mind is to the brain as software is to hardware. Thus the development of this particular piece of technology promised a resolution to an ancient philosophical problem, and much of the advancement of cognitive science as a discipline was fueled by the promise of computers, both as tools and as a model of the mind itself. Indeed the legacy of this technological metaphor still remains, in Chomsky's syntactical and computational linguistics, for example, and in classical cognitive models of information processing generally (e.g., Pylyshyn 1999).

3 Mind and Simulation, Simulation and Mind

This pattern extends into present conceptions of cognition. Current theories of the mind, with their emphasis on embodiment and interactivity, arguably reflect, to some extent, the increasingly interactive and virtual nature of our media environment, including the technology of simulation. We might inquire at this point, however, as to what counts as an instance of simulative technology. If we take flight simulators as a paradigm example of human-in-the-loop simulation, then the following characteristics might be considered central: interactivity, as specifically responsive in real-time; some virtual re-creation of an environment, with varying degrees of realism and refinement, ranging from the highly realistic to the more abstract; and an overall design directed towards training, with the aim of preparing for some *real*, non-simulated target activity. While these features certainly do not constitute necessary and sufficient conditions for simulation, they do point towards the widespread *simulative* nature of much of our everyday technology, in its increasing interactivity and real-time responsiveness, as well as the ubiquity and sophistication of simulated or *virtual* realities, and their manipulability in real time. That is to say, our culture and environment are thoroughly permeated by interactive technologies that may well be said to be *simulative* in some sense or another. Even the introduction of the simple Graphical User Interface (GUI) represents a significant move away from an older mode of computer use based on text or coding prompts towards a more visually simulative and perceptually responsive mode of computer interaction. Indeed the *Windows* operating system is itself a simulation of a physical office space, replete with the iconography of *desktops*, *files*, *folders*, and so on. Notably, though, the simulative nature of *Windows* is no longer directed toward any "real" non-simulated target activity; it is not a preparation for engagement with typical physical office items, but a *replacement* of them. This severance of the simulative from the real characterizes more and more of our virtual and media interactions. Simulation has detached itself and become evermore self-sufficient and self-serving. We are immersed, it seems, in immersion, as an end onto itself.

This development in our general interaction with computers parallels the development of cognitive science broadly speaking. The typical use of computers in the past, involving a step-wise procedure of inputting data, which the computer then processes, and then outputs the outcome, is akin to the information processing model of cognition, in which sensory inputs feed into a central processing mechanism, conceived as essentially symbolic in nature, which processes the inputs in terms of the syntactic manipulation of symbols, and which then feeds out the resulting output, whether expressed externally, as a motor or speech act, or manifested internally as an explicit representation. The shift to the flexibility and responsiveness of current interactive technology parallels the shift away from the conception of cognition as fundamentally a matter of internal information processing towards a more embodied interactive approach, which emphasizes and foregrounds our *sensorimotor* interactions with the environment. Gibson (1977), for instance, demonstrates this development with his theory of *affordances*, in which perception is conceived *ecologically*, as a function of our ongoing interaction with our environment, rather than the merely passive reception of information from it. Instead of the stepwise procedure of perception first feeding into a central cognitive processor, which then computes and outputs a response, our perception of an object is intrinsically informed by the capacities for interaction that that object *affords*. That is, perception itself comes geared for action, and is more properly understood as an aspect of our *two-way engagement* with the environment, rather than the merely one-way intake of information. This evolution in the conception of cognition is also represented in the evolution of approaches to Artificial Intelligence, in the move from centralized computation, in which cognition was modeled as a disembodied computer functioning fundamentally in isolation, to embodied robotics, in which artificial models of cognition and intelligence are built from the ground up, in terms of embodied activity within an environment. Thus what was once marginalized as detachable input and output modules feeding into and out from a centralized symbolic processor has now been positioned front and center: the mind, and our cognitive capacities generally, is increasingly conceived as a *simulator* of sorts, with thinking consisting in the *simulation* of situations, of sensorimotor scenarios of interaction.

Decety & Grezes, in "The power of simulation: Imagining one's own and other's behavior" (2006), detail this view of cognition-as-simulation. In the following passage they approvingly paraphrase Hesslow's (2002) *Simulation Hypothesis*:

> The simulation hypothesis states that thinking consists of simulated interaction with the environment and rests on the following three core assumptions: (1) simulation of actions: we can activate motor structures of the brain in a way that resembles activity during a normal action but does not cause any overt movement; (2) simulation of perception: imagining

perceiving something is essentially the same as actually perceiving it, only the perceptual activity is generated by the brain itself rather than by external stimuli; (3) anticipation: there exist associative mechanisms that
enable both behavioral and perceptual activity to elicit other perceptual
activity in the sensory areas of the brain...a simulated action can elicit
perceptual activity that resembles the activity that would have occurred if
the action had actually been performed. (Decety & Grezes 2006, p. 5)

Thinking here is a matter of simulation, of covertly, i.e. inwardly and sometimes
non-consciously, engaging in interactive scenarios. Neural overlap between actual
and imagined actions has been demonstrated by numerous functional magnetic re-
sonance imaging (fMRI) studies, where for instance similar patterns of brain activation are exhibited in pianist subjects both when playing a piece of music on a
keyboard and imagining playing the same piece (Meister et al. 2004). Furthermore, the work of Lakoff and Johnson on metaphor gathers convincing evidence
that even our most seemingly abstract forms of thought are ultimately grounded in
our physical and experiential interactions (e.g. *Philosophy in the Flesh*, 1999).

At this point a distinction should be drawn between explicit and implicit simulative processes. On the one hand, simulation is evoked during the explicit consideration of multiple possibilities in the course of high-level decision-making; here,
simulation seems more or less synonymous with *imagination* or *reenactment*, with
various scenarios explicitly represented and run through by the subject (why the
technical term *simulation* is used instead of *imagination* is a question to which I
will return). By contrast, an implicit version of simulation is invoked in theoretical accounts of the execution of actions. Even in simple acts like reaching to
grasp a cup, simulations in the form of forward motor plans are posited to run in
parallel with the actual act itself (Desmurget & Grafton 2000). These simulations
generate anticipated sensory consequences of the act, and are continuously updated against the progress of the action as it is performed in real time. The simulated outcome is then compared to the outcome of the unfolding action, in order to
both monitor and control the action as it proceeds and to confirm that the act was
in fact completed as planned. Significantly, these motor simulations supposedly
occur at the *subpersonal* neuronal level (at a level beneath and inaccessible to personal reflection and awareness): they are not the sort of processes that can be
made available to consciousness (Hurley 2008). This subpersonal status poses
problems that I will return to shortly.

4   Simulation and Social Cognition

The concept of simulation also figures prominently in theories of social cognition,
in accounts of how we come to understand others. One approach, called *Simulation Theory*, asserts that we use our own minds to create *simulations* of the minds
of others, in the attempt to position ourselves within their perspective, to view the
world from their point of view (see *Theories of Theory of Mind*, Carruthers &
Smith, ed. 1996). That is, in order to attribute mental states to another, we attempt
to imitate, or simulate, their mental activity by means of our own mental processes

and resources. So, when observing the behavior of another, we imagine, or *simulate*, behaving or acting in the same or similar way; this simulation, of course, is *covert* and internal, and does not result in overt outward action (Goldman 2002). These simulations can be either predictive, essentially asking the question: if I were to have these beliefs and desires, what would I then decide to do? Or they can be retrodictive, beginning with observed behavior and inferring which mental states would have led to them (Gallese & Sinigaglia 2011).

Additionally, the function of *mirror neurons* is conceived by some in terms of simulation, as a system that generates automatic and implicit simulations of the actions of others (Gallese & Goldman 1998) (Gallese & Sinigaglia 2011). Mirror neurons, first discovered by single-cell recordings in the F5 area of the brain in monkeys (Rizzolatti et al. 1996), fire both when the monkey performs a particular action as well as when the monkey observes another perform that same action. Mirror neuron activity has been demonstrated in humans as well (Mukamel et al. 2010), and is proffered as a possible neuronal mechanism underlying simulations of the actions of others, in the course of observing and understanding them. Put simplistically: to understand the action of someone reaching for food, or the meaning behind a smile, we, as it were, internally perform or simulate that act of reaching, or the facial expression of a smile, in order to realize the intention or emotion driving those acts. Thus we must, in a sense, perform, or simulate, the other's action, in order to then understand it. There is much controversy however over the appropriate interpretation of the role of mirror neurons, whether or not they do indeed underlie simulations of the actions of others. Nevertheless, the excitement surrounding them, and the prevalence of the simulation interpretation of them, indicates the organizing influence of the concept of simulation in current cognitive theorizing.

5 Cognition and Simulation, Critically Evaluated

Critics such as John Searle have bemoaned the distortions of metaphorical fixation, the trap of treating metaphors as reflections of the real thing. However, in response to these criticisms we might in turn observe that these forms of technology are themselves the results and expressions of our creative capacities. We are, after all, tool-making and tool-using creatures, and so the creation and development of technologies can be seen as deeply indicative of our intelligent engagement with the world (Clark 2003). Thus the direction of influence is in-to-out as well as out-to-in: technologies arise from the mind, from its creative and adaptive capacities, which in turn influence our conceptions of the mind.

Nevertheless, there is a way in which these metaphors profoundly deform our views of our minds and selves, by imposing concepts of control transposed from technology. The exertion of technological dominance over the world seems to seep into theoretical descriptions of our basic cognitive capacities. According to classical cognitive science, for example, the world is *taken up* into the mind and transformed into symbols, which are then computationally operated upon in the course of cognitive processing. The picture, then, is of the world internalized, devoured as it were, subject to complete manipulation under the command of the mind. Thus the mind here is a kind of inner sanctum, a sovereign inner space where its power reigns supreme. Once converted into internal code, the world itself, with its contingencies and unpredictability, is safely set aside, with the mind left alone in complete control of its own material.

This image of dominion, derived from technological mastery, appears under different guises in different theoretical contexts. Simulation theories, despite their apparent consonance with embodied interactive approaches, still implicitly appeal to this picture. That simulations seem to traffic in interaction with the environment is understandable, since simulations are themselves virtual re-creations of such interactions; yet even though these simulations are experiential in nature, and not an abstract symbolic code, they nevertheless occur internally, running on the mind's own resources, and thereby reiterate a view of the mind as fundamentally on its own, separate and independent. This is not to deny that we sometimes imagine scenarios to ourselves, *in our heads* as it were, when thinking through alternatives, but these explicit imaginings are distinct from the sorts of simulations that simulation theorists claim are basic to mental operations. Indeed, the fact that the term *simulation* is often used when *imagination* would seem to do just fine, in cases for instance of straightforwardly explicit "simulating" or reenacting, is itself instructive. Perhaps *simulation* is chosen for its aura of technical specialization, as opposed to the more ordinary and everyday *imagination*, which again speaks to the imposition of technological specification onto human mentality. And simulation as a technology is immersive to the point of deception, cloaking its fundamentally mediated status, making it all the more seductive as a metaphor, and hence all the more mistakable for the real thing (Turkle 2009).

With this general critical stance in mind, I will now briefly survey some of the potential problems with simulation theories. For example, embodied simulationist explanations of reading comprehension and linguistic meaning (e.g. Glenberg et al. 2004, Speer et al. 2009) assert that internal simulations underlie our understanding of language. However, whether purely internal representations are sufficient to ground linguistic meaning has been deeply controversial; indeed much 20th philosophy of language was dedicated to demonstrating that linguistic meaning cannot be secured internally, and that a public, external check on the use of the language is required in order for words to stably mean what they do (e.g. Wittgenstein 1953). Furthermore, assuming embodied simulations do occur during reading, presumably these simulations, if they appropriately and accurately reflect the content of the text being read, would be generated *because* the words being read already have meaning. That is, reference to simulations cannot be used to explain linguistic meaning, since the words would have to mean something to begin with in order for the appropriate simulations to come about. Therefore, while imaginings and reenactments may certainly serve to facilitate and flesh out linguistic comprehension, they cannot be said to *constitute* linguistic meaning itself (Weiskopf 2010).

Another problematic aspect is the supposedly *subpersonal* nature of certain simulations. Again, the previously mentioned forward motor simulations that monitor and regulate the performance of actions occur at the subpersonal neuronal level, and so are necessarily inaccessible to reflective awareness. These simulations though are sometimes said to *predict* the sensory consequences of planned actions, and it is questionable, perhaps even nonsensical, to speak of subpersonal processes making predictions. *People*, certainly, are capable of making predictions, of describing possible states of affairs and proposing that they will happen sometime in the future, proposals that can turn out to be right or wrong. However whether the same can be said of neuronal activity, or even the brain itself, is a matter of debate, one relating to the controversy concerning the role of representations in basic neurocognitive processes (e.g. Hutto 1999). These concerns arise as well in *Bayesian* models of the brain, in which the brain is a kind of Bayesian machine constantly making predictions and testing them against the environment; perception, then, is always informed by expectation, by what the brain expects to happen in particular circumstances (Knill & Pouget 2004). But here again we may ask if it makes sense to say that the brain, as opposed to the *person*, can make predictions about the environment, or claims about the world generally. To raise these concerns is not to cast doubt on the success of these models, but rather to question and clarify their interpretation. Instead of talk of *prediction*, these neuronal and brain processes might be described in terms of highly sensitive associations, with sensory and motor regularities correlated neuronally with a high degree of responsiveness and precision. If a particular kind of motor act is regularly followed by a particular set of sensory consequences, this regular co-occurrence would result in corresponding neuronal co-activations; so, when the motor act starts to occur, "anticipatory" activity in sensory cortex would occur as well, due to the established association. This admittedly simplistic sketch avoids questionable reference to neuronal processes making so-called predictions.

Problems with the subpersonal level are also evident in simulation theories of social cognition. For instance, if mirror neuron activity underlies embodied simulations of the actions of others, then these processes presumably are capable of supporting pretense, since simulation implies the creation of pretend "as if" mental states which are then attributed to others. Yet the question then is: does it make sense to apply the notion of pretend states to subpersonal neuronal processes? (Gallagher 2007) Again, *people* are capable of simulating the mental states of others, of generating mental states *as if* they were one's own in that situation. Even advocates of embodied simulation theories of mirror neuron activity that claim to repudiate the application of pretense nevertheless frequently slip into *as if* descriptions of mirror neuron functionality (e.g. Gallese & Sinigaglia 2011), which demonstrates the difficulty of applying even supposedly stripped-down versions of simulation at the neuronal level.

Furthermore, simulation theories of social cognition would seem to have trouble with cases in which the other being understood is in a state entirely different from one's own. Gallagher (2007) offers the example of someone enthusiastically and confidently handling a snake, while a snake-fearing observer looks on with repulsion and horror. Clearly the observer, though fearful and repulsed himself, would be able to recognize that the person handling the snake is enjoying herself. However, on the simulation account the observer must in some sense undergo and experience what the observed other is going through in order to understand the other. Thus simulation theories risk conflating the act of *spectating* with the act of actual *doing*, practically collapsing spectator and actor into the same experiential state. Finally, simulationist conceptions of social cognition tend to recapitulate a form of dualism with regard to the mind, in which the minds of others are conceived as otherworldly private realms, by their very essence hidden from view. Hence the need to use our own minds, to which we have direct access, to simulate the minds of others, which are necessarily inaccessible. An alternative is the phenomenological notion of *direct perception*, which rejects the need to simulate or infer the mental states of others in most basic cases, and affirms the perceptual availability of their affective and intentional states; that is, when we see someone smile or cry, we don't need to undergo a simulation of those behaviors in order to understand them: we simply *see* their emotional state directly manifested in and through their expression (Gallagher 2005). Such an approach restores our basic, immediate, intimate responsiveness to others.

6 Modeling and Simulation, Critically Evaluated

The core critique above, concerning ideas of control transposed from technology, applies to the use of modeling and simulation itself, particularly in research. Focusing for the moment on computer-automated simulations as research tools, as opposed to human-in-the-loop simulations (I will take up human-centered simulation in the next section, on the use of simulation in cognitive research and therapy), an obvious contrast with traditional scientific experiments is evident: a typical experiment engages *reality*, not a model thereof. A scientific experiment, in the usual sense, is run against the constraints of an independently existing reality, whereas a model or simulation is subject only to its own constraints, which may or may not be comparable with reality. Though empirical experimentation may well involve technological complexity and sophistication, allowing for very precise and sensitive manipulation of the environment, such control ultimately is for the sake of investigating the existence and nature of entities beyond our creation and contrivance (this may be safely said at least for the "hard" natural sciences; "soft" sciences such as psychology are a somewhat different matter). Furthermore, experiments control for the actual object of interest, and the experiment itself does not occur outside the causally closed physical universe, and so is subject to the same basic causal laws that govern the universe generally; simulations on the other hand operate with models, models designed and constructed beforehand by us, which then have to be related to the target phenomena of interest (Grüne-Yanoff & Weirich 2010). This contrast parallels the criticisms of cognitive theories above, in which a conception of cognition as engagement in and with the world is opposed to a view based on the manipulation of representations of the world: a view of the mind as responsive to the world, acting and acted upon by the world, as opposed to a god-like manipulator of a model or representation of the world, in dominion over its creation. Similarly, when we conduct actual experiments, we interact with the world of which we ourselves are part, whereas simulations run on a model of the world, rather than engaging with the world itself.

However, the fact that models are human constructions does not guarantee they will be completely comprehensible or predictable. While a simple model may be amenable to an analytic solution, which characterizes the general behavior of a model through all circumstances, this method is prohibitively complicated for complex models. More complex dynamic models may be solved by simulation, by plugging and running the numbers through an iterative or temporal process, yielding a *numerical* as opposed to analytic solution. But a solution by means of simulation is "epistemically opaque" in Humphreys' (2009) phrase: it lacks the cognitive clarity and generality of analytic or derivational solutions. Thus one has to find out how the model will behave under different conditions; prior to the results, the model itself is effectively opaque, incomprehensible and unpredictable. The intractability of the calculations underlying simulations might appear to justify their experimental status, as if we were interacting with something strangely independent of us, at once our creation yet beyond our control (indeed the same may be said about our relation to mathematics generally). And yet, while we may not be able to predict or anticipate all the consequences of the underlying mathematical models (especially if elements of randomness are introduced) the models nevertheless remain our construction, defined beforehand by us. So although there may be an element of psychological novelty or discovery, logically speaking there is no additional information to be discovered that isn't already there to begin with, predetermined by the constraints of the model.

These questions of epistemic opacity arise with agent based simulations as well. Agents operate autonomously, their behavior governed by rules of local interaction: they perceive and react to their environment and interact with other agents. These local interactions can result in high-level complex patterns of interactivity, patterns that can be very difficult to predict. So the question again is: by running the simulation, are we finding out something we didn't in some sense already know? Are we synthetically extending our knowledge, or analytically clarifying what we've previously defined and determined?

These are fundamental philosophical questions having to do with distinctions between the mathematical and the empirical, between analytic and synthetic knowledge, questions of whether and when information may be said to be gained, our base of knowledge extended, questions that I can only raise but not resolve here. And of course these are questions about the nature of simulations themselves. What are simulations? What kind of knowledge can we acquire with them? And how can, or should, we act on this knowledge? What are simulations for? What are the conditions of their proper use? These questions of ontology, epistemology, teleology, and methodology are ultimately bound together. This section in effect has been a preliminary inquiry into the epistemological implications of the ontological status of models and simulations, which in turn has consequences for their employment and application, and their incorporation into methodologies of various disciplines. Here I don't venture any absolute statements about the status of simulations; again I only suggest a parallel between the critique I offered earlier, of notions of technological control imposed upon cognition, and the use of modeling and simulation as a research tool. In other words, I wish to express caution at a possible loss of a sense of engagement with the world in the face of our increasingly sophisticated technological mastery and mediation; a call to keep in mind the difference between our direct interactions with reality versus our models and representations thereof, in both our scientific theory and practice. An over-privileging of simulation seems to me akin to a view of the mind that sets the world aside in order to manipulate representations of it: both work at a controlled remove from the world rather than dynamically interacting with it. Our degree of engagement with an independently existing reality, then, provides a clear standard by which to distinguish experiments and simulations as different kinds of activities, yielding different kinds of results.

However, the fact that they may be epistemically distinct does not necessarily privilege one over the other. Indeed the mediating status of models and simulations, poised between theory and reality, endows them with a methodological status all their own. For instance, models have a flexibility, a *prototypicality*, that coherent theories lack. Theories are under the constraint to explain and integrate, whereas models can be tried out and tested more freely. Models and simulations hence may be viewed as *transitional objects* of sorts (Wastell 1999), as means of generating and elaborating ideas, and as testing grounds for potential explanations. Furthermore, although natural scientific experiments interact directly with reality, their interpretation is underdetermined by the evidence they generate (Quine 1975). That is, experimental data may be interpreted in a variety of ways depending on the theoretical framework. Thus models and simulations may be used in conjunction with experimentation to work through various interpretations and possible explanations of data, again as a kind of transitional space between evidence and established theory.

7 Human-Centered Simulation in Cognitive Research

In this section I address the use of human-in-the-loop simulations in cognitive and psychological research and therapy. To a certain extent, the artificiality of simulation poses less of a problem for psychological research, since the typical experimental conditions under which human subjects behave are themselves highly contrived and controlled sets of circumstances. Hence, with the exceptions of ecological studies and certain cases involving deception, simulation technology may be fairly fluently incorporated into the methodology of cognitive research. A lot depends of course on the purposes of the experiment: if, for instance, simulation is being used to assess sensitivity to facial expressions, the details may need to be particularly convincing in order to avoid the problem of the uncanny valley. Thus far, however, results of the use of simulation technology have been promising, revealing our susceptibility to virtual simulative interactions. And although I've taken a critical stance toward some of the claims of simulation theories, the concept of simulation does speak to our experiential engagement with the world. Indeed much of the recent empirical and theoretical work on simulation has advanced the paradigm of embodied cognition, and has helped to reveal the extent of our experiential and imaginative openness to interactions with the physical and social environment.

For instance, the *Rubber Hand Illusion* provides an example of the plasticity and manipulability of our embodiment (see figure 1) (Botvinick & Cohen 1998) (Capelari, Uribe, & Brasil-Neto 2009). The participant places one hand behind a partition, out of view. A rubber hand (either a left or right hand depending on which hand of the participant is placed out of view) is placed in view in front of the participant. If both the rubber hand and the real hand are stroked simultaneously, many subjects begin to visually locate the sensation of being stroked in the rubber hand. This experiment is akin to Ramachandran's famous studies of phantom limbs (e.g. Ramachandran et al. 1992). Many amputees feel pain located in their "phantom" amputated limbs, often associated with a feeling of intransigent clenching that cannot be relieved. Ramachandran et al. positioned a mirror such that the reflection of the other limb is roughly placed where the phantom limb would be: the visual impression of voluntary movement then relieves the phantom pain. Thus the visual input tricks the brain into "believing" that the phantom limb is present and under voluntary control. This finding indicates the susceptibility of our sense of embodiment to virtual imagery, and indeed others (e.g., Giraux & Sirigu 2003, Murray et al. 2007) have begun to design various virtual versions and extensions of the original phantom limb experiment, with Gaggioli et al. (2010) adapting an immersive virtual system specifically for bilateral amputees.

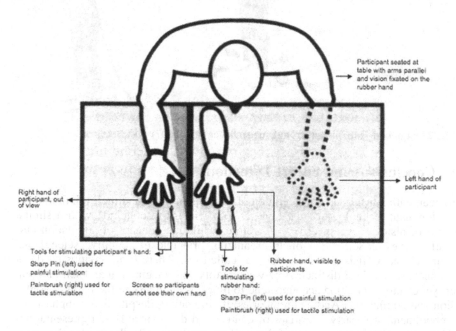

Fig. 1 Rubber Hand experiment (from Capelari et al. 2009)

Strikingly, Blanke et al. (2005) and Lenggenhager et al. (2007), with the aid of virtual reality headsets, extended the rubber hand illusion to the entire body (figure 2). Subjects are video recorded from behind, with the image then fed into the virtual reality headset that they're wearing: they thus see an image of their own body from behind. When stroked or prodded with a stick, a certain number of subjects report locating the sensation of being touched in the body they see before them. And when moved back several steps and then asked to walk to the spot where they thought they originally stood, those that reported the effect tended to walk past where they actually stood towards the location of the virtual body. Blanke has related these findings to the study of the phenomenon of "out of body" experience, and his research provides a prime example of the use of virtual and simulation technology in the investigation of embodied cognition, again exemplifying the degree to which our sense of our own embodiment is susceptible to virtual imagery.

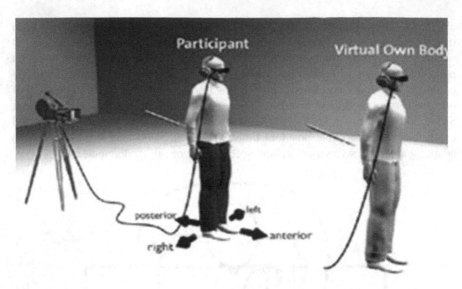

Fig. 2 Out of body experiment (from Lenggenhager et al. 2007)

8   Conclusions and Future Directions

I began with a historical consideration of theories of mind and their relation to the predominant technology of the time, and so will try to conclude with a similar sense of historical perspective. Therefore, while the concept of simulation currently serves as something of an organizing principle in cognitive science and neuroscience, criticism of the concept is called for. There is an obvious need, for instance, for a robust distinction between reality and imagination and self and other: yet certain simulation accounts tend to blur this boundary to the point of confusion and conflation. And again theories of simulation, despite their emphasis on embodiment, still carry overtones of control and dominance over representations of the world, to the detriment of dynamic engagement with the world. Notwithstanding these criticisms and qualms, however, simulation, both as a theory and technology, if understood and employed properly, holds promise in the fields of cognitive science and neuroscience.

Potential future directions include the use of simulation in new areas of research. Froese & Gallagher (2010) suggest a novel application of simulation technology to philosophical research, as an aid to imaginative variations of the environment in phenomenological investigations. Furthermore, simulation may be used to facilitate thought experiments, an important method in philosophy. In social cognitive experiments, simulated environments might be used to vary and study social dynamics that extend beyond the individual or dyad. Also, the use of simulation to support decision-making in real, time-sensitive operations is a particularly pressing issue in simulation research and development (Tolk 2009). Adopting principles and findings from embodied cognition may serve to inform the design and development of simulation technology: for example, multifaceted and multimodal sources of information could be fluently integrated into experientially cohesive simulative systems, facilitating the delivery of information and generation of alternatives in support of time-pressured decision-making (Morrow et al. 2011). These possibilities point to the development of technologies and virtual systems that at once exemplify and exploit the embodied simulative processes involved in human cognition and social interaction.

Acknowledgments. Thank you to Shaun Gallagher, Peter Kincaid, Stephen M. Fiore, and Patricia Bockelman Morrow for their commentary and encouragement. This work was supported by *TESIS: Towards an Embodied Science of InterSubjectivity*, a Marie Curie Initial Training Network (FP7-PEOPLE-2010-ITN, 264828).

References

Blanke, O., Mohr, C., Michel, C.M., Pascual-Leone, A., Brugger, P., Seeck, M., Landis, T., Thut, G.: Linking out-of-body experience and self processing to mental own-body imagery at the temporoparietal junction. The Journal of Neuroscience 25(3), 550–557 (2005)

Botvinick, M., Cohen, J.: Rubber hands 'feel' touch that eyes see. Nature 391, 756 (1998)

Capelari, E.D.P., Uribe, C., Brasil-Neto, J.P.: Feeling pain in the rubber hand: Integration of visual, proprioceptive, and painful stimuli. Perception 38(1), 92–99 (2009)

Carruthers, P., Smith, P.K. (eds.): Theories of Theory of Mind. Cambridge University Press (1996)

Clark, A.: Natural-born cyborgs: Minds, technologies, and the future of human intelligence. Oxford University Press (2003)

Decety, J., Grezes, J.: The power of simulation: Imagining one's own and other's behavior. Brain Research 1079, 4–14 (2006)

Desmurget, M., Grafton, S.: Forward modeling allows feedback control for fast reaching movements. Trends in Cognitive Science (4), 423–431 (2000)

Draaisma, D.: Metaphors of memory: A history of ideas about the mind. Cambridge University Press (2000)

Froese, T., Gallagher, S.: Phenomenology and Artificial Life: Toward a Technological Supplementation of Phenomenological Methodology. Husserl Studies 26(2), 83–106 (2010)

Gaggioli, A., Amoresano, A., Gruppioni, E., Verni, G., Riva, G.: A myoelectric-controlled virtual hand for the assessment and treatment of phantom limb pain in trans-radial upper extremity amputees: a research protocol. Studies in Health Technology and Informatics 154, 220–222 (2010)

Gallagher, S.: How the body shapes the mind, Oxford, New York (2005)

Gallagher, S.: Simulation trouble. Social Neuroscience (2), 353–365 (2007)

Gallese, V., Goldman, A.: Mirror neurons and the simulation theory of mind-reading. Trends in Cognitive Sciences 2(12), 493–501 (1998)

Gallese, V., Sinigaglia, C.: What is so special about embodied simulation? Trends in Cognitive Sciences 15(11), 512–519 (2011)

Gibson, J.: The theory of affordances. In: Shaw, R., Bransford, J. (eds.) Perceiving, Acting, and Knowing: Toward an Ecological Psychology, pp. 67–82. Lawrence Erlbaum Associates, Hillsdale (1977)

Giraux, P., Sirigu, A.: Illusory movements of the paralyzed limb restore motor cortex activity. Neuroimage 20, 107–111 (2003)

Glenberg, A.M., Gutierrez, T., Levin, J.R., Japuntich, S., Kaschak, M.P.: Activity and imagined activity can enhance young children's reading comprehension. Journal of Educational Psychology 96(3), 424–436 (2004)

Goldman, A.: Simulation theory and mental concepts. In: Dokic, J., Proust, J. (eds.) Simulation and Knowledge of Action. John Benjamins, Amsterdam (2002)

Grüne-Yanoff, T., Weirich, P.: The philosophy and epistemology of simulation: a review. Simulation & Gaming 41(1), 20–50 (2010)

Hesslow, G.: Conscious thought as simulation of behavior and perception. Trends in Cognitive Sciences 6(6), 242–247 (2002)

Humphreys, P.: The philosophical novelty of computer simulation methods. Synthese 169, 615–626 (2009)

Hurley, S.L.: The shared circuits model (SCM): How control, mirroring, and simulation can enable imitation, deliberation, and mindreading. Behavioral and Brain Sciences 31(1), 1–58 (2008)

Hutto, D.D.: Cognition without Representation. In: Riegler, A., Peschl, M., von Stein, A. (eds.) Understanding Representation in the Cognitive Sciences: Does Representation Need Reality?. Kluwer Academic/Plenum Publishers, New York (1999)

Knill, D.C., Pouget, A.: The Bayesian brain: the role of uncertainty in neural coding and computation. Trends in Neuroscience 27(12), 712–719 (2004)

Lakoff, G., Johnson, M.: Philosophy in the flesh: The embodied mind and its challenge to western thought. Perseus Books, New York (1999)

Lenggenhager, B., Tadi, T., Metzinger, T., Blanke, O.: Video ergo sum: manipulating bodily self-consciousness. Science 317, 1096–1099 (2007)

Meister, I.G., Krings, T., Foltys, H., Boroojerdi, B., Müller, M., Töpper, R., Thron, A.: Playing piano in the mind – an fMRI study on music imagery and performance in pianists. Cognitive Brain Research 19, 219–228 (2004)

Morrow, P.B., Elias, J., Streater, J., Ososky, S., Phillips, E., Fiore, S., Jentsch, F.: Embodied cognitive fidelity and the advancement of human robot team simulations. Proceedings of the Human Factors and Ergonomics Society Annual Meeting 55(1), 1506–1510 (2011)

Mukamel, R., Ekstrom, A.D., Kaplan, J., Iacoboni, M., Fried, I.: Single-neuron responses in humans during execution and observation of actions. Current Biology 20, 750–756 (2010)

Murray, C.D., Pettifer, S., Howard, T., Patchick, E.L., Caillette, F., Kulkarni, J., Bamford, C.: The treatment of phantom limb pain using immersive virtual reality: three case studies. Disability & Rehabilitation 29(18), 1465–1469 (2007)

Pylyshyn, Z.W.: Computers and the Symbolization of Knowledge. In: Morelli, Anselmi, Brown, Haberlandt, Lloyd (eds.) Minds, Brains and Computers: Perspectives in Cognitive Science and Artificial Intelligence. Ablex (1993)

Pylyshyn, Z.W.: What's in your mind? In: Lepore, E., Pylyshyn, Z. (eds.) What is Cognitive Science. Blackwell, Oxford (1999)

Quine, W.V.: On empirically equivalent systems of the world. Erkenntnis 9(3), 313–328 (1975)

Ramachandran, V.S., Rogers-Ramachandran, D., Stewart, M.I.: Perceptual correlates of massive cortical reorganization. Science 258(5085), 1159–1160 (1992)

Rizzolatti, G., Fadiga, L., Gallese, V., Fogassi, L.: Premotor cortex and the recognition of motor actions. Cognitive Brain Research 3, 131–141 (1996)

Searle, J.: Mind: A Brief Introduction. Oxford University Press, New York (2004)

Speer, N.K., Reynolds, J.R., Swallow, K.M., Zacks, J.M.: Reading stories activates neural representations of visual and motor experiences. Psychological Science 20, 989–999 (2009)

Tolk, A.: Using Simulation Systems for Decision Support. In: Abu-Taieh, E.M.O., El Sheikh, A.A. (eds.) Handbook of Research on Discrete Event Simulation Environments: Technologies and Applications, pp. 317–336. IGI Global, Hershey (2009)

Turkle, S.: Simulation and Its Discontents. The MIT Press, Cambridge (2009)

Wastell, D.G.: Learning dysfunctions in information systems development: Overcoming the social defense with transitional objects. MIS Quarterly 23(4), 581–600 (1999)

Weiskopf, D.: Embodied cognition and linguistic understanding. Studies in History and Philosophy of Science 41(3), 294–304 (2010)

Wittgenstein, L.: Philosophical Investigations. Blackwell Publishing, Malden (1953)

On the Value of a Taxonomy in Modeling

Roger Smith

Florida Hospital Nicholson Center
Orlando, FL, United Sates

Abstract. Though modern science and business have created and adopted classification schemes, taxonomies, and operating rules that can be applied almost universally, the practice of building models and simulations remains unbounded by
science. Like the arts, each practitioner has the freedom to create a model in any
form that appears to offer a solution to a specific problem. A Periodic Table of
modeling has not emerged. Practitioners do not rely on a framework of established, tested, and accepted modeling techniques to guide their work. Conversely,
there are also no known poor methods for structuring a model which are not acceptable and which would bring censure from the professional community.
The unbounded nature of the current practice of modeling is supportive of an
artistic approach to modeling that encourages creative freedom in imagining and
building a unique new model. The environment is also convenient to modeling as
a service in which a customer is allowed to direct the construction of a model in
almost any direction that will address the problem, with few restrictions applied
from known best practices. As expedient as these advantages are, they also allow
inaccurate and inefficient approaches to be used without an objective or historic
"model-of-modeling" as a reference. The current practice of modeling allows almost any approach while its measure of correctness is determined solely by the
usefulness of the resulting product. This chapter is an attempt to begin the construction of a model-of-modeling which can serve as the Periodic Table for our
profession.

1 Aristotle's Metaphysics

The Greek philosopher Aristotle published the first known classification system
for the universe in his metaphysical works [1]. His five basic elements – Earth,
Water, Air, Fire, and Aether - seem simplistic today, but at the time such an
organized view of the world was quite groundbreaking and very useful to
philosophers.

Examining the world around him, Aristotle classified items that were "cold and
dry" as belonging to the "earth" element. Today we most often refer to these items
as "solids". He classified items that were "cold and wet" as the "Water" element,
or a liquid. The "Air" element was anything that he considered "hot and wet", or
gases. "Fire" was "hot and dry", what we might consider heat or energy. Finally,
there was the divine element of "Aether" which made up the heavenly spheres.

A. Tolk (Ed.): Ontology, Epistemology, & Teleology for Model. & Simulation, ISRL 44, pp. 241–254.
springerlink.com   © Springer-Verlag Berlin Heidelberg 2013

Today we know that these are composed of the same elements found here on Earth and our spiritual understanding has moved from the field of science to that of religion [1].

He further classified the interactions between these elements by assigning a "natural place" for each element. He said that earth had a natural tendency toward the center of the universe (i.e. center of our planet). Water tends toward a sphere that surrounds the center of the planet. Air tends toward a sphere surrounding the water. And fire tends toward the lunar sphere. These tendencies then explained why a rock would fall through water to seek its natural place in the center, air bubbles in water rise to achieve their sphere above the water, and fire rises through the air in seeking the lunar sphere. Finally, Aristotle's heavenly spheres were separate and moved in "the perfection of circles", which we now understand to be their elliptical orbits.

Though these five elements are Aristotle's most famous classification scheme, he did not stop there. He continued to create schemes for classifying the causes of activity in the world, methods of change and movement, types of objects, and a ladder of life forms (Table 1). It seems he was obsessively seeking a general model of his universe. He wanted to understand what existed and how it changed according to universal principles. Aristotle was a modeler of his universe and he understood how a consistent and simple model of the world would help him to understand the behaviors that were occurring all around him. He may also have realized that a general model could predict the behaviors of objects before he had even discovered them.

Humans have continued this search for structure, organization, and generalization and in the process have created the sciences that we know today along with our view of an organized, clockwork universe [2].

Table 1 Aristotle's Classification Schemes

5 Elements	4 Causes	3 Changes	2 Types	Ladder of Life
Earth	Material	Growth	Universal	Angelic
Water	Formal	Locomotion	Particular	Humanity
Air	Efficient	Alteration	-	Animals
Fire	Final	-	-	Plants
Aether	-	-	-	Minerals

Collected from "Aristotle" entry on Wikipedia.

1.1 Chemistry

One of the most powerful and notable classification schemes in science is the periodic table of chemical elements. This table began to take shape in 1789 when Antoine Lavoisier published a list of 33 chemical elements. Lavoisier grouped the elements into gases, metals, nonmetals, and earths. This was a place to begin and chemists spent the following century searching for a more precise classification scheme. In 1829, Johann Wolfgang Döbereiner observed that many of the elements could be grouped into triads based on their chemical properties. Lithium, sodium, and potassium, for example, were grouped together as being soft, reactive metals.

The modern version of the table emerged when Russian chemistry professor Dmitri Ivanovich Mendeleev and German chemist Julius Lothar Meyer independently published their periodic tables in 1869 and 1870, respectively. They both constructed their tables by listing the elements in a row or column ordered by atomic weight, and starting a new row or column when the characteristics of the elements began to repeat. The success of Mendeleev's table came from two intentional decisions that he made in constructing the table. The first was to leave gaps in the table when it seemed that the corresponding element had not yet been discovered. The second decision was to occasionally ignore the order suggested by the atomic weights and switch adjacent elements, such as cobalt and nickel, to better classify them into chemical families (Figure 1). With the development of theories of atomic structure, it later became apparent that Mendeleev had listed the elements in order of increasing atomic number (Figure 2) [3].

ОПЫТЪ СИСТЕМЫ ЭЛЕМЕНТОВЪ.

ОСНОВАННОЙ НА ИХЪ АТОМНОМЪ ВѢСѢ И ХИМИЧЕСКОМЪ СХОДСТВѢ.

			Ti = 50	Zr = 90	? = 180.
			V = 51	Nb = 94	Ta = 182.
			Cr = 52	Mo = 96	W = 186.
			Mn = 55	Rh = 104,4	Pt = 197,4.
			Fe = 56	Rn = 104,4	Ir = 198.
		Ni = Co = 59	Pl = 106,6	O- = 199.
H = 1			Cu = 63,4	Ag = 108	Hg = 200.
	Be = 9,4	Mg = 24	Zn = 65,2	Cd = 112	
	B = 11	Al = 27,4	? = 68	Ur = 116	Au = 197?
	C = 12	Si = 28	? = 70	Sn = 118	
	N = 14	P = 31	As = 75	Sb = 122	Bi = 210?
	O = 16	S = 32	Se = 79,4	Te = 128?	
	F = 19	Cl = 35,6	Br = 80	I = 127	
Li = 7	Na = 23	K = 39	Rb = 85,4	Cs = 133	Tl = 204.
		Ca = 40	Sr = 87,6	Ba = 137	Pb = 207.
		? = 45	Ce = 92		
		?Er = 56	La = 94		
		?Yt = 60	Di = 95		
		?In = 75,6	Th = 118?		

Д. Менделѣевъ

Fig. 1 Mendeleev's Periodic Table of 1869
(Note 90 degree rotation from modern layout.)

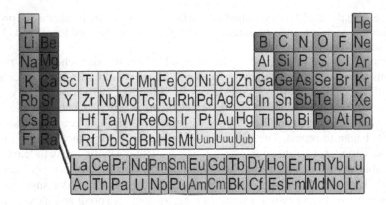

Fig. 2 Modern Periodic Table of Elements

1.2 Biology

Carolus Linnaeus carried Aristotle's classification scheme to a more elaborate form in his work, *Systema Naturæ* first published in 1735. In this work, he divided nature into three kingdoms: mineral, vegetable and animal. These then had five ranks: class, order, genus, species, and variety. The modern form of this ranking has eight major levels of classification beginning with "life" and proceeding through domain, kingdom, phylum, class, order, family, genus, and species (Figure 3). This attempts to provide a structure for all living things, but focuses on abstraction rather than order as is dominant in the periodic table.

A biological phylogenetic tree is much closer in structure and function to the periodic table. It focuses on specific items and attempts to define the relationships between them. It uses a branching diagram or tree to show the inferred evolutionary relationships among various biological species based upon similarities and differences in their physical and genetic characteristics. The "taxa" joined together in the tree are implied to have descended from a common ancestor (Figure 4). This hierarchy improves understanding and enables predictions of behavior and characteristics before they have been observed and measured.

Most of the sciences have a core touchstone similar to chemistry and biology which organizes and defines what belongs in that science, how everything is related, and how systems function.

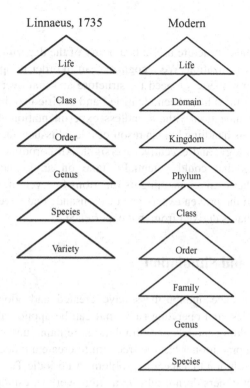

Fig. 3 Hierarchy of Biological Classification

Phylogenetic Tree of Life

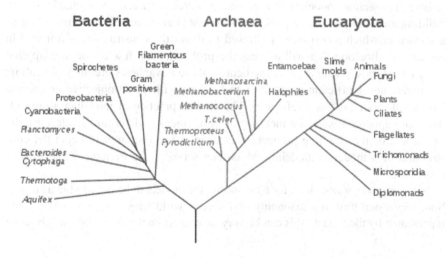

Fig. 4 A Phylogenetic Tree of Life

1.3 Business

Even modern business structure uses a taxonomy of the departments and functions that make up the organization. We recognize than in order to optimize the performance of a company, there is a need for structure and that over time some of the most effective structures have been identified and codified so that they can be applied to a new organization without endless experimentation. We recognize the need for departments handling human resources, legal issues, facilities, operations, information technology, and executive oversight. All corporate organizations are conceptual entities that could potentially take on any structure that can be imagined. But, rather than reimagining its functions for every new company that is created, we rely on the proven experience of thousands that have come before and adopt a structure that is derived from that experience.

2 Modeling and Simulation

Though modern science and business have created and adopted classification schemes, taxonomies, and operating rules that can be applied almost universally, the practice of building models and simulations remains unbounded by science. Like the arts, each practitioner has the freedom to create a model in any form that appears to offer a solution to a specific problem. A Periodic Table of modeling has not emerged. Practitioners do not rely on a framework of established, tested, and accepted modeling techniques to guide their work. Conversely, there are also no known poor methods for structuring a model which are not acceptable and which would bring censure from the professional community.

The unbounded nature of the current practice of modeling is supportive of an artistic approach to modeling that encourages creative freedom in imagining and building a unique new model. The environment is also convenient to modeling as a service in which a customer is allowed to direct the construction of a model in almost any direction that will address the problem, with few restrictions applied from known best practices. As expedient as these advantages are, they also allow inaccurate and inefficient approaches to be used without an objective or historic "model-of-modeling" as a reference. The current practice of modeling allows almost any approach while its measure of correctness is determined solely by the usefulness of the resulting product. This chapter is an attempt to begin the construction of a model-of-modeling which can serve as the Periodic Table for our profession.

When a modeler does look for a taxonomy for models to be included in a simulation, they often turn to a taxonomy of the real world target domain that is being represented by the model. This can be very helpful when the simulation will be used only within that domain. It can significantly improve the ability of the modeler to explain the resulting simulation to the potential customer; and validation can be easier to achieve when the models closely follow the domain taxonomy. However, this approach also inherits the limitations of the domain itself. More abstract modeling efficiencies that may be valid and powerful are not used because they are not familiar to customers in the target domain, though these techniques may be a proven approach in the modeling field. When adopting domain taxonomies, their differences create incompatibilities between models and simulation systems, as when attempting to combine models of humans, machinery, and energy.

2.1 Model Taxonomy: Entities, Actions, Relationships

A taxonomy of the entities, actions, and relationships that can be used as a framework for guiding all model construction may be expressed using Entity-Relationship or UML diagrams, but the need is not for a notational standard; rather we need a base framework which can guide the connections and interactions that are necessary and possible within and across models and simulation systems [4].

The useful structures in a model seem to include entities (objects), actions (events), and relationships (connections). Given a set of entities that have a set of action capabilities, the relationships between the entities would indicate where the actions could be applied (Figure 5).

At this level, the specific attributes or parameters of the entities, actions, and relationships are not of interest. These play an important role in structuring our understanding of the models, but at a finer resolution. Just as Aristotle did not give the detailed universal characteristics of earth, air, fire, and water, the model attribute problem should be tackled separately, after a solution to the larger problem has begun to emerge.

Since a taxonomy in modeling is at its beginning stages, perhaps the best approach is to return to Aristotle and create a classification scheme based on observation and experience, with the understanding that such a scheme will prove to be naive and simple after it has motivated the creation of a more rigorous and complete definition of the problem. We must allow ourselves to follow Aristotle in beginning with our experience and using that to grow into a larger and more complete work.

2.2 Entities

Typical modeling approaches involve making a list of the objects in the world that the customer loves and then assigning them the actions that they have in the real world. This "realistic" approach to modeling insures that each model begins in a unique place and without reference to underlying principles that have been previously created or discovered in the artistic and scientific profession of modeling.

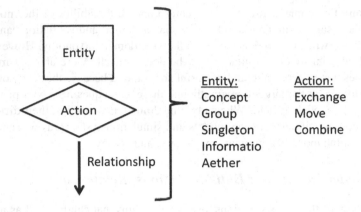

Fig. 5 Naive Taxonomy of Entities, Actions, and Relationships

When Picasso begins to paint a bowl of fruit it is certain that the resulting picture will present the items in a very unique and artistically brilliant way. But, even he begins the task with a mental model of the objects, the light, and the function of those objects in his mind. This links the painting to the real world, but still allows abstract representation based on Picasso's own view of the world and his understanding of how to communicate to his audience.

If there is only one way to represent an apple, then there would only be one painting of an apple. Similarly, if there is only one way to model an airplane, then we would only need a single model of that airplane. The real world properties of the airplane are just one input into the best way to model it. The real world taxonomy of an object is just one set of information that is useful and valuable in modeling something from the real world. There are many ways to model the airplane, just as there are many ways to paint a bowl of fruit. Methods for modeling and painting emerge from experimentation with new ideas, and some of those experiments will lay the groundwork for a new school of art or a new method of modeling. The effective methods should make up the science of modeling and allow the creation of a taxonomy to support that science.

The fruit bowls of Picasso, Monet, and Cezanne cannot be translated point-by-point from one artist into another because each incorporates such a unique style of representation. But, the viewer can identify corresponding features and appreciate the similarities and differences between the works. In modeling, we do not seek a perfect taxonomy of the real world from which our models are drawn. But rather an abstraction that allows us to identify where one entity or action belongs and how it is related to other entities and actions. We may see little functional difference between apples, pears, oranges, and plumbs. But we are very interested in the relationship between apples and airplanes, and understanding what such different classes of entities have in common that would allow them to interact with each other. We seek a codification of this understanding which will allow us to automate interactions so they can be handled by a model in a computer, without constant human intervention.

At first blush one might assume that there is little connection between an apple and an airplane. Certainly the two naturally inhabit different physical spaces. But apples may be cargo to airplanes. Is there a structure that can identify that this is possible and which can guide the construction of a cargo-bearing aircraft? Fuel poses a more complicated relationship. This liquid may be the cargo in an aircraft which remains unchanged over time, or it may be the energy source that powers the airplane and is consumed during an interaction. How can the fuel e represented so that it can take both forms and be converted naturally from one to the other?

There are millions or billions of different entities in the real world which could be encoded into a model. We have traditionally expected the human modelers to create unique, useful, and correct representations of every combination. There has been no guiding structure which carries the experience of thousands of prior modelers that can be used to guide new models. How are these billions of objects organized and related so that a generalized representation can be handled by a computer?

A naïve Aristotelian structure for entities is proposed in Table 2. The contents attempt to separate concepts, groups, single items, and information from each other. Non-modelers tend to see entities as the individual and independent objects that they experience in their daily lives. Entities like "car" appear to be the natural representation of the world. But experienced modelers have learned that when many cars come together, they may be better represented as a group which functions together. Or conversely, when the car is the subject of maintenance, it may itself be the aggregation of many smaller parts which need to be represented to support the replacement and repair that is required to return the car to operation.

Abstraction and aggregation are two powerful tools for identifying a model's representation of the world. Abstraction creates hierarchy, while aggregation internalizes or eliminates hierarchy.

A "group" may emerge through the aggregation of multiple objects which are all the same (homogeneous) or which are different (heterogeneous). The techniques that support homogeneous aggregation are usually simpler than those called for in heterogeneous aggregation. A group may also be formed simply through linking, where each object remains uniquely and independently identified, but is linked to other objects so that an action or relationship to the group can be applied to each individual entity.

More general than a group is the "concept", something which exists as an idea or identity, but has no physical expression. Governments, religions, and societies are made up on many entities, each of which has the ability to take action. But it is their self-identification as belonging to the concept that brings the concept into existence. The individual entities may choose which actions to take based on their identification with one of these concepts.

Finally, there may be a separate class for entities which form the background to support the other more dynamic objects. These are the new aether. In some cases, the earth itself plays an important role in modeling actions, but is itself not explicitly represented or changed by the actions. The gravity and atmosphere provided by a planet are often taken for granted in a model and applied, not as a separate interacting object, but as a universal presence and force that becomes hard-coded into the model.

This derivation of entity classes or structure must certainly be too naïve to be the best and final solution to our problem. But perhaps it is a place to begin.

Table 2 Naive Entity Taxonomy

Concept	Group	Singleton	Information	Aether
Gov't Religion Society Plan	Homo Hetero Linked	Simple Composed	Bit Packet Message	Weather Earth

2.3 Actions

How many actions can there be? Are there an infinite number, few, or just one? Can all actions be derived from some generic action?

"Exchange" is a general form of action that we find in many models (Table 3). Weapons are a unilateral exchange of an object that acts upon the recipient's state variables, especially those related to health and physical structure. Communication is a bilateral exchange in which both parties can receive information. The information is copied from sender to recipient. However, the energy that carries the information is a unilateral exchange just like a weapon. In a surgical simulation, the instruments have an effect very similar to a controlled weapon. The objective is often to reach a point at which a unilateral exchange can happen where a piece of tissue is removed from the patient and taken by some outside system. When this

Table 3 Naive Action Taxonomy

Exchange	Move	Combine
One Way	Continuous	Merge
Two Way	Discrete	Separate
Multi Way	Sequential	Link

is achieved, then the similarity to a weapon changes to something more like construction or repair, returning the patient object to a more desirable state.

Sensing and communicating are actions very commonly found in models and simulations. But both of these seem to lend themselves to a specialization of the "exchange" action. Passive sensing is the exchange of the target's signature with the sensor. Active sensing is the two-way exchange of energy from the sensor, to the target, and back to the sensor.

Movement appears to be a unique and near-universal form of action. It is different from "exchange" in that an entity can perform this action on itself without reference to or effect on other entities.

Combination may be an action which creates a new form of aggregation. This may join, link, or separate groups of entities.

This list of root actions is very short, but it is difficult to propose an action that is not subsumed into the "exchange" category.

2.4 Relationships

Does potential action define a relationship, or does a relationship define potential action? In a world with a single object can there be any action? Action is a change with respect to a prior state. So action is significantly limited without multiple entities and the relationships that exist between them. Without a relationship, it appears that an object has "potential action" which cannot be realized as "dynamic action" until the relationship is defined.

At the most general level, the relationships that exist between entities would seem to address their hierarchy (rank), intention (affiliation), location (locality), and connectivity (Table 4).

Table 4 Naive Relationship Taxonomy

Rank	Affiliation	Locality	Connectivity
Superior	Friend	x,y,z, t	Physical
Subordinate	Opponent		Frequency
Partner	Neutral		
Indeendent	Null		

Hierarchical relationships provide the structure necessary to identify the flow of information and action among cooperating and competing objects. Affiliation determines whether the actions between entities are supportive, competitive, or neutral. Locality can support relationships that change as entities move through space and acquire or lose neighbors to be acted upon based on nearness in space. Connectivity identifies a means for sharing or exchanging information.

3   Model: Entity Is; Action Does; Relationship Can Do

Entity, action, and relationship define a model through time and space. An Entity "is", it exists now and for some amount of time in the past and future. Like a boulder on a plain, it is a distinct feature of the world which can be referenced and awaits action.

An Action "does". It exists for a short period and potentially creates a change to the world. The action itself may come and go very quickly, but its effect may persist through changes to an entity's state.

A Relationship "can do". It defines potential action and lines through which actions and change can be applied.

All three of these can change at any time. An object's state changes all the time as a result of incoming actions. Actions can change based on the originator, environment, recipient, and time. Relationships can change in response to actions, just as actions change entities. Actions must be defined which follow relationships, change entities, and change the relationships themselves.

4   Art, Science, or Service?

Is modeling a science, an art, or a service? It is a service when performed as a work-for-hire for a customer who defines the product and process. This form is similar to hiring an artist to paint a picture for the living room. The client has a great deal of influence over the model and how it is constructed. Such models are designed as much by the novice, inexperienced client as they are by the professional modeling team that has been hired to do the work. This client-modeler relationship can significantly limit or undermine the use of proven modeling techniques or the exploration of new concepts. Such works typically repeat known practices to accomplish a specific objective.

Modeling is an art when modelers search for ways to express themselves and their capabilities. The goal is unique expression, just as now famous artists created entirely new approaches to expression by being free to explore new paths in their works. This can lead to the discovery of new methods or schools of thinking as Pablo Picasso, Monet, Jackson Pollack, Roy Lichtenstein, and Andy Warhol created new genres in art. But when following the artistic path to creativity, for every successful new method, there are hundreds or thousands of unsuccessful methods. This prevalent failure is a primary reason that clients do not encourage or support an artistic approach to modeling. Given their limited funds and business needs, they cannot gamble on being the one out of one hundred new methods that will succeed. Hence, most commissioned models are very similar to one of the successful predecessors.

Modeling is a science when we attempt to organize what has been created and discovered in the field in an attempt to create a working and valuable abstraction of that field. The goal is to create or discover laws, structures, and guidelines that provide valuable forms of representation. Like the artistic approach, these experiments do not lend themselves to a project's timelines and deliverables. Therefore, the science of modeling must be relegated to academic and research organizations that have the latitude to explore, fail, and publish discoveries, rather than delivering a working product.

Most modeling is currently done as a service, just as most new buildings are commissioned for practical usage. Models are occasionally commissioned with the hope and intention of expanding the science of modeling. The artistic approach to modeling is typically relegated to independent modelers working on their own time and initiative, and measured by the unique and satisfying products that result more than by the usefulness of the product. Occasionally these are adopted and applied to service projects.

The existence of a taxonomy (modeling science) may guide and reorient projects (modeling service) in their choice of models and may create tools for self-expression (modeling art).

5   Validation

When simulation is provided as a service, validation often proceeds based on the opinions of the sponsor or their designated representative. As a work-for-hire, the customer applies their own level of rigor to the acceptance of the product, rather than referencing the best practices in the field for validation.

This approach places the opinions of subject matter experts side-by-side with the scientific approach to modeling. It gives equal credibility to subjective and objective methods, without calling for a validation of the validation method. This subjective validation is a confounder. It corrupts, confuses, and obscures a scientific approach to modeling. Subjective methods of validation can assign negative reviews to very valid modeling methods, and positive reviews to ad hoc and unsubstantiated methods. At its core, this implies that there are no objectively valid or scientifically-based foundations for the practice of modeling and that any method which arrives at a convincing product may be acceptable.

Without rigorous and objective validation methods, it is difficult to establish, build, and defend rigorous methods for modeling. In the current service environment, modeling is robbed of its status as a scientific process constructed from years of experimentation and experience, and becomes only a subjective and expedient method of representing what a customer wants to see.

6   Conclusion

This chapter is meant to motivate a deeper exploration of that nature of modeling through a taxonomy of the scientifically supportable methods and structures that have been created and discovered over many years of research and practice. Aristotle's categorization of the world into five substances provided a very useful structure for future scientists. Just as important, it illustrated the need for a classification scheme to understand the world and its operations. Aristotle's five substances were to naïve, but they started the process of scientific classification. The minimal taxonomy of model entities, actions, and relationships that are presented here are similarly naive, but they are offered as a place to begin the process of creating a more scientific taxonomy of modeling methods to support the practice of modeling and simulation.

References

[1]   Aristotle (350BCE) Physics (Books 1-6). MIT Classics web site,
      http://classics.mit.edu/Aristotle/physics.1.i.html
      (accessed April 2, 2012)
[2]   Dolnick, E.: The Clockwork Universe: Isaac Newton, the Royal Society, and the
      Birth of the Modern World. Harper, New York (2011)
[3]   Gordin, M.: A Well-Ordered Thing: Dmitrii Mendeleev and the Shadow of the Periodic Table. Basic Books, New York (2004)
[4]   Turnitsa, C.D., Padilla, J.J., Tolk, A.: Ontology for modeling and simulation. In: Proceedings of the 2010 Winter Simulation Conference, Baltimore, MD, pp. 643–651 (December 2010)

Resources

The following books are not referenced in the chapter, but they are valuable resources for scholars and students seeking more information.

Banks, J.: Handbook of Simulation. John Wiley & Sons, New York (1998)
Booch, G.: The Unified Modeling Language User Guide. Addison-Wesley, Uppersaddle River (1990)
Davis, P.K., Bigelow, J.H.: Experiments in Multiresolution Modeling. RAND Report (1998),
      http://www.rand.org/pubs/monograph_reports/2007/MR1004.pdf
Davis, P.K.: Distributed Interactive Simulation in the Evolution of DoD War-fare Modeling and Simulation. Proceedings of the IEEE 83(8) (1995)
Fishwick, P.A.: Simulation Model Design and Execution: Building Digital Worlds. Prentice Hall, New York (1995)
Law, A., Kelton, W.D.: Simulation Modeling & Analysis. McGraw Hill, New York (1991)
Nance, R.E.: A History of Discrete Event Simulation Programming Languages. History of Programming Languages II. ACM Press, New York (1997)

A Bayesian Approach to the Validation of Agent-Based Models

Kevin B. Korb[1], Nicholas Geard[2], and Alan Dorin[1]

[1] Monash University
   Melbourne, Victoria, Australia
[2] University of Melbourne
   Melbourne, Victoria, Australia

Abstract. The rapid expansion of agent-based simulation modeling has left the theory of model validation behind its practice. Much of the literature emphasizes the use of empirical data for both calibrating and validating agent-based models. But a great deal of the practical effort in developing models goes into making sense of expert opinions about a modeling domain. Here we present a unifying view which incorporates both expert opinion and data in validating models, drawing upon Bayesian philosophy of science. We illustrate this in reference to a demographic model.

1 Introduction

Agent-based models (ABMs) are computer simulations of numerous, heterogeneous "agents". The models' microbehavior is determined by explicitly programmed rules, while their macrobehavior is not, instead emerging from the collective behavior of the population of agents, usually in very complex ways. This kind of simulation has grown from early efforts in ecology and artificial life into one of the most widely applied computer methods across the sciences today. Nevertheless, skepticism about the interpretation and epistemological standing of these models remains widespread and will do so until at least the fundamentals of ABM validation are agreed upon. Here we present and defend a Bayesian approach to ABM validation.

As many have remarked, the theory of how to validate ABMs is vastly underdeveloped compared to its practice [e.g., Klein and Herskovitz, 2005; Kleindorfer et al., 1998]. This is unsurprising given how rapidly ABMs have grown from a niche computer application in the 1980s to a leading research technology for ecology [Grimm and Railsback, 2005], economics [Tesfatsion and Judd, 2006], epidemiology [Auchincloss and Diez Roux, 2008] and dozens of other sciences (see http://jasss.soc.surrey.ac.uk/JASSS.html for a full range of examples).

In this paper we take some of the principles of Bayesian theories of scientific method and develop them into an account of validation practice for simulation

A. Tolk (Ed.): Ontology, Epistemology, & Teleology for Model. & Simulation, ISRL 44, pp. 255–269.
springerlink.com   © Springer-Verlag Berlin Heidelberg 2013

science. The Bayesian approach to philosophy of science explicitly recognizes the distinction between the current understanding of the behavior of a system (prior belief) and the data (likelihood), which provides us with a framework for integrating both qualitative and quantitative approaches to validation. In essence, our prior belief about the model is updated in the light of experimental data gathered from our simulations.

Bayesian inference tends to accord well with an Ockham-like favoritism for simplicity (e.g., Wallace, 2005). By contrast, both the systems under study and the ABMs themselves tend to be complex, nonlinear and high dimensional. This complexity raises some special epistemological questions about Ockham's Razor, which we address in §3.

After developing a Bayesian approach to validation in the abstract, we illustrate it in reference to the demographic submodel of an epidemiological ABM.

2 Bayesian Philosophy of Science

Klein and Herskovitz [2005] have presented a case that Karl Popper's falsification-ism [Popper, 1959] be made the basis for the epistemology of simulation. Popper's account of methodology has many virtues, which have made his name prominent throughout the sciences and perhaps even seem synonymous with philosophy of science. For example, Popper's emphasis on "severely testing" theories — pitting them experimentally against an alternative, such that one or the other must become falsified — is very agreeable to the empirical spirit. Likewise his emphasis on the fallibility of scientific method agrees with both the history of science and traditions in scientific education. Regardless, there are many difficulties standing in the way of a Popperian theory of method. Kuhn [1962], Lakatos [1970] and Feyerabend [1975] all demonstrated with numerous historical examples how in a great many cases unexpected results were rationally held to be *anomalous*, rather than *falsifying*, demanding, not rejection of the theory under test, but instead the discovery and elaboration of new auxiliary hypotheses which could explain the discrepancies between theory and observed reality.

In view of the importance they place on accounting for the accumulation of scientific knowledge, more troubling for Klein and Herskovitz [2005] will be the fact that Popper never gave any reasonable account of the *growth* of knowledge. His reliance strictly upon falsification left any account of support, confirmation or growth of things *known* at best open. To be sure, Popper talked much of "corroboration", even developing a measure of degrees of corroboration. That is, theories that have survived more severe tests are meant to have higher degrees of corroboration than theories that have survived less severe tests. This was supposed to fill the vacuum left by an epistemology exclusively reliant upon refutations, but a vacuum filled with a fictional aether is still just a vacuum. Popper insisted not just that all such corroborated theories were lacking any empirical support, but also that they were, in point of fact, *false*. On Popper's repeated account, all synthetic universal hypotheses are false [e.g., Popper, 1959, Appendix *vii] and simply waiting for their refutations to be found![1]

2.1 Bayesian Confirmation Theory

Bayesian philosophy of method has grown from the ashes of Popperianism. Bayesianism has been propelled by numerous factors: in artificial intelligence and statistics by the development of new methods for exact inference (in Bayesian networks; e.g., Pearl, 1988; Korb and Nicholson, 2010) and approximate inference (in MCMC simulation; e.g., Friedman and Koller, 2003); in cognitive neuroscience [Glimcher, 2004]; and generally across many sciences through the explosive growth of the accessible computational power needed for these kinds of analyses.

In philosophy a driving force for Bayesianism has been a string of successful Bayesian re-analyses of Popperian insights into method, combined with an approach that supplies what Popper could not: a theory of theory confirmation.

All of this originates in Bayes' theorem (Bayes, 1763), which simply describes the posterior probability of a hypothesis (conclusion) and in terms of its prior probability and likelihood. In particular,

$$P(h|e) = \frac{P(h) \times P(e|h)}{P(e)} \tag{1}$$

This is an analytic theorem. Bayesian confirmation theory goes well beyond it: it asserts that the proper way to assess confirmation is to adopt the probabilities conditional upon the available evidence — as supplied by Bayes' theorem — as our new posterior probabilities. This move to a posterior distribution is called *Bayesian conditionalization*.

Given this view, the simplest way of understanding the concept of the confirmation or support offered by some evidence is as the difference between the prior and posterior probabilities of a hypothesis; that is, e supports h just in case $S(h|e) = P(h|e) - P(h) > 0$ (cf. Howson and Urbach, 1993, p. 117). A second measure of support, the ratio of likelihoods e given h over e given not-h, is equally defensible [Good, 1983]:

$$\lambda(e|h) = \frac{P(e|h)}{P(e|\neg h)}.$$

It is a simple theorem that the likelihood ratio is greater than one if and only if $S(h|e)$ is greater than zero. $\lambda(e|h)$ (or, simply, λ) can be understood as a degree of support most directly by observing its role in the odds-likelihood version of Bayes' theorem:

---

[1] This was Popper's extreme skeptical "solution" to Hume's problem of induction: stop inducing! And never mind that his statement itself is a synthetic universal. Popper was no more bound by the petty hobgoblin Consistency than any inductivist!

$$O(h|e) = \lambda O(h).$$

This asserts that the conditional odds on h given e should equal the prior odds adjusted by the likelihood ratio. Since odds and probabilities are interconvertible $(O(h) = P(h)/P(\neg h))$, support defined in terms of changes in normative odds measures changes in normative probabilities just as well as $S(h|e)$. λ has a significant advantage over $S(h|e)$ however: it is easier to calculate. Since hypotheses often describe how a system functions given initial conditions, finding the probability of the evidence assuming h is often straightforward. What a likelihood ratio reports is the normative *impact* of the evidence on the posterior probability, rather than the posterior probability itself (which would require also the *prior* probability of h). However, confirmation theory is concerned with accounting just for rational *changes* of belief, and so λ turns out to be the best tool for understanding confirmation, as we show now with two examples.

(1) Likelihood ratios make clear why Karl Popper's (1959) insistence that scientific hypotheses be subjected to severe tests makes sense. Intuitively, a severe test is one in which the hypothesis, if false, is unlikely to pass; that is, whereas the hypothesis predicts some outcome e, its competitors do not. Since the hypothesis predicts e, $P(e|h)$ must be high; since its competitors do not, $P(e|\neg h)$ must be low. Together these imply that the likelihood ratio is very high. So, a severe test will be highly confirmatory if passed and highly disconfirmatory otherwise — providing the most efficient approach to testing a hypothesis, as Popper pointed out.

(2) Another example is the preference which experimental scientists exhibit *ceteris paribus*, when confronted by two possible tests of a theory, for that test which is most different from one previously passed. For example, Eddington had two alternatives to testing Einstein's general theory of relativity (GTR) in 1919: either repeating Einstein's analysis of the precession of Mercury's perihelion or checking the predictions which GTR made of a "bending" of starlight by the mass of the sun, observable during a total eclipse. Despite the fact that astronomical observations of the motion of Mercury are cheaper and simpler, Eddington famously chose to observe the starlight during the eclipse over the Atlantic. Intuitively, we can say that this was because a new result, as opposed to a repeated experiment, offers a more severe test of the theory. For formal Bayesian analyses of this case, see Franklin [1986] and Korb [2004].

More comprehensive accounts of Bayesian method can be found in Howson and Urbach [1993] and Korb [1992]. For our purposes here, it suffices to point out that λ provides a tool for understanding the direction and degree of confirmation or disconfirmation, allowing guidance for validation techniques even when a full probabilistic account is unavailable. We now proceed to a qualitative account of Bayesian ABM validation.

2.2 Bayesian Validation

The goal of empirical validation of computer simulation — its central epistemological question — is to determine whether the simulation is telling us the truth about some target process in the world — whether the theory which it instantiates is true or false.

Some researchers take an unnecessarily narrow view of this process. For example, Windrum et al. [2007] suggest that in order for a validation process to be *empirical* it must directly involve data, which may become an excuse to downplay expert opinions and intertheoretic relations between the theory behind the simulation and related science. But, while empirical knowledge ultimately rests upon sensory experience, it does not have to *directly* rest upon it. We see empirical validation as encompassing both statistical tests using data and expert opinion, which itself (hopefully) derives largely from experience. Bayes' Theorem, in fact, provides a natural form in which to combine these: expert opinions are readily interpreted as providing prior probabilities of a model being correct, while statistics can be used to measure the fit of the model to the data — i.e., its likelihood. This division does not perfectly divide model validation activities, but it does work roughly and, more importantly, serves to reinforce the importance of *combining* expert- with data-oriented validation methods.

ABM simulation is widely understood to involve a tripartite relation:

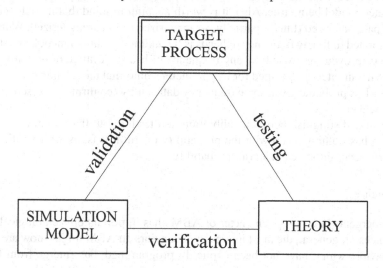

The central epistemological question can be answered once we know the status of any two of these relations [Mascaro et al., 2010, Chap 3].

Verification

As the goal of validation is to determine the representational accuracy of a model, the process logically begins with the construction of the model. Assuming that a simulation strictly reflects some underlying scientific theory, then whatever probability that theory has must be shared with its simulation. So, activities which contribute to the confirmation of theories, together with activities which verify that a simulation is true to some theory, also are properly considered an aspect of validation and are most naturally accommodated as contributing to an assessment of the model's prior probability.

Calibration

In a similar vein, calibrating a model contributes to its probability of being true. This may be an uncommon observation, which is a natural consequence of the calibration of *some* models being trivial. For example, calibrating a binomial model to fit the observed tosses of a coin is trivial, and it also doesn't obviously contribute to the binomial model being true, since *whatever* the bias of the coin, the binomial model could have been appropriately calibrated. Falsificationism suggests that a model which can accommodate *any* data cannot even be tested, let alone be regarded as a true (or good) scientific theory. However, this suggestion is misleading. For one thing, if there were ever any models which could *not* have been calibrated to fit the data, then successful calibration rules them out. Whatever probability *those* models started with must be redistributed, raising the probability of a successfully calibrated model being true. Also, it is worth keeping in mind that models may be either parameterized (fitted), partially parameterized or unparameterized. What can be calibrated to fit any frequency of heads is the unfitted binomial model, and it cannot be disproved (or proved) by any frequency of heads. A fitted, or partially fitted (e.g., with an interval specified for its parameter), binomial model, however, will be more or less probable given some frequency data, and so confirmed or disconfirmed by those data.

We may distinguish between calibration and testing, but that is not to say calibration has nothing to do with the probability of truth. It is, as with verification, properly accommodated in a prior probability.

Emergence

The emergent (bottom-up) character of ABMs has important epistemological consequences. In general, the most interesting behaviors an ABM might show are macrobehaviors which have not been explicitly programmed, but emerge from lower level rules which have been explicitly programmed. Normally, the higher level behavior could be realized (at least qualitatively) in multiple distinct ways at the lower level. In philosophical terms this relation is supervenience: higher-level behavior supervenes on one or more lower-level supervenience bases.[2]

As the higher level, emergent behavior is often the target of interest, the behavior we should like to predict or explain, it is also the behavior we should like to validate our model against. So, wherever possible, it makes sense to preferentially calibrate with lower level data and validate or test with higher level data.

---

[2] This is often, and wrongly, characterized as micro-reduction. On supervenience, see McLaughlin and Bennett [2011].

3   Simplicity or Complexity?

The complexity of ABM models raises the question of the status of Ockham's Razor. Edmonds and Moss [2005] have argued influentially that ABM modeling, contrary to the usual methodological advice, should start out complex and devolve towards simplicity. Ockham's Razor, of course, suggests the opposite: that we should add complexity only in the face of some evidential setback — specifically, that where two theories do equally well with the data, the simpler is to be preferred (Keep It Simple, Stupid). Edmonds and Moss claim that this rule has little or no merit in ABMs and, more specifically, that simplicity confers no epistemological virtue to a model. ABMs are aimed at understanding complex phenomena, and, according to them, should aim to represent them in the "most straightforward way possible", meaning as descriptively detailed as possible (and so their "Keep It Descriptive, Stupid").

Striking just the right balance between simplicity and fit to the data — what Grimm et al. [2005] call finding the "Medawar zone" — is always going to be difficult. And it may well be that many overemphasize simplicity to their own disadvantage. But we disagree that simplicity confers no epistemological advantages.

Undoubtedly, Edmonds and Moss's starting point, the presence of so much complexity in the systems being modeled, can seduce people into over-specifying their models, but that's a danger, not an essence nor a virtue of ABMs. Methodological simplicity, on the other hand, has a number of real, if modest, virtues:

1. The KISS approach is at least a *possible* inductive strategy. Adding complexity where required by evidence is a possible path to the truth, as Reichenbach's (1949) vindication of induction argument suggests. The inverse approach in most domains, where complexity is unbounded, doesn't even begin to make sense, since there is no beginning. And choosing one model from the multitude having complexity comparable to some target system can hardly be justified at the *start* of a research program.
2. Starting out with a complex model implies having a large parameter space. This is not only operationally inconvenient, it is hugely methodologically suspect, since over-specified models fit noise and fail to generalize.
3. As a simple model with features added only as needed, a KISS model is far more promising as a vehicle for the consilience of induction, i.e., we can try to adapt it to new and related domains. For example, a KISS measles model might well be usable in a pertussis problem with minimal (and motivated) changes. A KIDS measles model will always only be a measles model.

Perhaps the main virtue that has been put forward for Ockham's Razor, and the one Edmonds and Moss [2005] contest most vigorously, is that simplicity *ceteris paribus* corresponds to higher probability. While widely regarded as true, by both Bayesians and their opponents, this would be exceedingly hard to prove — or to disprove. We don't have any convenient, unbiased collection of examples for testing it. As the probability of simplicity is an exceedingly complex matter, and the advantages of simplicity above are independent, we pass over it (but see Wallace, 2005, for a Bayesian defence of simplicity).

Modeling is not much different from theorizing, as its epistemology shows. It's simply cognitively impossible to start out with a theory that is as complex as the phenomenon. One starts out with a central idea or two, which then get enhanced. Furthermore, the *goal* is to end up with a theory that is at least somewhat simpler than the phenomenon at issue, that can explain it, rather than simply reproduce it. Ockham's Razor is methodologically inevitable.

4   ABM Validation Methods

Here we present a number of recognized types of validity for ABMs, characterizing them in terms of both prior and posterior considerations. We suggest that, as in the case of Bayesian analyses of scientific methods mentioned above, Bayesian analyses of these validation methods can be made and may well improve their usage.

We don't propose that each kind of validity considered here needs to be adopted in every ABM study, however these varieties will generally be worth considering. Here we consider them in the abstract; in §5 we consider some of them relative to our own simulation.

1. **Expert opinion (prior).** This is the usual starting point for constructing and refining computational models. We suggest that this covers most kinds of validation which do not directly involve data, corresponding to what Pitchforth and Mengersen [2012] call **nomological validity:** establishing that the model fits within its wider scientific context. Some of the terminology comes from psychology, by way of Pitchforth and Mengersen [2012]. That study focuses upon Bayesian network simulations, however the concerns of simulation epistemology are strikingly similar across ABMs and Bayesian networks.

   a. **Face validity:** Does the model look right to an expert? While face validity is a weak kind of test of a model, it is nevertheless central to most modeling endeavors. Models that look wrong are often abandoned without further ado, something which often causes headaches with machine learned models, since learning algorithms rarely incorporate any kind of aesthetic sense. Face validity should be examined throughout the modeling process, analogously with agile software development processes, where end users provide continuing feedback on the adequacy of software.

   Aside from a holistic assessment of a model, all the other forms of validation under *expert opinion* are similarly subjective assessments. Frequent reviews from different experts provide an opportunity for those with varying assumptions about both the model and the domain to provide feedback. Such reviews are also an opportunity to negotiate validity criteria, perhaps including exemptions, when unrealistic aspects of model structure or behavior are deemed less relevant for validation.

b. **Content validity** considers whether the most important factors and relationships between variables noted in the literature are present in the model. Expert opinion will be the primary guide here, but focused reviews of the literature will also be useful.

c. **Case analysis** takes specific instances and examines how the model deals with them. This shares conceptually again with software engineering, where "use cases" are often applied to review software usability, etc. The specific instances may (should) include both normal and extreme cases. They also may be constructed from setting specific initial conditions (as in a historical case study) or from setting parameters that govern relations between individual roles within the simulation (e.g., reproductive or immigration rates) or, of course, from both.

A thorough validation might take further inspiration from software engineering and do an equivalence partitioning of initial conditions to generate a suite of cases that looks at all (or many) varieties of normal and abnormal conditions. Since the results need to be judged by an expert, the value of this depends also upon the patience of the experts available.

d. **Internal validity** examines whether variation in the model's variables is reasonable [Sargent, 2010]. This could specifically consider covariation between sets of variables, to determine whether changes in some variable either cause or are codependent with changes in others, in ways which are judged sensible by experts; this is generally called **sensitivity analysis**. The inverse process of **robustness analysis** aims to identify features of the model that are resistant to varying initial conditions [Grimm and Railsback, 2005, Sec 9.7].

2. **Data (likelihood).**

a. **Predictive validity** is the primary way of validating in many discussions. If we were to take "prediction" literally, then even the use of historical data not employed in calibrating the model would be (improperly) excluded (what has been called "retrodiction").

Measuring the fit to data of a model — i.e., predictive accuracy — is again often the only way considered of assessing predictive adequacy. However, predictive accuracy has limitations; see, e.g., Korb and Nicholson [2010, Sec 7.5] for a discussion.

Regardless of the measure used, testing picks up wherever the calibration left off. Reusing data used to calibrate a model to "test" it is generally just an error, since what is then being tested is only the ability of a model (with tuned parameters) to remember what was used to train it. A possible approach to getting the most out of a finite pool of data would be to adapt cross validation methods from machine learning, e.g., using randomly selected splits of the data to repeatedly calibrate with one split and test with the other. The difficulties of calibrating ABMs may limit the utility of this approach, however.

For any measure, some account must be made of the *degree* of accuracy required of the model. It may be that the model is intended to fit data to some precisely specifiable degree of tolerance. Perhaps more common is a requirement that some qualitative aspects of the data be matched, what in economics are called **stylized facts** [Kaldor, 1961, p. 178] and Grimm et al. [2005] call **patterns**. An example from economics would be the positive dependency between support for public education and GDP per capita (e.g., Barro, 2002). This is well established for industrial societies, so a model of modern economies allowing for exceptions would be reasonable, but a model showing no such tendencies would not.

3. **Other.** Not every technique cleanly falls into data or expert opinion, but has aspects of both.

   a. **Convergent validity:** how similar are the model structure, discretization and parameterization to other models that are intended to describe a similar system? Where divergence between models in their assumptions or methods suggests a divergence in results, then we have **discriminant validity**.

   The judgment of the similarity (and relevance) of other models and their features will have to be made by experts, but may well be made in part on the basis of statistical features of data generated by those models.

   b. **Visualization; traces; animation.** Different ways of visualizing the results of simulations may support expert judgments of convergent validity, sensitivity analyses, etc.

   c. **Fruitfulness.** As with the assessment of scientific theories themselves, the fruitfulness of a simulation, its successful adoption by other researchers in application to related problems, is an indirect measure of its validity. In particular, a model which is widely and successfully (re)applied in related problem areas cannot be an entirely wrong-headed model across these domains.

5 Validating a Model of Household Demography

We now briefly illustrate how the validation techniques discussed above might apply in a real ABM, using as a case study a model of household demographics, developed as a component of a larger epidemiological simulation. This model is relatively simple, exhibiting emergence of household and population-level dynamic patterns from individual-level demographic processes.

Households are an important focus of disease transmission with a special relevance for childhood diseases, with the probability of transmission known to be affected by family size and composition [Viboud et al., 2004]. Existing models typically assume a static household distribution. However, this is inappropriate when dealing with long-term patterns of disease and immunity for endemic diseases like measles or pertussis [Glass et al., 2011], during which dramatic shifts in underlying demographic rates may occur. However, accounting for the variety of household types and the transitions between them in a mathematical model would be extremely difficult; hence our ABM.

The primary requirements of our model were that it capture the composition and dynamics of households containing children in a plausible fashion over extended periods of demographic change, and that it be amenable to calibration using data from a variety of different developed and developing countries, allowing for international comparison.

Our model represents a population of individuals, defined by their ages, sexes and the households to which they belong. At each time step, depending on their current attributes, individuals can experience one or more of the following demographic events: death, birth of a child, leaving their family home, forming or breaking a couple with another individual.

For some parameters of our model, such as mortality and fertility, age and sex specific rates were directly available [Australian Bureau of Statistics, 2010a,b]. However, for other parameters, such as the probability of leaving home, and the formation and separation of couples, data were not readily available. We adopted relatively simple rules for estimating the probabilities of these events occurring, which we subsequently adjusted by calibrating simulation performance against the data that was available. For example, to determine parameters for couple formation we tested our model's output against survey data on the percentage of people at particular ages who had never been in a couple [de Vaus, 2004]. This process involved adjusting the age at which an individual becomes eligible for forming a couple with another individual as well as the probability of an eligible individual forming a couple. A similar procedure was used to calibrate parameters corresponding to couple dissolution.

Having calibrated our model using statistics concerning individual-level events, our validation exercise focused primarily on population structure and household dynamics. The quantity of data against which we could validate varied according to country and year, so a broader approach than just data comparison was required. Space precludes a complete description of our validation methods and results (a paper on this is in preparation); instead we describe how each of the categories in Section 4 could be applied to our model. Note that some of the validation processes were more straightforward than others, and, in general, any one validation process may be more or less relevant depending on the particular model.

1. **Expert opinion (prior).**

   a. **Face validity.** To some extent we are all familiar with the varied dynamics of population and households and our own intuitions provided a first point of contact for face validity. The field of demography (via both expert researchers and literature) provided more specialized perspectives on what constitutes a model that looks 'right'. One important point is that experts from different disciplines may judge the same model differently. This validation process therefore provided an opportunity to negotiate an appropriate set of criteria for further validation, as well as to identify 'exemptions' — aspects of model behavior that may be unrealistic, but are deemed unimportant in the context of the research question. For example, our model does not currently allow for the existence of 'group households' (e.g., student share households); however, as these types of household typically do not contain young children (the focus of our research question), this was considered an acceptable omission. In the context of a different research question (e.g., the epidemiology of sexually transmitted diseases), this design choice may render the model invalid.

b. **Content validity.** As mentioned above, engagement with domain experts and literature provided the check on the completeness (or reasoned omission) of factors and relationships in our model. Particularly helpful were documents such as the Australian Institute of Families report [de Vaus, 2004], which aggregated and contextualized census and survey data on households under chapter headings that matched the types of individual life transitions we wanted to capture in our model (e.g., chapter titles include "Marriage and remarriage", "Transition of young people to adulthood" and "Lone parent families").

c. **Case analysis.** During the development and verification of our model we used Australian data collected in the last decade. Despite keeping our calibration (individual level) and validation (household level) data sets separate, we were aware of the possibility that we could consciously or unconsciously be 'designing' our model to reproduce a very specific pattern of behavior. To guard against this, subsequent to final development, we validated model behavior against two new cases, using previously unused data sets: historical Australian data from 1921 and Zambian data from 2000. Both of these populations differed from the modern Australian population data along several dimensions. For example, the average household size in Australia in 1921 was 4.3 individuals, as compared with 2.6 in 2000. The success of our model in passing validation tests on this data, without requiring new adjustments to the underlying mechanics, strengthened our confidence in the general model.

d. **Internal validity.** We took two approaches to assessing the internal validity of our model. First, we re-collected output data on distributions of the individual events whose probabilities we had calibrated. As calibration was performed on individual model components, comparing these output distributions against the calibration data provided a straightforward way of checking that interactions between components were not producing any unexpected side-effects in the combined model. Our second approach was to conduct a sensitivity analysis on the input parameters governing household formation and dissolution (i.e., leaving home and the formation and separation of couples). Compared with the easily available mortality and fertility rates, these parameters required more indirect estimation from available data. Therefore, assessing the sensitivity of our model output to these parameters provided an indication of how critical these values are and how successful our estimation had been.

2. **Data (likelihood).**

a. **Predictive validity.** The general principle we adopted in separating calibration and validation data was to calibrate the probabilities of individual events (birth, death, couple formation, etc.) and validate against higher-level properties of households. Data available for validation included the distribution of household sizes, distributions of household types occupied by individuals of given ages (couple households with/without children, lone person households, etc.), and household transition matrices, mapping the proportion of individuals in a household of type X who had been in a household of type Y at some point in the past [Wilkins et al., 2011]. Each of these constituted a set of data that was clearly distinct from our calibration data, against which model output could be compared in a quantitative fashion.

6 Conclusion

Our data-directed and expert validation efforts have shown that the demographic model is doing a reasonable job of recreating long-term demographic patterns in our target population (currently Australia), supporting our planned use of it as a platform for developing epidemiological simulations.

The simple Bayesian message we would like to finish with is that a validation process that concentrates on expert consensus to the exclusion of collecting statistics from data, or, equally, one which tests against data but ignores expert opinion, is incomplete. It is only by combining prior probabilities with likelihoods that we obtain a balanced picture of the empirical merits of a model.

Acknowledgements. This work was in part support by ARC grant DP110101758. We would like to thank our project partners Jodie McVernon, James McCaw, Kathryn Glass and Emma McBryde for their support. Alessio Moneta, Roman Frigg, Jay Pitchforth and referees also provided helpful comments.

References

Auchincloss, A.H., Diez Roux, A.V.: A new tool for epidemiology: The usefulness of dynamic-agent models in understanding place effects on health. American Journal of Epidemiology (1), 1–8 (2008)

Australian Bureau of Statistics, Births, Australia, 2009, cat. no. 3301.0 (2010a) (viewed August 18, 2011)

Australian Bureau of Statistics, Life Tables, Australia, 2007–2009, cat. no. 3302.0.55.001 (2010b) (viewed August 18, 2011)

Barro, R.J.: Education as a determinant of economic growth. In: Lazear, E.P. (ed.) Education in the Twenty-First Century, pp. 9–24. The Hoover Institution, Palo Alto (2002)

de Vaus, D.: Diversity and change in Australian families: Statistical profiles. Australian Institute of Family Studies, Melbourne (2004)

Edmonds, B., Moss, S.: From KISS to KIDS – An 'Anti-simplistic' Modelling Approach. In: Davidsson, P., Logan, B., Takadama, K. (eds.) MABS 2004. LNCS (LNAI), vol. 3415, pp. 130–144. Springer, Heidelberg (2005)

Feyerabend, P.: Against Method. Verso (1975)

Franklin, A.: The Neglect of Experiment. Cambridge University (1986)

Friedman, N., Koller, D.: Being Bayesian about network structure: A Bayesian approach to structure discovery in Bayesian networks. Machine Learning 50, 95–125 (2003)

Glass, K., McCaw, J., McVernon, J.: Incorporating population dynamics into household models of infectious disease transmission. Epidemics 3, 152–158 (2011)

Glimcher, P.W.: Decisions, Uncertainty, and the Brain: The Science of Neuroeconomics. MIT Press (2004)

Good, I.J.: Good Thinking: The Foundations of Probability and its Applications. University of Minnesota (1983)

Grimm, V., Railsback, S.: Individual-based Modeling and Ecology. Princeton University Press, Princeton (2005)

Grimm, V., Revilla, E., Berger, U., Jeltsch, F., Mooij, W.M., Railsback, S.F., Thulke, H., Weiner, J., Wiegand, T., DeAngelis, D.L.: Pattern-oriented modeling of agent-based complex systems: Lessons from ecology. Science 310, 987–991 (2005)

Howson, C., Urbach, P.: Scientific Reasoning: The Bayesian Approach, 2nd edn. Open Court (1993)

Kaldor, N.: Capital Accumulation and Economic Growth. Macmillan (1961)

Klein, E.E., Herskovitz, P.J.: Philosophical foundations of computer simulation validation. Simulation & Gaming 36, 303–329 (2005)

Kleindorfer, G.B., O'Neill, L., Ganeshan, R.: Validation in simulation: Various positions in the philosophy of science. Management Science, 1087–1099 (1998)

Korb, K.B.: A Pragmatic Bayesian Platform for Automating Scientific Induction. Ph. D. thesis, Indiana University (1992)

Korb, K.B.: Bayesian informal logic and fallacy. Informal Logic 24 (2004)

Korb, K.B., Nicholson, A.: Bayesian Artificial Intelligence, 2nd edn. CRC Press, Boca Raton (2010)

Kuhn, T.: The Structure of Scientific Revolutions. University of Chicago, Chicago (1962)

Lakatos, I.: Criticism and the Growth of Knowledge. Cambridge University (1970)

Mascaro, S., Korb, K.B., Nicholson, A.E., Woodberry, O.: Evolving Ethics: The New Science of Good and Evil. Imprint Academic (2010)

McLaughlin, B., Bennett, K.: Supervenience. In: Zalta, E.N. (ed.) The Stanford Encyclopedia of Philosophy (Winter 2011)

Pearl, J.: Probabilistic Reasoning in Intelligent Systems. Morgan Kaufmann (1988)

Pitchforth, J., Mengersen, K.: A proposed validation framework for expert elicited Bayesian networks. Decision Support Systems and Electronic Commerce (submitted, 2012)

Popper, K.: The Logic of Scientific Discovery. Translation of Logik der Forschung. Basic Books, New York (1934)

Reichenbach, H.: The the pragmatic justification of induction. In: Feigl, H., Sellars, W. (eds.) Readings in Philosophical Analysis, pp. 305–327. Appleton-Century-Crofts, New York (1949)

Sargent, R.G.: Verification and validation of simulation models. In: Proceedings of the 2010 Winter Simulation Conference (WSC), pp. 166–183 (2010)

Tesfatsion, L., Judd, K.L. (eds.): Handbook of Computational Economics, Volume II: Agent-Based Computational Economics. Elsevier, Amsterdam (2006)

Viboud, C., Boëlle, P.Y., Cauchemez, S., Lavenu, A., Valleron, A.J., Flahault, A., Carrat, F.: Risk factors of influenza transmission in households. British Journal of General Practice 54, 684–689 (2004)

Wallace, C.S.: Statistical and Inductive Inference by Minimum Message Length. Springer (2005)

Wilkins, R., Warren, D., Hahn, M., Houng, B.: Families, Incomes and Jobs, Volume 6: A Statistical Report on Waves 1 to 8 of the Household, Income and Labour Dynamics in Australia Survey. Melbourne Institute of Applied Economic and Social Research, Melbourne (2011)

Windrum, P., Fagiolo, G., Moneta, A.: Empirical validation of agent-based models: Alternatives and prospects. Journal of Artificial Societies and Social Simulation 10 (2007)

A Framework for Modeling and Simulation of the Artificial

Scott A. Douglass[1] and Saurabh Mittal[2]

[1] Air Force Research Laboratory
WPAFB, OH, United Sates
[2] L-3 Communications
WPAFB, OH, United Sates

Abstract. *Artificial systems* that generate contingency-based teleological behaviors in real-time, are difficult to model. This chapter describes a modeling and simulation (M&S) framework designed specifically to reduce this difficulty. The described Knowledge-based Contingency-driven Generative Systems (KCGS) framework combines aspects of SES theory, DEVS-based general systems theory, net-centric heterogeneous simulation, knowledge engineering, cognitive modeling, and domain-specific language development using meta-modeling. The chapter outlines the theoretical and technical foundations of the KCGS framework as realized in the Cognitive Systems Specification Framework (CS2F), a subset of KCGS. Two executable models are described to illustrate how models of autonomous, goal-pursuing cognitive systems can be modeled and simulated in the framework. The technical content and agent descriptions in the chapter illustrate how the *M&S of the artificial* depends critically on *ontology*, *epistemology*, and *teleology* in the KCGS framework.

1 Introduction

This chapter describes the Cognitive Systems Specification Framework (CS2F), as a subset of Knowledge-based Contingency-driven Generative Systems (KCGS) framework; a modeling and simulation (M&S) framework designed to support the M&S of models, agents, and cognitive systems capable of autonomously designing their own behavior in real-time. The CS2F framework is based on advances made in System Entity Structure (SES) theory [43], the Discrete Event Systems (DEVS) Unified Process [18], and large scale cognitive modeling (LSCM) research initiative [5, 20, 21].

A. Tolk (Ed.): Ontology, Epistemology, & Teleology for Model. & Simulation, ISRL 44, pp. 271–317.
springerlink.com   © Springer-Verlag Berlin Heidelberg 2013

The chapter begins with a discussion of a modeling problem the framework is intended to solve. The problem boils down to the present difficulty of modeling and simulating autonomous cognitive systems. The CS2F framework is intended to decrease this difficulty. Section 2 describes *artificial systems* and the artificial phenomena they produce. This section argues that autonomous models and agents are difficult to specify because: (1) they produce artificial phenomena; phenomena that reflect contingencies, choice, and teleology; (2) current modeling frameworks lack comprehensive support for the formal specification of relationships between contingencies, choices, and goals. Section 3 presents the CS2F framework and discusses the modeling formalisms and net-centric simulation technologies that constitute it. This section describes the framework as a componentized environment in which artificial phenomena can be readily modeled and simulated. After describing the framework, the chapter illustrates how models of artificial phenomena are actually specified and executed. Section 4 describes two agents: one that learns to adjust its behavior to match the probability structure of the environment; and another that uses abduction, a type of inference that refines knowledge, to make sense of its situation. While these agents are simple to facilitate exposition, they clearly demonstrate how the CS2F framework is used to model and simulate artificial phenomena. In Section 5, the broader theoretical background and ambition of the KCGS framework are presented. In Section 6, the chapter finally argues that the framework's effectiveness can be traced to the integration of aspects of *ontology*, *epistemology*, and *teleology* into modeling and simulation.

1.1 The Problem

Cognitive scientists employing computational process modeling in their research consider cognitive activity to be a product of an open system that interacts with the environment [41]. This perspective has motivated many cognitive scientists, especially those in the information processing psychology and cognitivist research traditions, to study *cognitive architecture*, the structural and behavioral system properties underlying cognitive activity that remain constant across time and situation. The Adaptive Character of Thought-Rational (ACT-R) is a theory of human cognition in the form of a cognitive architecture [2]. While cognitive modeling frameworks such as ACT-R allow cognitive scientists to explain/predict activities and processes occurring within an invariant architecture, they say little about how the influences of situational contingencies in which cognition occurs should be formally captured and related to behavior.

Let a *program* be sequence of instructions that perform a task when executed. Let an *agent* be something that perceives and acts. Agents can act on the basis of: (1) contingencies; (2) built-in knowledge; or (3) a combination of contingencies and built-in knowledge. Let an *autonomous agent* be one that bases its actions on its contingencies. Let *acting rationally* be given beliefs, acting so as to achieve goals. Cognitive scientists are struggling to develop cognitive models that behave more like autonomous rationally acting agents than programs. To be autonomous, an agent must base its actions on the constraints and affordances of its situation. A key consequence of this dependence on situational factors is that the contingencies an autonomous agent acts in (factors outside the invariant architecture) play as important a role in determining its actions as its invariant architecture. Autonomous rational agents are difficult to develop because the broader system in which contingencies and cognitive activity mesh must be represented and processed by the agent.

1.2   The Solution

AFRL efforts to resolve the above problem have produced a Cognitive Systems Specification Framework (CS2F), as a subset of the proposed KCGS framework. CS2F combines modeling formalisms (or Domain Specific Languages) in which models of systems that produce artificial phenomena can be specified. The KCGS framework provides a metamodel-based computing infrastructure wherein the CS2F modeling formalisms can be formally anchored in DEVS component-based systems specifications and ultimately simulated. The following aspects of the KCGS framework execute in concert to computationally realize the modeling and simulation of artificial systems as realized in CS2F:

1. Domain specific languages (DSLs) allowing modelers to formally specify the structure of knowledge related to; the environment, agent behaviors, states, goals, and domain theories. These DSLs allow domain experts to provide collaborative input in a larger systems context wherein heterogeneous components and multiple implementation platforms are the norm.
2. DSLs that use hierarchy to manage complexity in systems consisting of a large number of entities. These DSLs capitalize on formal properties of DEVS related to closure under coupling and the formal systems specification of hierarchy.
3. DSLs that use domain abstractions to limit specification and computational complexity during the specification and simulation of artificial systems.
4. Model-to-model transformation technologies that formally transform model component specifications in DSLs into executable DEVS systems-models. These executable models combine modeled aspects of both agents and their environments.
5. Knowledge processing mechanisms that refine knowledge while the system is in operation. These processes occur during simulation and allow agents to generate effective action and learn.
6. Capabilities based on variable structure modeling that support structural change in the system in operation. These capabilities change an agent's behavioral repertoire so that it reflects the dynamism and contingencies of the environment.
7. Capabilities based on event-based modeling that inject new knowledge into an autonomous agent at runtime. These capabilities support the learning and adoption of new knowledge.

The CS2F framework is designed to allow modelers to combine state, goal, and domain knowledge in cognitive domain ontologies (CDO) and simulate artificial phenomena in DEVS. These abilities allow cognitive scientists to model and simulate cognition as an artificial phenomenon contingent on situational factors, guided by a library of cognitive capacities (or behavior repertoire) and runtime constraints that operate between the agent and its environment.

2 Artificial Systems

In a review of embodied cognition, Wilson [41] proposes that science should study systems that are essentially permanent in structure; systems whose behaviors are invariant across situations. Underlying Wilson's proposition is a notion that if science is to understand and predict systems and processes in nature it must focus on, and model, fundamental principles of organization and function, not the behaviors of systems in specific situations. Put another way, when the specification of a scientific model appeals to situation-specific factors, scientists cannot predict the behavior of the model when the situation changes. Wilson illustrates the importance of focusing on the invariant aspects of studied systems by pointing out how scientific understanding of hydrogen is based on fundamental understanding of atoms, not understanding of how hydrogen behaves in a large number of contexts. Wilson suggests that scientists working to understand how cognition is *situated* or *embodied* are straying from a preferred focus on system behaviors that are invariant across situations. Put another way, rather than studying cognitive activity in specific situations, cognitive scientists should study cognitive architecture, the invariant structural and behavioral system properties underlying cognitive activity that remain constant across time and situation.

2.1 Artificial Phenomena

The "Achilles' heel" of Wilson's proposition is the obvious fact that human behavior is quite different from the behavior of hydrogen. Simon [35] draws out this difference by contrasting natural phenomena with artificial phenomena. *Natural phenomena* (for example, the behavior of hydrogen) are based on necessity; the behaviors of systems producing natural phenomena are subservient to natural law. *Artificial phenomena* (for example, the goal-pursuing actions of a human) are based on contingency; the behaviors of systems producing artificial phenomena are improvisational and reflect choices and requirements. Cognitive scientists endeavoring to model autonomous models and agents should develop theories and methods that enable them to understand cognition as an artificial phenomenon profoundly influenced by situational factors and contingencies. This chapter illustrates how knowledge about situational factors and contingencies can be represented in ontologies and processed by teleological (goal-pursuing) agents in order to refine their knowledge and gain real-time behavioral autonomy.

2.2 State and Process Descriptions

Simon [35] argued that two representations of knowledge underlie artificial human behavior: *state description* knowledge and *process description* knowledge. Humans generate/design effective behaviors by posing and solving problems that link their goals to the actions they can take. They pose the problems by clarifying state descriptions of their goals and then solve the problems by discovering sequences of actions or processes that produce their goal states. We propose that for an agent to act autonomously, it must be able to convert state descriptions (representations of goals) into process descriptions (representations of actions). Specifically, an agent must be able to determine: (1) *what* actions are likely to achieve goals; and (2) *how* to perform these actions.

While it may be straight forward to specify how an agent is to perform actions using sequences of instructions (a program), it is extremely difficult to specify how an agent is to autonomously determine what it should do in its circumstances. This difficulty is partly due to the way answering the *what* question takes place *in situ*; in the confluence of situational constraints, current goals, perceived possible actions, cognitive limitations, preferences, etc. It is virtually impossible to specify all the ways contingencies shape actions in a model consisting of built-in rules. To develop autonomous agents that pursue goals in unpredictable environments, cognitive modelers need modeling formalisms and frameworks with which they can specify and execute models and agents that "soft-assemble" their actions. These formalisms and frameworks must allow modelers to separate the *what* and *how* concerns so that answers to the *what* question can be used to assemble sequences of instructions or rules that answer the *how* question *in situ*. This chapter describes formalisms and an execution framework meeting these requirements.

2.3 Modeling Artificial Systems

Modeled as intelligent artificial system, autonomous agents pursue goals while interacting with the environment and dealing with their contingencies. Such an autonomous agent must be able to situate itself in the environment, perceive the constraints and affordances of the moment, and relate contingencies to goals in order to take effective action. In order to design such an autonomous artificial system, the Modeling and Simulation discipline must be able to formally specify the structure of the domain of the agent. Unless we formally specify the entire agent/contingency environment, the behavior specification of the agent will be fragile in the face of events occurring in the environment not anticipated by pre-defined rules. This implies that we should model the *whole* environment and a *rich* behavior representation of such an agent situated in it. This certainly is computationally prohibitive and better methods of managing such information are being developed by the chapter authors.

3   CS2F Framework for Modeling and Simulating Artificial Systems

This section describes technologies and a framework for specifying, modeling and simulation of artificial systems. It addresses concepts like model interoperability, model transparency, domain specific languages, formal ontology representation, knowledge engineering, search mechanisms and their integration aspects. We begin this section by describing *meta-modeling* as a necessary aspect of model interoperability. Meta-models are abstract models of the domain of interest and result in a set of rules that define a "domain-model". Performing model transformations at the meta-modeling layers paves way for model integration and collaborative development. In the next subsection, we will discuss how meta-models are transformed to the DEVSML stack to make them executable. It should be noted that the models are not "executable" by design at the meta-modeling layer. They have to be transformed to a framework (DEVS in this case) that takes models and makes them executable (as a simulation). This separation of concerns is a central theme in DEVS based modeling and simulation that separates the modeling and simulation layers using a simulation relation.

3.1   Foundations of the CS2F Framework

3.1.1   Meta-modeling

A domain specific language (DSL) is a dedicated language for a specific problem domain. For example, HTML is a DSL for web pages, Verilog and VHDL are DSLs for hardware description. A DSL can be can have a textual, graphical, or a hybrid concrete syntax and is essentially a meta-model of allowable specifications. A DSL exploits abstractions so that the respective domain experts can specify their problem without paying much attention to the general purpose computational programming languages such as C, C++, Java, etc. which have their own learning curve. In our efforts, we employ the Generic Modeling Environment (GME) as a DSL development framework. GME is a highly configurable meta-modeling environment developed by the Institute for Software Integrated Systems (ISIS) at Vanderbilt University [13, 14, 28, 38]. The GME is essentially a tool for creating and refining domain-specific modeling and program synthesis environments. GME meta-models are specified using a graphical/textual notation resembling UML class diagrams.

GME has been used by the authors to develop the Cognitive Systems Specification Framework (CS2F), a composition of domain-specific languages tailored to the requirements of specifying models and agents that base goal-pursuing behaviors on contingencies. The DSLs currently composed in CS2F are:

CS2F/DM   A specification language based on the OWL [17] ontology standard used to specify an agent's propositional or declarative knowledge.

CS2F/CDO   A specification language based on SES theory used to specify models of domain knowledge combining aspects of agents and the situational factors or contingencies constraining their behavior.

CS2F/BM   A specification language based on behavior models (predicated nondeterministic finite state machines) used to specify an agent's behavioral repertoire.

These three CS2F specification languages have been composed into a single metamodel defining an integrated authoring environment in which a modeler can specify the declarative, domain, and procedural knowledge of an agent.

3.1.2 Representations of State, Goal and Domain Knowledge

System entity structure (SES) theory is a formal ontology specification framework that captures system aspects and their properties [43]. In the past, SESs were used in design and simulation environments to formally capture configurations of systems that achieve a common design objective [11, 31–33, 42, 44].

In the early 1990s, researchers working at the overlap of artificial intelligence and modeling and simulation began to design and implement environments that automated the process of design space exploration [31] to solve engineering problems [32]. SESs were used to represent system configuration alternatives in these environments. The SES was primarily used to specify the relations between these entities [33]. In addition to capturing aspects, entities, taxonomic relationships, variable values, and structural/configuration alternatives, these SES included information about how entities in the SES could be realized in the DEVS formalism and composed into an executable model. To systematically explore design spaces in these environments: (1) rule-based search processes were used to derive all valid pruned entity structure (PES) captured by the SES; (2) information based on entities and aspects in each PES was used to compose an executable model using DEVS components stored in a model repository; (3) each composed model was simulated; and (4) simulation results were analyzed in order to identify the most desirable design alternative. The Solutions set is determined by the pruning process on the SES and the optimal solution was determined by simulation of each of the designs in Solutions set.

Rather than being used to capture system alternatives to be explored through DEVS-based modeling and simulation, SES are currently being used to formally capture structural and relational information about domains. SES are being used to specify entire ontologies rather than just system configurations that solve engineering problems [15, 16, 43]. This current use exploits similarities between SES and general ontologies. Current research and modeling and simulation activities utilizing SES demonstrate that extraordinarily diverse domains can be formally captured and related to each other through formal structures such as domain ontologies, pragmatic frames, and overlapping pruned entity structures (PES).

The CS2F framework described in this chapter uses cognitive domain ontologies (CDOs), a theoretical extension of SES to represent *spaces of behavior* as if they were system configurations. Situational/agent properties, aspects and constraints can be formally captured in CDOs. CDOs are processed by an agent to determine *what* it should do. The framework constitutes a modeling architecture that explicitly supports the representation and processing of CDOs. This capability allows modelers to separate the *what* and *how* concerns and specify agents that generate process descriptions by using answers to the *what* question to identify and "soft-assemble" knowledge into contextually appropriate process descriptions.

3.1.3 DEVSML 2.0 and the DEVS Unified Process

Discrete Event System Specification (DEVS) [46] is a formalism which provides a means of specifying the components of a system in a discrete event simulation. The DEVS formalism consists of the model, the simulator and the experimental frame as shown in Fig. 1. The Model component represents an abstraction of the source system using the modeling relation. The simulator component executes the model in a computational environment and interfaces with the model using the simulation relation or the DEVS simulation protocol in the present case. The Experimental Frame facilitates the study of the source system by integrating design and analysis requirements into specific frames that support analyses of various situations the source system is subjected to.

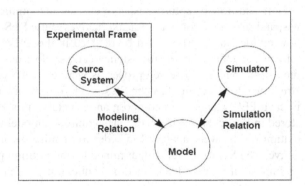

Fig. 1 DEVS Framework elements

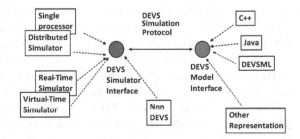

Fig. 2 Standardizing the Model and Simulator interfaces

While historically models have been closely linked to the platform (such as Java, C, C++) in which the simulator was written, recent developments in platform independent modeling and transparent simulators [25–27, 30] have allowed the development of both the models and simulators in disparate platform. To facilitate interoperability, integration and composability, a layered DEVS Modeling stack was proposed that executes on Service oriented Architecture (SOA) [24, 26]. Current efforts are focusing on a standardization process [39] wherein the simulation relation can be standardized for further interoperability [27, 45].

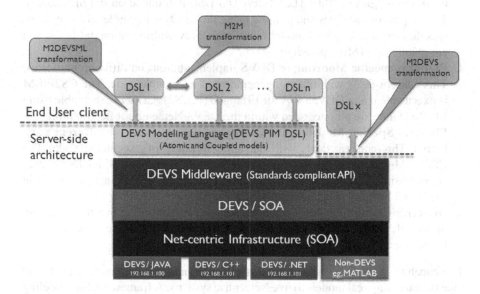

Fig. 3 DEVSML 2.0 stack enabling model and simulator interoperability

The latest version of this stack, shown in Fig. 3, was proposed as a part of Air Force Research Laboratory's Large Scale Cognitive Modeling (LSCM) initiative [5, 20, 21]. While the earlier version of the DEVSML stack was designed to provide XML interoperability and the netcentric transparent simulation to the DEVS models, the current version was designed to enhance scope and model interoperability [22]. Models specified in the new DEVSML 2.0 stack are specified in domain specific languages (DSLs) and then through transformations are taken to the DEVS framework. The idea of accommodating suites of DSLs at the top layer of the stack is a major addition in the DEVSML 2.0 stack.

The DEVSML stack has been an integral component of the larger DEVS Unified Process (DUNIP) [18, 23]. DUNIP is a universal process and is applicable to multiple domains. However, the understated objective of DUNIP is to incorporate discrete event formalism as the binding factor at all phases of this development process. The important concepts, the processes within DUNIP and how they relate to CS2F are listed below:

1. **Requirements and Behavior specifications using Domain Specific Languages (DSLs)**: We mentioned CS2F/BM, CS2F/DM, and CS2F/CDO as DSLs that are designed to support a very specific objective. Similarly, any DSL designed specifically for requirements specification is positioned here.
2. **Platform Independent modeling at lower levels of systems specification using DEVS DSL**: This step involves the development of M2DEVSML or M2DEVS transformations to yield DEVS and/or DEVSML models from CS2F/BM specifications.
3. **Model Structures at higher level of System resolution using Cognitive Domain Ontologies (CDO)**: The CS2F/CDO DSL is founded on the SES theory. This step allows analysis and pruning using the CDOs at higher levels of systems specification and employs model-based repository within the model integrated computing [38] (MIC) paradigm.
4. **Platform Specific Modeling** or DEVS implementations on different platforms: This concept deals with the autogeneration of executable code. The CS2F/BM is executable using DEVSJAVA, or Erlang/OTP. CS2F/CDO is executable using LISP and they are all integrated within the DEVS Netcentric infrastructure.
5. **Platform Specific Modeling i.e. DEVS implementations on different platforms**: This concept deals with the autogeneration of executable code. The CS2F/BM is executable using DEVSJAVA, or Erlang/OTP. CS2F/CDO is executable using LISP. These DSL execution capabilities are all integrated within the DEVS Netcentric infrastructure.
6. **Net-centric execution in a distributed setup**: This concept allows the execution of any DEVSML model in a Netcentric environment where the simulation can be executed in a local-centralized or a remote-distributed setting.

The capabilities defined above allow us to specify any kind of domain models and take the executing real models to live Netcentric systems. A framework for modeling and simulation of the artificial must have these basic capabilities.

3.2 Technical Description of the CS2F Framework

The CS2F framework consists of three components implemented as net-centric services: (1) soaDM, an associative memory based on the declarative memory system of ACT-R; (2) soaCDO, a domain ontology processing application based on a non-deterministic constraint solver; and (3) the DEVSML Stack, a DEVS-based agent execution framework. Models and agents simulated in the framework base their behavior on: declarative knowledge; cognitive domain ontologies; and behavior models. The CS2F DSLs and framework components used to represent and process these aspects of models and agents are summarized in Table 1.

Declarative, or factual knowledge in a propositional form, is maintained in soaDM. Net services provided by soaDM provide agents with an associative memory through which they can retain and retrieve knowledge. Domain knowledge, or cognitive domain ontologies (CDOs) capturing goals and behavioral design objectives, are maintained in soaCDO. Net services provided by soaCDO provide agents

Table 1 Framework DSLs and the net-centric components in which they are processed

DSL/Formalism	Representation Specialization	Framework Component
CS2F/DM	Declarative Knowledge	soaDM
CS2F/CDO	Domain Knowledge	soaCDO
CS2F/BM	Procedural Knowledge	DEVSML Stack

executing in the KCGS framework with an ability to choose *what* to do on the basis of contingencies. Behavior models are predicated non-deterministic finite state machines capturing procedural knowledge in sub-assemblies. Behavior models are maintained and executed in the DEVSML Stack. The DEVSML Stack interacts with the other framework components and realizes the low-level behavior of the agent. Fig. 4 shows the component diagram of CS2F and its Netcentric implementation.

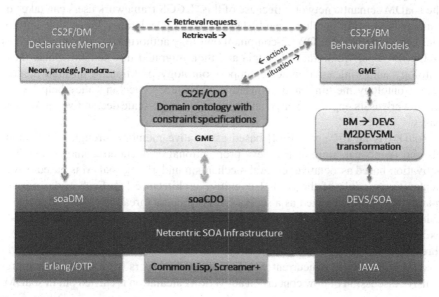

Fig. 4 SOA components in CS2F

3.2.1 CS2F/DM: OWL Ontologies Capturing Declarative Knowledge

Human memory is a part of a quintessential *artificial* system that learns and acts in the world. Human behavior is as flexible as it is because we know lots of things and can use what we know to craft contextually appropriate and effective actions in many different circumstances. Humans know a great deal and can quickly cull through all that they know in order to retrieve and apply the right knowledge given their circumstances. Behavioral flexibility is enabled by a memory system that: (1) provides access to vast amounts of knowledge; and (2) tunes this knowledge to match the information structure of the environment through learning. The ACT-R cognitive architecture [1, 2] includes an associative memory system providing these properties and capabilities. ACT-R's declarative memory system is based on a set of equations explaining the sub-symbolic calculation, learning and utilization of activations and associative strengths [1, 2]. soaDM, a net-centric component of the KCGS framework based on work described by Douglass and Myers [6], utilizes the equations underlying ACT-R's declarative memory system.

Declarative ontologies represent knowledge that models and agents can acquire through experience and retrieve when relevant. CS2F/DM, the DSL with which modelers specify declarative knowledge in soaDM, is based on the OWL ontology standard [17]. CS2F/DM declarative knowledge ontologies describe the classes, class properties, object properties, data properties, and instances constituting a domain. Declarative knowledge ontologies specified in CS2F/DM are translated into files that configure a semantic network in soaDM. Any consistent OWL-compliant ontology can be translated into CS2F/DM and subsequently be used to configure the soaDM semantic network. Because of this, KCGS framework users can take advantage of existing ontologies and RDF databases. Declarative knowledge ontologies can be authored in OWL2-compliant ontology authoring environments such as NeOn, Protégé, Wandora, or Ontopia and then migrated into soaDM. Since these ontology authoring environments support ontology partitioning through namespaces, ontology merging, and knowledge consistency checking, they help KCGS framework users engineer, verify, and understand large-scale declarative knowledge bases.

soaDM is an Erlang/OTP [4] based associative memory through which models and agents can store and retrieve propositional or declarative knowledge. The activation-based associative retrieval mechanism underlying soaDM is based on the declarative module of the ACT-R cognitive architecture [1]. Each node in a semantic network is realized as a separate OTP process thread in Erlang. Activation calculation spreads in soaDM semantic networks as messages are asynchronously exchanged between the process threads constituting their nodes. Since process threads in Erlang execute concurrently, activation-based associative retrieval in soaDM is massively concurrent. See Douglass and Myers [6] for a more comprehensive discussion of how concurrent activation calculation is carried out in soaDM.

3.2.2 CS2F/CDO: Cognitive Domain Ontologies Capturing Contingencies

To be capable of generating autonomous rational action, a model or agent must be able to transform *state descriptions* into *process descriptions*. Transformations of this sort link high-level goals (states) to low-level actions (processes). The vast majority of contemporary cognitive models are built up from productions, rules, or procedural descriptions that combine information about goals and actions. On the surface, this mixing of state and process description knowledge seems to be a natural way of combining the translation of a state description (goal) to a process description (set of actions). A problem with this approach surfaces when it is employed by a modeler specifying a large model that must act autonomously in a complex and dynamic environment: it's almost impossible to specify all the required procedural descriptions combining goals and actions over large spaces of environmental contingencies. In order to express a rich knowledge set that includes environment, contingencies, resources, possible actions and much more, we need a framework that allows us to represent knowledge in many facets or dimensions. The soaCDO is a net-centric component of the framework the uses CS2F/CDO to represent knowledge in such a way.

CS2F/CDO, the DSL in which domain models integrating knowledge related to goals, requirement, situational factors, and possible actions can be specified, is based on System Entity Structure (SES) theory [43]. Cognitive domain ontologies specified in CS2F/CDO are translated into constraint networks in soaCDO. The distinguishing feature between a CDO and an SES representation is the inclusion of constraint language in a CDO. While the SES theory lays the foundation of specifying constraints and how they operate, the CDO constraint language is a formal specification and is an integral part of domain ontology. CS2F/CDO has been developed within the GME, a meta-modeling environment in which domain-specific modeling languages and multi-paradigm modeling frameworks can be formally specified.

The most efficient way to describe CS2F/CDO is to describe its underlying meta-model. Fig. 5 shows the portion of the CS2F/CDO meta-model formally describing correct cognitive domain ontologies (CDOs). The entities, concepts, and relationships constituting the abstract syntax of a DSL are expressed in UML class diagrams in GME. Concepts and entities are represented as classes. Connections terminating with a solid diamond indicate containment relationships between classes. The

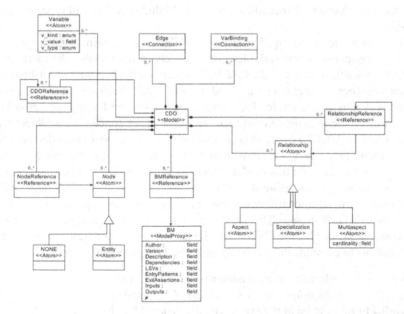

Fig. 5 GME class diagram specifying the portion of the CS2F/CDO meta-model related to valid entities, variables, and relationships in CDO

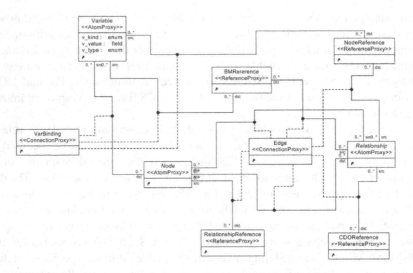

Fig. 6 GME class diagram specifying the portion of the CS2F/CDO meta-model related to valid connections between entities, variables, and relationships in CDOs

cardinalities of containment relationships are displayed at their source. Connections terminating with arrows indicate reference relationships. For example, a reference relationship in the Fig. 5 indicates that "NodeReference" entities are allowed to refer to "Node" entities. Triangles denote inheritance; in the figure below a triangle indicates that "Aspect", "Specialization", and "MultiAspect" are all types of "Relationship".

The meta-model in the Fig. 5 specifies that CDOs can contain: nodes/entities; relations, edges/connections; variables; and a variety of references. The meta-model in Fig. 6 shows the portion of the CS2F/CDO meta-model formally describing connections between these elements allowed in valid CDOs. Each allowable connection is represented as a dark circle. The source, destination, cardinality, and associated connection type constraints in the meta-model work in concert with Object Constraint Language (OCL) constraints (not shown) to enforce axioms underlying SES theory. Entity properties, containment and reference relationships, and constraints in the meta-model ensure that models specified in the CS2F/CDO DSL are "correct by construction" and therefore do not violate axioms critical in SES theory. GME meta-models can additionally define the concrete syntax or appearance of a DSL. Fig. 7 shows a GME-based CDO authoring environment presenting a graphical/textual concrete syntax for the CS2F/CDO DSL. The DSL's concrete syntax enables CDO authors to combine the following elements in a graphical workspace:

Entity	domain entity (concept) denoted by '<>'
Aspect	decomposition (is made up of) denoted by '|'
Specialization	can be of type (is a type of) denoted by '||'
Multi-Aspect	decomposition into similar type denoted by '|||'
Variable	variables attached to entities with ranges or values denoted by '~'

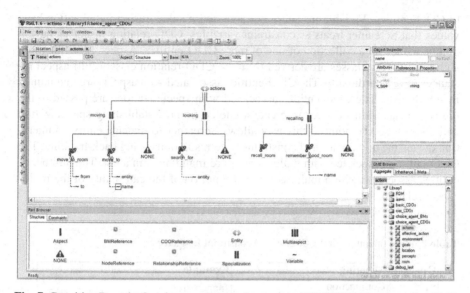

Fig. 7 Cognitive Domain Ontology under development in CS2F/CDO. Note how the GME interface explicitly supports the specification of CDOs

The concrete syntax of CS2F/CDO allows KCGS framework users to graphically specify CDOs containing entities, aspects, specializations, multi-aspects, attached variables, and domain-specific constraints. User actions and choices violating SES axioms during CDO specification are either not allowed by the CS2F/CDO meta-model or generate error messages. This real-time meta-model conformance checking process ensures that CDO are correct by construction.

soaCDO is written in Common Lisp [37]. Non-deterministic programming capabilities based on the Screamer [36] and Screamer+ [40] Common Lisp extensions are used by soaCDO. Screamer adds two basic mechanisms to Common Lisp: (1) a non-deterministic special form called *either* that takes zero or more lisp expressions as arguments; and a deterministic function called *fail* that takes no arguments. The *either* special form non-deterministically evaluates one of the expressions passed to it, returns the value of the evaluation, and establishes a choice point. The *fail* function triggers back-tracking to the most recent choice point. If un-evaluated expressions are encountered at the choice point, the next value is returned. If no additional expressions are encountered at the choice point, then back-tracking continues to another choice point. Screamer+ extends the functionality of either/fail and allows non-deterministic programming to take advantage of complex data types and Common Lisp Object System [10].

CDOs are computationally realized as structured sets of Common Lisp Object System objects. The root object of a CDO is an instance of a CDO-entity class. CDO-entity instances have a unique name, a collection of zero or more attached CDO-variables, and a collection of zero or more CDO-relations. CDO-variable instances have a unique name and a value. CDO variables can only be connected to CDO entities. Variable instances not assigned a value during initialization have values that Screamer treats as constraint variables. CDO-relation instances have a unique name and a collection of one of more CDO-entities. Aspect, specialization, and multi-aspect classes are derived from the CDO-relation class. Multi-aspect instances have a cardinality. The CDO-entities associated with aspects are maintained in simple lists. The CDO-entities associated with specializations are passed to the *either* special form in order to create a choice point. Establishing specialization entities as a choice point in this way allows Screamer to manage entity enumeration during the computation of constraint system solutions using back-tracking. The CDO-entities associated with multi-aspects are maintained in a list. The number of entities associated with a multi-aspect is a function of the cardinality of the multi-aspect relation.

Table 2 Basic operators in the CS2F/CDO constraint language

Operator	Meaning	Example
`and`	Conjunction	`(and p q)`
`or`	Disjunction	`(or p q)`
`not`	Negation	`(notq)`
`==>`	Implication	`(==> p q)`
`<==>`	Biconditional	`(<==> p q)`
`false`	Logical Falsity	`(and (not p) q)`
`true`	Logical Truth	`(or p (not p))`
`e@`	Entity located in CDO	`(e@ musical_performance style)`
`v@`	Variable attached to CDO entity	`(v@ (actions moving move_to) name)`
`equale`	Entities are equal	`(equale ensemble small-group)`
`equalv`	Variable has a value	`(equalv weight 105)`

`let`	var/val Binding	`(let ((p its-raining)`
		`    (q groun-gets-wet)))`
		`    (==> p q))`

The CDO pruning process is cast as a constraint-satisfaction problem (CSP) in soaCDO. Constraint variables correspond to CDO-variable values and the entities connected to relations. The domains of constrain variables corresponding to variable values are a function of variable type. CDO-variables can currently be: integers, floats, strings, lists, vectors, non-numeric enumerated sets, integer ranges, float ranges, and Boolean values. The domains of constraint variables related to a CDO-relation are the set of all entities connected to the relation. Constraints relate sub-sets of the constraint variables and specify the domain values variables are allowed to assume. CS2F/CDO constraints are specified in a language based on first order logic (FOL). Constraints can employ universal and existential quantifiers, implication, bi-directional implication, conjunction, disjunction, negation, and a comprehensive set of non-deterministic functions. Table 2 lists important basic operators that can appear in constraints.

Constraints expressed in well-formed statements are translated into implicative normal form (INF). This translation reduces complex FOL-based statements to disjunctions of implications that are then mapped into Screamer. Appendix 1 lists important complex operators that can appear in constraints that have been translated into INF. The translation into INF produces the following basic implications:

1. Implications consisting of *conventional antecedents* and *conventional consequents*. These are mapped into Screamer as conditional constraints that use *assert!* to propagate constraints in CDOs. For example, the INF implication (==> p q) would be mapped into as (ifv p (assert! q))
2. Implications with *conventional antecedents* and *consequents equivalent to logical false*. These are mapped into conditional constraints that use fail to trigger back-tracking. For example, the INF implication (==> p false) would be mapped into Screamer as (ifv p (fail))
3. Implications with *antecedents equivalent to logical true* and *conventional consequents*. These are mapped into unconditional assertions. For example, the INF implication (==> true q) would be mapped into Screamer as (assert! q)

CDO pruning starts with a process that relates situational factors to corresponding entities in a CDO through the use of the *assert!* operator. These assertions combine with domain-specific constraints in a subsequent search process that finds CSP solutions using a non-deterministic search with chronological back-tracking. The search for CSP solution in soaCDO can: (1) find one solution: (2) find the ith solution: (3) find the "best" solution; (4) find all solutions; or (5) find a solution, present it to the user/agent, and then ask if another solution is required. The ability to obtain solutions from soaCDO while back-tracking over choice points means that CDO with significant combinatory complexity can still be effectively processed.

The example CDO, shown in Fig. 8, is based on an example System Entity Structure discussed in [43]. The CDO represents a set of *musical_performance* entities. Each *musical_performance* has *style* and *ensemble* characteristics; each of which is a specialization. A *musical_performance* can therefore have a style of *symphonic*, *folk*, or *jazz*. A *musical_performance* can also therefore have an ensemble of *orchestra*, *small_group*, or *soloist*. With no additional CS2F/CDO constraints, processing of this CDO in soaCDO would result in 9 CSP solutions generated by crossing all 3 styles with all 3 ensembles. Some of these solutions are clearly implausible. For example, "symphonic soloist" performances are obviously impossible. Implausible entities such as these can be removed from the set of *musical_performance* entities allowed by the CDO with CS2F/CDO constraints.

As previously mentioned, the CS2F/CDO DSL includes a powerful constraint language that can be used to incorporate domain-specific constraints into CDOs. The translation of these constraints into INF in soaCDO allows a rich constraint language based on FOL to be seamlessly integrated into a CDO processing infrastructure built upon non-deterministic search and chronological backtracking. Table 3 illustrates how CS2F/CDO constraints refining the entity relations in the *musical_performance* CDO translate into INF. Constraints are defined in the CS2F/CDO constrain language using a *define-constraint* macro. The first argument to

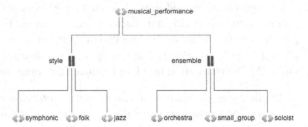

Fig. 8 CDO specifying a space of possible musical performances

define-constraint is a unique name to be assigned to the constraint. Assigning names to constraints allows KCGS framework users to simplify interactions with soaCDO. The second argument to *define-constraint* is the scope of the constraint. The scope of a constraint is the CDO entity that is to be treated at the root of all entity references in the constraint.

Table 3 Example constraints that refine the possible space of musical performance

Constraint	Implicative Normal Form

`(define-constraint m1`
`  :musical-performance`
`  (==> (equale (e@ style)`
`            symphonic)`
`       (equale (e@ ensemble)`
`            orchestra)))`
	`(orv`
	`  (ifv (equale (e@ style) symphonic)`
	`       (assert!`
	`         (equale (e@ ensemble)`
	`              orchestra))))`

`(define-constraint m2`
`  :musical-performance`
`  (==> (equale (e@ style) folk)`
`       (or (equale (e@ ensemble)`
`                small-group)`
`           (equale (e@ ensemble)`
`                soloist))))`
	`(orv`
	`  (ifv (equale (e@ style) folk)`
	`       (assert!`
	`         (orv (equale (e@ ensemble)`
	`                  soloist)`
	`              (equale (e@ ensemble)`
	`                   small-group)))))`

`(define-constraint m3`
`  :musical-performance`
`  (==> (equale (e@ style) jazz)`
`       (or (equale (e@ ensemble)`
`                small-group)`
`           (equale (e@ ensemble)`
`                orchestra))))`
	`(orv`
	`  (ifv (equale (e@ style) jazz)`
	`       (assert!`
	`         (orv (equale (e@ ensemble)`
	`                  orchestra)`
	`              (equale (e@ ensemble)`
	`                   small-group)))))`

In Table 3, the constraint *m1* is defined with a scope of *musical-performance* (the root entity in the example CDO shown in Fig. 8). Basing *m1* on this scope allows for the *style* and *ensemble* specializations to be clearly related in a conditional constraint requiring that when the style of the musical-performance is symphonic the ensemble must be orchestra. To ensure that constraints are as computationally efficient as possible, constraint authors should define the scopes of their constraints so that the CSP solution process can "push" constraints as far into the sub-structure of CDOs as possible. The last argument to *define-constraint* is an expression specifying entity/value requirements and variable assignments in the indicated scope. The *m2* and *m3* constraints in Table 3 provide additional domain-specific constraints that refine the domain knowledge captured in the *musical_performance* CDO. An additional constraint limiting the nature of musical performances is provided in Appendix 2. The *m4* constraint in Appendix 2 demonstrates how the transformation to INF allows constraint authors to specify groups of implications in a single constraint. Note how the negations in the consequents of *m4* translate into failure assertions in the implicative normal form. During the CSP solution process in soaCDO, these failure assertions trigger: (1) the elimination of CDO entity/variable assignments; and (2) chronological backtracking.

Table 4 Examples showing how CS2F/CDO constraints impact CSP in soaCDO

Constraints	style	ensemble
none	symphonic	orchestra
	symphonic	small-group
	symphonic	soloist
	folk	orchestra
	folk	small-group
	folk	soloist
	jazz	orchestra
	jazz	small-group
	jazz	soloist
m1, m2, m3	symphonic	orchestra
	folk	small-group
	folk	soloist
	jazz	orchestra
	jazz	small-group
m4	symphonic	orchestra
	folk	small-group
	folk	soloist
	jazz	orchestra
	jazz	small-group

Table 4 lists CSP solutions found by soaCDO under conditions when: (1) no additional domain-constraints were allowed to impact constraint propagation; (2) the simple *m1*, *m2*, and *m3* constraints are allowed to impact constraint propagation; and (3) the complex *m4* constraint defined in Appendix 2 is allowed to impact constraint propagation. Close inspection of the *m4* constraint reveals that it predominately impacts the constraint propagation process through fail-based backtracking. For example, when the ensemble constraint variable is bound to orchestra and the style constraint variable is bound to folk, an assertion of failure immediately eliminates the solution and initiates backtracking.

soaCDO is a Common Lisp based service through which models and agents can represent and process cognitive domain ontologies formally capturing the entities, constraints, and relationships constituting the requirements of the tasks they are performing. soaCDO translates cognitive domain ontologies specified CS2F/CDO into entity/relation networks that are processed with a non-deterministic constraint solver. The constraint-based search/pruning mechanism functions as a type of *cognitive control* allowing models and agents to match their goals to possible actions in such a way that its goals are achieved despite the vagaries of its situation. Cognitive domain ontologies represent knowledge that models and agents are able to process in order to determine *what* they should do. Executing in real time, this mechanism allows models and agents to generate behavior *in situ*.

3.2.3   CS2F/BM: Behavior Models Capturing Behavioral Sub-assemblies

In state-of-the-art cognitive modeling frameworks such as such as ACT-R [2], EPIC [3], and Soar [29], procedural knowledge is specified in productions or rules. Each production is essentially an association between antecedent context requirements and consequent actions. During model simulation, productions whose context requirements are met form a conflict set. Utility calculations or preference are typically used to select which production in the conflict set is allowed to exercise its consequent actions. Unless context is embellished with persistent information, individual productions are unaware of productions that precede or follow them during model simulation. This makes it very difficult to model complex behaviors based on sequences of productions. Modeling frameworks lacking a representation of behavior *above the production* require their users to carefully embellish context with state information if their models depend on behaviors based on sequences of productions. Behaviors based on sequences of productions also must be shielded from interruption. Failure to shield sequences of productions underlying complex behaviors frequently leads to model brittleness in complex dynamic environments.

In the KCGS framework, procedural knowledge is represented in CS2F/BM behavior models; formal structures that allow a modeler to represent behavior above the level of the production [5, 20]. Behavior models can be stored in repositories and used in different contexts. CS2F/BM allows a cognitive modeler to build models and agents from sub-assemblies (behavior models) that conceal complexity rather than large numbers of primitives (productions) that expose complexity. Behavior models are computationally realized as predicated finite state machines. Transitions in behavior models are functionally equivalent to productions; they have pattern-based guards that represent context requirements and side-effects that represent consequent actions. Transition pattern guards are compared to a set of events/facts maintained in a working memory. Behavior models are specified in CS2F/BM, a DSL developed and delivered in the Generic Modeling Environment (GME).

During model execution in the KCGS framework, an agent's behavior is determined by the set of behavior models currently in its *behavior repertoire*. While transition activity in behavior models is typically localized (transitions and generated actions are concurrent across behavior models), it is possible for them to interact or synchronize through the exchange of events or messages. This allows behavior models to be organized into hierarchies. Discussions of behavior models and their execution can be found in [5, 20].

Fig. 9 presents an example behavior model (BM) in a hybrid graphical/textual concrete syntax. The BM allows an agent to attempt to retrieve a room from soaDM, the associative memory component of the KCGS framework. Transitions in the BM are labeled with brief comments explaining or documenting the purpose of each

Fig. 9 Graphical representation of a behavior model in CS2F/BM

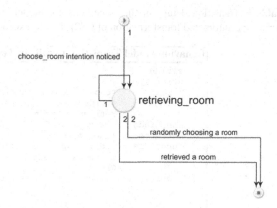

transition. The state in the BM has been assigned a name that also explains or documents the behavior captured by the BM.

While specifications in the graphical/textual concrete syntax effectively summarize the transitions and state changes underlying a BM, the formal details of the BM remain hidden. The formal details of states and transitions can be specified and edited in GME by selecting a state or transition in an editor and entering attributes in a set of text entry cells. This process is illustrated in [5, 20]. The automated model-to-model translation process that semantically anchors BMs specified in CS2F/BM produces a text-only intermediate description of each behavior model. An example of this textual CS2F/BM form is provided in Table 5.

As illustrated in Table 5, BM transitions can have the following attributes (priority, src, and dst attributes are required):

priority	resolve conflict when more than one transition is possible
label	a description of the function/purpose of a transition.
src	the state from which a transition originates.
dst	the state to which a transition leads.
pre_binds	"name=value" statements used to bind and compare locally scoped variables (LSVs).
patterns	predicate/event constraints that must be met.
functions	execute calculations involving LSVs and context pattern elements.
assertions	predicates/events added to working memory after a transition.
post_bindings	name/value pairs that overwrite LSVs maintained by a BM.

Predicates/events are represented in transitions as tuples delimited by curly-braces. For example, "{choose_room}" in the patterns of the first transition in Table 5 is a predicate representing an agent intention. In this transition, pre_binds and functions are used to assemble the sub-parts of an assertion that executes a retrieval through soaDM. The second transitions in Table 5 specifies how the sub-parts of a *room_chosen* assertion are to be assembled from properties of a successfully retrieved set of facts about a destination/room.

Table 5 Transition details of the same behavior model specified in a text form generated during the automated transformation of CS2F/BM to executable DEVSML

```
Behavior Model Transition Details in Textual CS2F/BM
transition {
       priority     1
       label        "choose_room intention noticed"
       src          startstate
       dst          retrieving_room
       pre_binds    w=W,context=C
       patterns     {choose_room}
       functions    Endo=[{type, destination}],
                    Exo=expand_context(C, W),
                    Cs=[]
       assertions   {execute_retrieval, Cs, Endo, Exo}
}
...
transition {
       priority     2
       label        "retrieved a room"
       src          retrieving_room
       dst          stopstate
       patterns     {retrieval_success, C, _},
                    {type, C, destination},
                    {name, C, CN}
       assertions   {room_chosen, C, CN}
}
```

In the previous section we saw how a DSL such as CS2F/BM can specify behaviors similar to those produced by sets of ACT-R productions. The approach proposed in this section takes the CS2F/BM meta-model in its entirety. The meta-model is semantically anchored in DEVS, which provides solutions to interoperability, extensibility, composability and scalability. CS2F/BM is a recast of our earlier described Research Modeling Language (RML) and detailed transformation is available in [5, 20]. The next subsection provides an overview of the methodology.

From structure perspective, any DEVS system is made up of three elements, the model components (atomic or coupled), the messages that flow between them, and the couplings that communicate these messages between components [46]. Both the atomic and coupled DEVS components transmit and receive messages. However, the capacity to interpret the message and use it to express the behavior is solely the characteristic of a DEVS atomic component. A new message originates exclusively within an atomic component per its behavior specification and is then placed at the output interface of the atomic component. The behavior of an atomic component is a function of superposition of two behaviors i.e. when an external message is received and when it is not. In order to specify the behavior, a state space is specified and the transitions between these states are defined with respect to an 'event' abstraction.

Describing the richness of DEVS atomic behavior is outside the scope of this paper. We will consider a subset of DEVS formalism known as Finite Deterministic DEVS (FDDEVS) [9]. FDDEVS implemented in the DEVSML 2.0 stack is called the DEVS modeling language [22] that abstracts the DEVS formalism. An automated transformation process using EBNF and Xtext Eclipse Modeling Framework

(EMF) is formally specified to preserve the true DEVS semantics. The platform independent DEVS modeling language, as illustrated in Fig. 3 is semantically anchored to the DEVS M&S framework through a middleware.

The notion of 'state' in DEVS is associated with occurrence of an 'event'. Now, looking at each of the transitions in Fig. 9, we find that each transition although specifies the source *src* and the destination *dst* state, has more going on inside it. For example, the *pre_binds*, *post_binds*, *patterns* and *assertions* elements. As per the CS2F/BM semantics, the model will expect the pre_bind variables to match up with the patterns, and if matched, will perform the post_bindings and assertions and will then finally move to the dst state. In DEVS semantics, this operation can be considered as two events, and consequently, two states. The first state being, beginOperation, wherein evaluation is being made per input patterns and the second state being, dst itself. On completion of first state, assertions (output) is being sent and the model then moves to dst state. While there is no problem in the CS2F/BM semantics, the DEVS formalism requires the specification of *output function* which is associated with a specific state. If we preserve the CS2F/BM state set then the point where two events happen together, ie. Incoming patterns and assertions, breaks the notion of discrete event in DEVS formalism. The DEVS semantics very clearly expresses this in the output function. Using the system homomorphism concepts [46] as shown in Fig. 10, by introducing a Zero time state, we not only preserve the CS2F/BM semantics but also transform the state machine into a DEVS state machine. Table 6 lists the mapping of CS2F/BM semantics into FDDEVS elements.

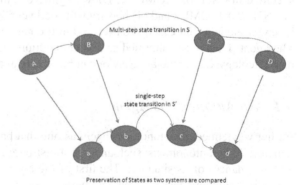

Fig. 10 Preservation of States as two systems are compared and M2DEVS transformation is performed

We have provided an overview on the execution of M2DEVSML transformation from one CS2F/BM DSL into another DSL (DEVSML) that is semantically anchored in DEVS. More details about the atomic behavior, coupling and structure of the transformed CS2F/BM model into DEVS atomic and coupled models can be seen in [20].

Table 6 Semantic mapping from CS2F/BM to FDDEVS

CS2F/BM Elements		FDDEVS Elements
Globals		
	states	S
Transitions		1. If patterns > 0, then each tuple in patterns is an incoming external message and be addressed in ext. The src state must transition to beginDst state in zero time.
		2. If *assertions* > 0 then each tuple is an outgoing message and be addressed in in state beginDst.
		3. every beginDst state should internally transition to dst in 0*ms*. Every dst must match the *ta* = 50*ms* of CS2F/BM state and once elapsed should internally transition to passive.
	src	s in S
	dst	s in S
	patterns	X
	assertions	Y

4 Modeling in the CS2F Framework

Two agents will be described in the following section. Discussions of how these agents are specified and executed in the KCGS framework illustrate: (1) how declarative knowledge is specified in CS2F/DM (Protégé); (2) how behavior models are specified in CS2F/BM (GME); (3) how cognitive domain ontologies are specified in CS2F/CDO (GME); and (4) how transformed versions of declarative ontologies, behavior models, and CDOs are executed in the net-centric simulation framework. The agents have been simplified so that connections between ontology, epistemology, teleology, and artificial behavior can be clearly and effectively made.

4.1 An Autonomous Agent

Earlier we defined an autonomous agent as one that bases its actions on its contingencies. To be autonomous, such an agent must base its actions on the constraints and affordances of its situation. The first of the agents that will be discussed navigates in a synthetic task environment while searching for a reward item. Through instance-based learning enabled by an associative memory [8], this agent adapts is behavior over time in order to match the information structure of the environment. This agent represents *what* it should do in a CDO, *how* it should behave in a set of BMs, and facts it knows and learns in a declarative memory. Rather than basing its behavior on pre-specified rules, the autonomous agent: (1) assigns aspects of its contingencies to entities and variables in a CDO: (2) processes the CDO using a constraint-satisfaction process in order to determine *what* it should do: (3) determines *how* it should behave by determining which entities in a CSP solution correspond to BMs: and then (4) effectively acts by incorporating these BMs into its behavioral repertoire.

The agent acts in a virtual environment consisting of four rooms. A centrally located room is known as the *home_room*. The other rooms are known as *room1*, *room2*, and *room3*. When the agent moves to a *trigger_plate* in the home room, a reward (small item) randomly appears in *room1*, *room2*, or *room3*. After triggering this event, the agent chooses a room (by retrieving a memory corresponding to it from declarative memory), moves to the room, and then searches the room for the reward. If the reward item is visible in the chosen room, the agent: (1) enters the room: (2) collects the reward; (3) strengthens the activation of the room in declarative memory by making a mental note of the room; and (4) navigates back to the *home_room*. If the reward item is not visible in the chosen room, the agent does nothing. Having collected the reward or not, the agent then returns to the *trigger_plate*.

Initially, the agent has no preference for *room1*, *room2*, or *room3*; the three pieces of declarative knowledge in memory corresponding to the rooms all have the same level of activation. If the appearance of the small item is truly random across the three rooms, then the agent will effectively never come to prefer one room over the others. If the small item is allowed to appear with different probabilities across the rooms, then finding and collecting the item leads to the agent preferring one room over the others. With time and trial repetition, the agent's room preferences adapt to match the reward probabilities of the rooms as mental notes about rooms lead to activation changes in relevant pieces of declarative knowledge.

4.1.1   CS2F/DM – The Agent's Declarative Knowledge

The autonomous agent requires little declarative knowledge to be effective. Appendix 3 summarizes the initial configuration of the agent's declarative memory (maintained by soaDM). Declarative information is represented as nodes in the soaDM semantic network. Edges in the semantic network represent relations between nodes and other nodes (object properties) or numbers/strings (data properties). Properties are always arity/2 (relate 2 things) and have domain and range restrictions. For example, the agent initially knows that *room1*: (1) is of type *destination* (is somewhere is can consider as destination goal); (2) is connected to *home_room*; and (3) has a string name of "room1". An inspection of *door1* reveals that it is both a *way_in* and *way_out* of both *room1* and *home_room*. In other words, nodes and relations represent knowledge that the agent can navigate from *home_room* to *room1* through *door1*.

4.1.2   CS2F/CDO – The Agent's Domain Knowledge

The agent uses a single CDO to determine **what** it needs to do in order to achieve its goal of finding and acquiring the small item. The top-level entity in this CDO represents an *effective_action* (or space of behaviors that achieve the design objective of a particular goal). The primary decomposition of *effective_action* is shown in the Fig. 11.

Fig. 11 Top-level entities and relations in the *effective_action* CDO. Note that the percepts, actions, goals, and environment entities are actually reference to previously specified entities in a repository

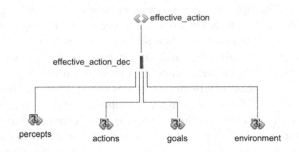

The CDO formally captures the space of *effective_actions* by decomposing them (through aspect decomposition) to *percepts*, *actions*, *goals*, and the *environment*. This decomposition specifies that *percepts*, *actions*, *goals*, and the *environment* are aspects of *effective_actions*. In the CDO, *percepts*, *actions*, *goals*, and *environment* are actually references to additional CDOs held in a CS2F/CDO repository maintained in GME. References can be expanded when the modeler wishes to view or modify the details of the referred-to entities. The ability to use entity references in CDOs encourages the re-use of CDOs and significantly reduces the visual complexity of large CDOs.

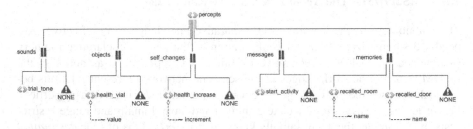

Fig. 12 Component CDO specifying the percepts the autonomous agent can comprehend

Fig. 12 shows that *percepts* entities have characteristics related to *sounds*, *objects*, *self_changes*, *messages*, and *memories*. Each of these characteristics is actually a specialization. The sounds specialization for example, specifies that a percepts sound can either be *trial_tone* or *none*. Instances of the percepts entity correspond to events/facts the agent is able to perceive. A *health_vial* corresponds to the small item the agent is seeking to locate and acquire. A *health_increase* corresponds to an event/fact generated by the act of collecting the reward item. This percept type is used by the agent to recognize when it has successfully collected the reward item. A *start_activity* corresponds to a message provided to the agent by an operator or experiment frame in order to initiate an agent's behavior. Lastly, the *recalled_room* and *recalled_door* percepts correspond to events/facts retrieved from declarative memory.

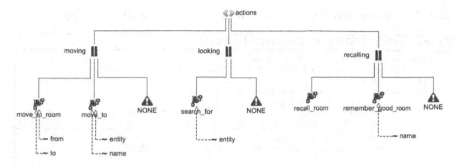

Fig. 13 Component CDO specifying the actions the autonomous agent can initiate

Fig. 13 shows that *actions* entities have characteristics related to *moving, looking*, and *recalling*. These characteristics are specializations. The *move_to_room, move_to, search_for, recall_room*, and *remember_good_room* entities in the actions CDO are actually **references to behavior models**. When the agent processes the *effective_action* CDO *in situ*, instances of these references in CSP solutions will be used to dynamically reconfigure the behavior repertoire of the agent.

Fig. 14 shows that *goals* entities are a combination of a part_task entity expressing the sub-goal underlying an *effective_action* and a desired destination. The destination specialization specifies that goals can involve a desired destinations related to a *room* or *location*. The room entity can be room1, room2, room3, or home_room. The location entity can be door1, door2, door3, trigger_plate, or none.

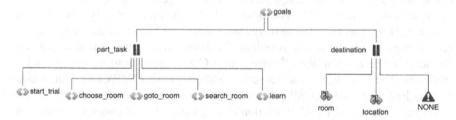

Fig. 14 Component CDO specifying the goals the autonomous agent can maintain

Table 7 illustrates how CDO and domain constraints capture what an agent should do *in situ*. The table shows three constraints that allow the autonomous agent to determine what it should do upon perceiving a *trial_tone* sound in a virtual environment.

Table 8 shows how the CSP solution process provided by soaCDO can use a CDO and additional contingency-based assertions to help an agent determine **what** it should do *in situ*. To simplify explication, a direct interaction with the CSP infrastructure of soaCDO is shown. The top part of Table 8 consists of a call to the soaCDO function "soaCDO-solutions". This primitive initiates a non-deterministic

Table 7 Constraints allowing the autonomous agent to respond to a sound percept

Examples of the CS2F/CDO Constraint Language

```
(define-constraint p_hear_trial_tone
    ;; If the trial_tone is heard, then choose a room.
    :effective_action
    (==> (equale (e@ percepts sounds) trial_tone)
        (and (equale (e@ goals part_task) choose_room)
            ;; The trial_tone can only be heard in the home_room.
            (equale (e@ environment current_room room room_spec) home_room))))
```
```
(define-constraint c_choose_room
    ;; Limits the context in which choose_room is applicable.
    :effective_action
    (==> (equale (e@ goals part_task) choose_room)
        (and (equale (e@ percepts objects) none)
            (equale (e@ percepts self_changes) none)
            (equale (e@ percepts messages) none)
            (equale (e@ percepts memories) none)
            (equale (e@ actions looking) none)
            (equale (e@ actions moving) none)
            (equale (e@ goals destination location location_spec) none))))
```
```
(define-constraint g_choose_room
    ;; To act out choose_room, recall a room from memory.
    :effective_action
    (==> (equale (e@ goals part_task) choose_room)
        (equale (e@ actions recalling) recall_room)))
```

CDO search that returns the first configuration of entity and attached variable assignments meeting a CDO's structural constraints and additional domain constraints expressed in the CS2F/CDO constraint language. In Table 8, one CSP solution from the *effective_action* CDO is being requested. The call to soaCDO-solutions includes one additional assertion that maps properties of the agent's contingencies to entities in the *effective_action* CDO. Assertions such as this are essentially function calls accepting two arguments. The first argument is the CDO scope of the assertion. The second argument is the actual assertion. The assertion in Table 8 indicates that in the scope of *percepts* in the *effective_action* CDO, sound is to be constrained to *trial_tone*. This contingent-based assertion and additional CS2F/CDO domain constraints lead to the single CSP solution shown in the bottom part of Table 8. The displayed summary of the CSP solution clearly indicates that under the contingencies expressed by the assertion, the autonomous agent should pursue the action of recalling a room (recall_room).

Listings in Tables 7 and 8 illustrate how the autonomous agent acts effectively after perceiving a *trial_tone* sound. The critical thing for the reader to remain aware of is that in the KCGS framework, the agent is determining *what* it should do by mapping aspects of its contingencies to a CDO and then using a constraint-satisfaction process to determine how it should use BMs to achieve its goals. The framework allows modelers to exploit an abstraction layer between BMs and CDOs.

The autonomous agent described in this section illustrates how knowledge about contingencies, possible actions, and goals can be represented in CDOs and processed by agents in order to achieve a form of real-time behavioral autonomy. Under these circumstances, CDOs are used to formally relate high-level goals or behavioral

Table 8 Example showing how CSP in soaCDO results in a CDO solution or prune determining what action(s) the agent should take in order to choose a room

Examples of the CS2F/CDO Constraint Language

```
(soaCDO-solutions
  (effective_action '(assertion :percepts (equale (e@ sounds) trial_tone)))
  :one)

Percepts: sounds/trial_tone, objects/none, self_changes/none, messages/none,
          memories/none
Actions: move/none, look/none, recall/recall_room
Goals: part_task/choose_room, destination/none
Environment: room/home_room, location/none
nil
```

design objectives (state descriptions) and low-level actions (process descriptions) in such a way that the agent's behavior is not simply a function of pre-specified rules. CDOs represent connections between contingencies in which agents must act, the agent's goals, and behaviors the agent might utilize to achieve these goals. The key to processing the domain knowledge captured in a CDO under these circumstances is a search process that finds configurations of entities and variables that: (1) meet structural constraints expressed through the aspect, specialization, and multi-aspect relationships in a CDO; and (2) satisfy domain-specific constraints expressed in the CS2F/CDO constraint language.

4.1.3 CS2F/BM – The Agent's Procedural Knowledge

The autonomous agent uses 7 behavior models to generate effective action in the synthetic task environment. The behavior models enable the agent to perform fundamental behaviors necessary for it to act in the task environment. The behavior models are as follows:

1. **assess_separation:** Enables the agent to track the separation distance between itself and a destination location. The separation is reported using the qualitative categories *separated* and *close*.
2. **find_item:** Enables the agent to determine the location of an item by: seeing it directly in percepts; scanning for it in a 90 degree rotation to the left; and scanning for it during a final 180 degree rotation to the right.
3. **move_to:** Enables the agent to either see or recall the location of a named entity and move to it. If the location information is neither visible nor recallable, then the agent rotates until the location information is visible.
4. **move_to_room:** Enables the agent to use a retrieved doorway and the behavior model *move_to* to move an agent from one room to another. If the agent is unable to remember a doorway that leads from its current room to the desired room, the agent initiates a new trial.
5. **recall_room:** Enables the agent to either: retrieve a room that has provided a high frequency of reward items in the past; or randomly choose a room.

6. **remember_good_room:** Enables the agent to increase the activation of declarative knowledge corresponding to a room in which the small item was collected.
7. **search_for:** Enables the agent use the *find_item* and *move_to* behavior models to locate and collect the reward item.

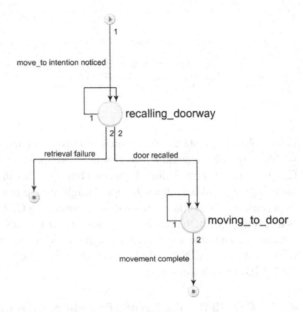

Fig. 15 Graphical representation of a behavior model in CS2F/BM specifying how the autonomous agent can achieve the objective of moving to a room

The *move_to_room* behavior model is shown in Fig. 15. The graphical rendering of the BM shows that the overall behavior involves recalling a door and moving to it. When an agent intends to move to a room, it remembers which door leads to the room and then navigates to the door. When the agent reaches the door, its intended movement is completed and the behavior model transitions to an end state.

Table 9 shows the details of two transitions in the *move_to_room* behavior model. To move to a room, the agent must either visually locate or recall the location of a door that leads from the current room in which it is located and the room it considers its destination. The first of the transitions allows the agent to retrieve a door from declarative memory. Retrieval constraints require that the retrieved door be a *way_out* of the room the agent is moving from and an *way_in* to the room the agent is moving to. The second transition allows the agent to actually move to the successfully retrieved door. The transition essentially: (1) verifies that information about a door has been retrieved; (2) obtains the name of the door; and (3) initiates a sub-goal to *move_to* the door. The "{assert_intention, {move_to, C, N}}" assertion in this transition initiates transition activity in the *move_to* behavior model. These two transitions demonstrate how a behavior model can interact with:

- soaDM in order to base behavior on retrieved declarative knowledge
- other behavior models in order to coordinate hierarchical behavior based on the execution of sub-goals

Table 9 A partial listing of the transition details of the same behavior model specified in a text form generated during the automated transformation of CS2F/BM to executable DEVSML

```
Behavior Model Transition Details in Textual CS2F/BM

transition {
    priority    1
    label       "move_to intention noticed"
    src         startstate
    dst         recalling_doorway
    pre_binds   context=C,w=W
    patterns    {move_to_room, From, To}
    functions   Constraints=[{type, door}, {way_in, To}, {way_out, From}],
                Endo=[{type, door}],
                Exo=expand_context(C, W)
    assertions  {execute_retrieval, Constraints , Endo, Exo}
    post_binds  from=From,to=To
}
...
transition {
    priority    2
    label       "door recalled"
    src         recalling_doorway
    dst         moving_to_door
    patterns    {retrieval_success, C, _}, {type, C, door}, {name, C, N}
    assertions  {assert_intention, {move_to, C, N}}
    post_binds  door=C,door_name=N
}
...
```

4.1.4 Summary of the Autonomous Agent's Runtime Behavior

When the autonomous agent is initially situated in the simulated task environment, it has: (1) declarative knowledge about rooms, doorways, and the *trigger_plate*. Initially, the agent has a single *central_executive* behavior model in its behavior repertoire. Transitions in the *central_executive* behavior model are sensitive to the following percepts/events:

1. A *start_activity* message originating from a modeler or experiment frame indicating that the agent should begin to perform the overall activity of trying to find and collect the reward object.
2. A *trial_tone* sound originating from the external environment indicating that the agent should initiate a single effort to find the reward object. This sound is produced when the agent stands on the trigger_plate.
3. A *recalled_room* retrieved memory originating from the agent's associative memory indicating which room the agent expects to find the reward object in.
4. A *search_room* goal originating from an internal intention indicating that the agent wants to search for the reward object in a room.
5. A *health_increase* perceived change originating from self-monitoring indicating that the agent has collected the reward object and should make a mental note (increase the activation) of the current room.

As percepts/events originating from outside or inside the agent trigger these transitions, the agent asserts entity and variable values from its contingencies into the *effective_action* CDO and initiates a CSP-based process in soaCDO that "prunes" the CDO. This process utilizes constraints similar to those presented in Table. 7. The agent then integrates any behavior models referred to in one of these CSP solutions into its behavior repertoire. These new, but contextually relevant, behavior models generate effective action until some future percept/event triggers another transition and precipitates another "prune" of the *effective_action* CDO.

The *central_executive* behavior model only transitions when percepts/events require that it re-assess *what* it should be doing in order to achieve its goal. Between these transitions, behavior models in the agent's behavior repertoire autonomously tell the story of *how* the agent should act in the moment. The process of using cognitive domain ontologies to determine *what* to do given contingencies and behavior models to determine *how* to act *in situ* allows an agent to translate state descriptions to process descriptions.

4.2 Agents That Use an Abduction-Based Inquiry Process

The agent described in the previous section demonstrates how constraint-based processing of CDO can inform an agent *what* it should do *in situ*. The agent described in this section demonstrates how CDO can additionally be used by an agent to systematically increase its understanding of its situation. The agent is capable of a type of sensemaking. Before describing this agent, sensemaking and abduction, the type of inference that enables sensemaking, will be defined.

Sensemaking is a process shown in Fig. 16 through which people attempt to understand complex and ambiguous situations so that they can make reasonable decisions and act effectively [12]. In context of this chapter, sensemaking will be defined as abduction-based inquiry.

Abduction can be thought of as a type of inference that plays a role in a process through which inquiry reduces doubt [7, 34]. As a person assesses and understands the context they are trying to act effectively in, they either: (a) find that it's "business as usual" and act according to routine; or (b) are surprised by unexpected events and observations and try to make sense of things through *designed inquiry*. A surprised person uses abduction, a type of inference from observations to likely explanations or causes, to generate new ideas (hypotheses) about their situation. Through deduction and induction, these hypotheses can be expanded and confirmed/disconfirmed. If necessary, follow-on actions can refine knowledge and hypotheses. The model described in this section uses domain knowledge captured in a CDO and abduction to generate knowledge and hypotheses. This model of abduction is intended demonstrate that an inference-based process (artificial phenomena) that increases the knowledge and autonomy of an agent can be modeled and simulated in the KCGS framework.

In addition to allowing an agent to determine *what* it should do *in situ*, CDO surprisingly allow agents to systematically increase their understanding of situations through an abduction-based inquiry process. The essence of this ability is a non-monotonic reasoning process through which agents: (1) assesses evidence about

Fig. 16 Central concepts, relations and constraints in a model of sensemaking as abduction-based inquiry

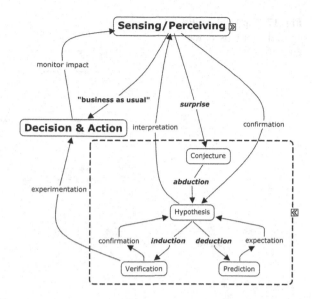

their situations; (2) assert this evidence and other related aspects of their situations into CDOs representing world knowledge; (3) use constraint propagation to process the CDOs; (4) treat the resulting set of CSP solutions as a hypothesis sets constituting explanations of their situation; and (5) design actions that will allow them to effectively reduce their hypothesis set.

When used this way, CDOs are not just matching contingencies to goals and indicating *what* the agent should do *in situ*. Rather, CDOs are being used to capture world knowledge that can be used to relate small-scale observations (evidence) to large-scale ontologies (explanations) in a process that, employing designed action, increases the epistemological quality of the agent's knowledge!

To illustrate this process in as simple an agent as possible, this section will describe an agent that pursues a singular goal of trying to discover the identity of an unknown person. The agent has some general world knowledge about individuals and facial characteristics. The agent knows that certain uniquely identifiable individuals have certain visual characteristics. When asked to guess the identity of an initially unknown person, the agent asks questions designed to constrain the identity of the person so as to systematically refine its knowledge about them.

4.2.1 CS2F/CDO – The Agent's Domain Knowledge

The following figures partially present the CDO used by the identity determination agent. The top-level *guess* CDO in Fig. 17 contains entity references that reduce the visual complexity of the CDO. In order to conserve space, only two of the entity references are presented in full detail in Fig. 18 and Fig. 19. Each entity reference is replaced by the details of the referred to CDO by soaCDO when CS2F/CDO specifications are translated into executable Common Lisp.

Fig. 17 Top-level entities and relationships in the guess CDO

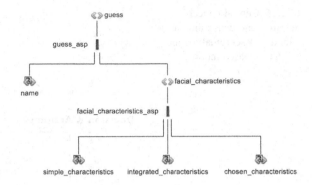

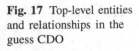

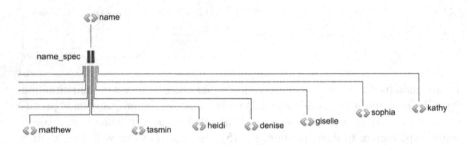

Fig. 18 Component CDO specifying the names guesses can be based on

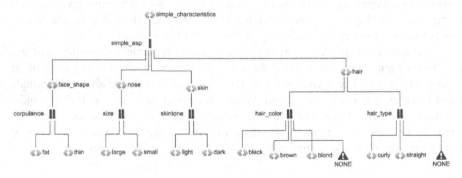

Fig. 19 Component CDO specifying the simple-characteristics guesses can be based on

The *guess* CDO specifies a space of identities. Without additional domain constraints, this CDO would produce a large number of solutions or prunes when processed by soaCDO. These solutions would combine names with constellations of simple, complex, and chosen characteristics. Without domain constraints specifying the unique characteristics of each guess name, multiple solutions based on each name would exist.

If constraints similar to those shown in Table. 10 are defined and incorporated into the CDO, then each *guess name* is constrained to have a specific set of characteristics. Under these circumstances, the *guess* CDO captures a set of named identities with fixed characteristics. With these domain constraints, only one solution for each *guess name* can exist.

Table 10 Example constraints refining the aspects of guesses. Note how these constraints "define" individuals by relating a set of characteristics to the *name* of a *guess*.

Examples of the CS2F/CDO Constraint Language

```
(define-constraint adam
  :guess
  (let ((<simple_asp> (n@ guess ... simple_asp))
        (<integrated_asp> (n@ guess ... integrated_asp))
        (<chosen_asp> (n@ guess ... chosen_asp)))
    (==> (equale (e@ guess guess_asp name name_spec) adam)
        (and (equale (e@ <simple_asp> face_shape corpulance) fat)
             (equale (e@ <simple_asp> nose size) small)
             (equale (e@ <simple_asp> skin skintone) light)
             (equale (e@ <simple_asp> hair hair_color) brown)
             (equale (e@ <simple_asp> hair hair_type) straight)
             ;;
             (equale (e@ <integrated_asp> countenance countenance_spec) smile)
             (equale (e@ <integrated_asp> gender_spec) male)
             (equale (e@ <integrated_asp> age age_spec) old)
             ;;
             (equale (e@ <chosen_asp> facial_hair mustache_spec) none)
             (equale (e@ <chosen_asp> facial_hair beard_spec) none)
             (equale (e@ <chosen_asp> headgear headgear_spec) hat)
             (equale (e@ <chosen_asp> eyeware eyeware_spec) none)))))
... Additional Constraints ...
```

Possessing world knowledge capturing information about the characteristics of named individuals, the identity determination agent is able to guess the unknown individual by systematically acquiring knowledge about his or her characteristics. To acquire knowledge about the characteristics of the unknown individual, the agent simply asks if they have a specific characteristic. Each answer to these queries becomes an assertion that reduces the number of subsequent CSP solutions found by soaCDO. If CDO solutions are considered hypotheses about the identity and characteristics of the unknown individual, then each assertion reduces the number of hypotheses. This process clearly uses inquiry to reduce doubt. The listing in Table. 11 shows how 4 assertions reduce the hypothesis space represented in the *guess* CDO to 2. The assertions indicate that previous questions led to knowledge that the individual: (1) is wearing a hat; (2) is not wearing glasses; (3) is female; and (4) is young. When these characteristics are asserted as requirements during the constraint propagation process, only 2 solutions are found. To continue to make sense of the identity of the unknown individual, the agent would note that the remaining hypotheses differ with respect to *hair_color* and ask "Does the unknown person have black hair?" The answer to this query would provide the last piece of evidence required to disambiguate the identity of the unknown individual.

Table 11 Example showing how CSP in soaCDO results in a set of CDO solutions constituting the hypothesis space resulting from abducing from evidence of characteristics to explanations based on named guesses meeting constraints based on the evidence

```
Example CSP Solution
(soaCDO-solutions
  (guess_ '(assertion :integrated_characteristics (equale (e@ age_spec) young))
          '(assertion :integrated_characteristics
                      (equale (e@ gender_spec) female))
            '(assertion :chosen_characteristics
                        (notv (equale (e@ eyeware_spec) glasses)))
            '(assertion :chosen_characteristics (equale (e@ headgear_spec) hat)))
  :print)

Name: sophia
Simple Aspects: face_shape/thin, nose/small, skintone/light, hair_color/blond,
               hair_type/curly
Integrated Aspects: expression/smile, gender/female, age/young
Chosen Aspects: facial_hair/(none, none), headgear/hat, eyeware/none

Do you want another solution? (y or n) >> y

Name: petra
Simple Aspects: face_shape/thin, nose/small, skintone/light, hair_color/black,
               hair_type/curly
Integrated Aspects: expression/smile, gender/female, age/young
Chosen Aspects: facial_hair/(none, none), headgear/hat, eyeware/none

Do you want another solution? (y or n) >> y
nil
```

The ambiguous situation the identity determination agent is trying to make sense of centers on the ambiguous identity of an individual. When the agent is initially asked to make sense of its situation, it is unable to discount any of the named individuals it has knowledge about in the *guess* CDO. The hypothesis space the agent seeks to reduce using an abduction-based inquiry process contains all named individuals. The agent reduces the hypothesis space by asking questions about characteristics distinguishing a subset of the hypothesis space. The agent uses a CDO and the assertion of accumulating evidence to refine its knowledge of the identity of the unknown individual. Using a CDO in this way is quite different than using one to determine what to do *in situ* since it enables an agent to refine its knowledge.

5 The KCGS Framework

In order to express a rich knowledge set that includes environment, contingencies, resources, possible actions and much more, we need a framework that allows us to represent knowledge in many facets or dimensions. While a cognitive rational agent uses all this knowledge to compose its immediate action, it is very difficult as a modeler to construct this knowledge-set if there is only one dimension. For a multi-dimensional and multi-resolutional knowledge representation, the knowledge framework must itself allow constructions of this kind of representation. Ontology, in technical terms is a graph of nodes and information is presented in the relations that exist between these nodes. Of course, it is a great step as the knowledge can now be presented in associative terms, more like a semantic network. It is now more amenable to data engineering efforts but it is essentially flat and not suitable for piecewise construction or layered methodologies for better manageability. The SES formal knowledge representation mechanism with its set of axioms and rules helps develop an ontology that can be constructed and deconstructed in piecewise manner through SES aspects and specializations. The latest work in SES ontology domain is an evidence of such efforts [43].

We have shown in our narrative earlier how an agent can have its description in multiple aspects and specializations. Such aspects and specialization can be added or removed incrementally and intuitively without changing other facets of the system and still understandable by the common modeler. In other words, the modeler is not overwhelmed by the influx of new knowledge as it builds upon the existing ones. This is important because in large systems, large knowledge-set often results in 'information paralysis' at the modeler end. Such aspects and specializations give ontology a multi-resolutional capability and can be called upon at real-time execution of the system. Also note here that adding such elements is piecewise isolated and it is the defined rules that create relationships between different SES elements at run-time thus managing complexity. It also implies that while the general structure of the proposed ontology remains intact, it is the defined rules that dictate the association and affordances of the entire system at run-time. These rules then become dynamic and dictate how the knowledge entities interact. This property of SES is a major way forward as compared to existing cognitive models where the rules are a function of the invariant architecture itself and any change in the architecture calls for major upgrades in the modeling system. The realization of these rules by DEVS formalism in a SES modeled system is much easier, manageable and formally verifiable at run-time.

Another advantage of this piecewise construction is partitioning of the expert knowledge in the domain of interest. It now becomes much more feasible to integrate the expert knowledge of other cognitive scientists as aspects of such ontology. Therefore, we attempt to construct and open the proposed Cognitive Domain Ontology for further input and contributions from the community at-large. Once the structure of these aspects is laid out, it is easier to define and modify rules that related different aspects of the ontology.

5.1 Putting CS2F in the KCGS Perspective

This chapter describes the CS2F framework in order to illustrate how *artificial systems* producing *artificial phenomena* can be modeled and simulated. The framework combines aspects of SES theory, DEVS-based general systems theory, cognitive architectures (ACT-R), and DSL development using meta-modeling to change the way artificial systems are formally specified and simulated. The presentation of the CS2F framework and example agents has been tailored to the objective of highlighting how ontology, epistemology, and teleology play roles in the realization of autonomous models and agents in the framework. The framework is significantly more than just a set of net-centric applications capable of executing a set of new DSLs though. This section will describe how CS2F is an instance of a larger KCGS framework.

The KCGS framework is based on three major areas with formal SES theory at their centers:

I   Ontology and data representation.
II   Knowledge engineering and parallel distributed computation search mechanisms.
III   DEVS Unified Process.

I deals with knowledge representation and how data interoperability is achieved between different ontologies using SES foundational framework. In its current state, basic programmatic pruning mechanisms are used. II deals with the entire knowledge engineering and data-mining aspect of executing the pruning process that transform data into information. This computational process has to align with the AI-based search mechanisms, and real-time execution capabilities that will lead to formal SES-based pruned SESs. Finally, III takes the formal PESs and using the DEVS M&S technology, provides the requirement traceability, platform independent M&S, Verification and Validation and various other capabilities such as SOA execution, and system component descriptions in DEVS Unified Process.

The capabilities of the KCGS framework as realized in CS2F allow us to specify many kinds of domain models and take the executing real models to live netcentric systems. A framework for modeling and simulation of the artificial must have these basic capabilities to incorporate large-scale heterogeneous systems. Table 12 lists some of the requirements of such a framework and Fig. 20 shows how CS2F and the larger KCGS framework address these requirements.

Fig. 20 also shows how different disciplines interact together and interface with the formal SES theoretical framework. CS2F/DM is a DSL that formally captures declarative knowledge in ontologies. CS2F/BM interfaces with DEVSML 2.0 stack through various transformations. CS2F/CDO works at the intersection of SES theory, constraint satisfaction problems and various other knowledge engineering measures as overlaid on SES theoretical framework. The DEVS Unified Process is a superset that incorporates formal DEVS System theory, platform transparent M&S layered framework called as DEVSML 2.0, requirements engineering at the intersection with formal domain ontology representations using SES and other methods.

Ultimately, the solution we are looking for is an ontological framework that lends itself seamlessly into the simulation-based component modeling framework. Fig. 21 presents the meta-meta-model of an autonomous system. It formally captures real-world facets like environment and resources and agent-based facets like goals and behavior. Constraints play a dual role in an Autonomous System's ontology [19]. There are two types of constraints. Type I constraints are physics based (hard truths) and Type II constraints are situation-based. While Type I constraints are hard constraints, the Type II constraints are soft constraints that are dynamic. The Type II constraints are the ones that are responsible for contingency based behavior and

Table 12 Mapping requirements for M&S Framework for the Artificial with CS2F and KCGS components

Framework Requirements	Technical Foundation	CS2F Component	KCGS Component
Based on General Systems Theory	DEVS System Theory		DEVS Unified Process
Facilitate model-based development and engineering	DEVS M&S Framework, SES Theory	CDO	SES Theory
Scalable and component-based	DEVS M&S Framework	BM, CDO	DEVSML 2.0, DEVS Unified Process
Manage Hierarchy and abstractions	DEVS Systems Theory	BM, CDO	DEVS Unified Process
Interoperable across implementation platforms	DEVS M&S Framework		DEVSML 2.0
Formal specification	DEVS M&S Framework	BM, CDO, DM	DEVS Unified Process, SES Theory
Domain and platform neutral	DEVS Systems Theory, SES Theory	CDO	Ontology and Knowledge Representation
Agile and persistent	DEVS Systems Theory, SES Theory	CDO, DM, BM	DEVS Unified Process, SES Theory
Interface with AI knowledge engineering methodologies	SES Pruning	CDO Pruning	SES Theory, Data Mining, Constraint Satisfaction Problem (CSP)

situated behavior. The pruning process will work on these Type II constraints to generate a CDO that is 'situated'.

Earlier research demonstrated that SES, rule-based search processes, and conventional simulation can be used to capture, search through, and evaluate system configuration spaces. These efforts demonstrated how a rule-based search or SES pruning process can derive system configurations that meet the design objectives explicit in the SES. Current research efforts have demonstrated that SES can capture domains other than physical system design spaces. The KCGS framework described in this chapter combines current and previous patterns of SES use in modeling and simulation by: (1) using CDOs to represent behavior configuration spaces consisting of agents (beliefs, goals, behavioral constraints) and situational contingencies (task requirements, action affordances, physical constraints); (2) searching through (pruning) CDOs at runtime in order to generate behaviors that meet the goals and contingencies; and (3) executing cognitive agents employing CDOs in larger M&S framework such as DEVS Unified Process [18]. Future work will refine the KCGS framework and explore the relationships between declarative, procedural,

Fig. 20 Putting CS2F in perspective of Knowledge-based Contingency-driven Generative Sytems framework

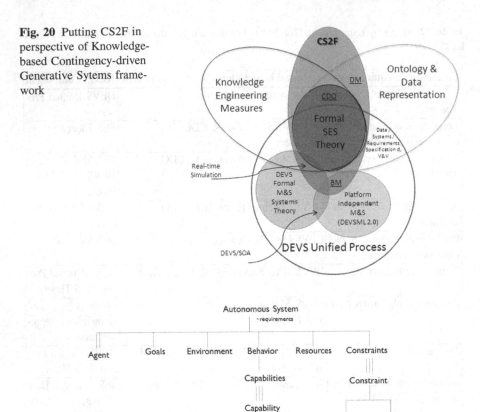

Fig. 21 Meta-meta-model for an Autonomous System

and domain knowledge in a formal modeling and simulation framework founded on the DEVS [46] Unified Process [18, 27]. Future work will also explore how large-scale autonomous (artificial) models and agents can be integrated into systems of systems and Human-in-loop solutions.

6 Concluding Remarks

The cognitive system specification framework (CS2F) is a composition of DSLs tailored to the needs of cognitive modelers. The abstract syntaxes of two of the DSLs composed in CS2F (CS2F/DM and CS2F/BM) are strongly influenced by the ACT-R cognitive architecture [1, 2]. The abstract syntax of CS2F/CDO is influenced by System Entity Structure (SES) theory [43]. The concrete syntaxes of CS2F/DM and CS2F/BM are designed so that a modeler with experience in ACT-R can specify behaviorally equivalent models at a high level of abstraction. The concrete syntax of CS2F/CDO is designed to allow modelers to rapidly specify theoretically sound CDOs regardless of their experience working with SES.

This chapter explored how a new modeling and simulation framework can allow a modeler to: (1) formally represent actions as behavior models in CS2F/BM; (2) formally represent goals and contingencies as cognitive domain ontologies in CS2F/CDO; and (3) let autonomous models and agents decide *what* to do on their own *in situ*. The framework consists of three major net-centric components each representing and processing a unique type of agent knowledge:

soaDM   an Erlang/OTP based associative memory through which models and agents can store and retrieve propositional or declarative knowledge.

soaCDO   a Common Lisp based service through which models and agents can represent and process cognitive domain ontologies formally capturing the entities, constraints, and relationships constituting the requirements of the tasks they are performing.

DEVSML Stack   a DEVS/Java based integration service through which models and agents can represent and process behavior models formally capturing actions they can perform.

Models and agents technically realized in the KCGS framework do not use predefined rules and knowledge that interleave state and process descriptions to act in anticipated circumstances. Instead, they are persistent computational entities that use CDO search/pruning *in situ* to re-configure their own behavioral repertoires to match their objectives and goals to their contingencies. The generative framework allows modelers to specify behavior above the level of the CS2F/BM behavior model. The 3 components of the generative framework discussed above are computationally realized in the modeling and simulation infrastructure developing in the LSCM initiative [21]. The DEVSML Stack translates behavior models specified in CS2F/BM into DEVS coupled models which are then executed in a net-centric realization of DEVS [20].The generative framework represents a significant modeling capability advancement that will add to, and leverage, the DSLs and M&S capabilities being researched and developed in the LSCM initiative.

6.1 Roles of Ontology, Epistemology, and Teleology in Artificial Systems

The KCGS framework discussed in this chapter supports the modeling and simulation of autonomous agents. These agents base their behaviors on their contingencies not just pre-specified rules. Autonomous agents modeled in the KCGS framework use three knowledge representations to gain autonomy: (1) procedural knowledge; (2) declarative knowledge; and (3) domain knowledge. These three knowledge representations allow KCGS framework users to model and simulate agents that exploit ontologies, produce artificial behaviors that are teleological, and refine their knowledge over time. The framework's effectiveness can be attributed to its exploitation of:

Ontology	Declarative and domain knowledge are represented using ontologies. Declarative knowledge is described in OWL ontologies, translated into semantic networks, and processed by agents in a soaDM associative memory component of the framework. Domain knowledge is described in CDOs, translated into constraint networks, and processed by agents in a soaCDO component.
Teleology	Agents use CDOs to determine *what* they should do *in situ*. Agents do this by mapping characteristics of their contingencies into CDOs and propagating them through: (1) constraints reflecting the aspects, specializations, and multi-aspects characterizing the CDOs; (2) domain constraints specified in a CS2F/CDO constraint language. When CDOs represent spaces of behaviors that achieve a behavioral design objective (goal), they allow an agent to generate goal-pursuing behavior in complex and dynamic environments.
Epistemology	Agents use CDOs to infer abductively from observed evidence to likely explanations. Under these circumstances, agents relate observations (in the form of asserted evidence) to what they know about the world (in the form of a CDO). By selecting actions that elicit additional evidence from the environment, agents can refine their situational knowledge.

Appendix 1: Complex Operators in the CS2F/CDO Constraint Language

Operator	Meaning	Example
assert!	Set value of a constraint variable	`(assert!` `  (equalv (v@ (weight) kg) 100))`

andv	Conjunction with variables	`(andv (equale (e@ aspect sport)`
		`          golf)`
		`    (equale (e@ aspect size)`
		`          small))`

orv	Disjunction with variables	`(orv (equale (e@ aspect size)`
		`          small))`
		`    (equale (e@ aspect size)`
		`          large))`

notv	Negation with variables	`(notv (equale (e@ ensemble)` `          soloist))`

ifv	Implication with variables	`(ifv (equale (e@ ensemble)`
		`          soloist)`
		`    (notv (equale (e@ style)`
		`              symphonic)))`

fail	Failure with backtracking	`(ifv`
		`  (andv (equale (e@ ensemble)`
		`          soloist)`
		`    (equale (e@ style)`
		`          symphonic))`
		`  (assert! (fail)))`

a-member-of	Non-deterministic selection from a list [creates choice point]	`(assert! v (a-member-of '(a s d f)))`
either	Non-deterministic selection from arguments [creates choice point]	`(either small medium large)`

an-integer-above
an-integer-between
an-integer-below
	Define integer ranges	`(equalv v (an-integer-above 10))`
		`(assert!`
		`  (equalv (v@ (weight) kg)`
		`          (an-integer-between 75 105)))`

a-real-above
a-real-between
a-real-below
	Define real ranges	`(equalv pi`
		`  (a-real-between 3.0 4.0))`
		`(ifv (notv (equalv pi`
		`              3.141592653589793))`
		`  (fail))`

>v, >=v, <v, <=v	Comparison functions that accept constraint variables	`(ifv (<=v v 10) (fail))`

Appendix 2: Example Constraints that Refine the Possible Space of Musical Performance

Constraint	Implicative Normal Form
| ```
(define-constraint m4
  :musical-performance
  (and (==> (equale (e@ style) jazz)
            (or (equale (e@ ensemble)
                        small-group)
                (equale (e@ ensemble)
                        orchestra)))
       (==> (equale (e@ style) folk)
            (not (equale (e@ ensemble)
                         orchestra)))
       (==> (equale (e@ style) symphonic)
            (not
             (or (equale (e@ ensemble)
                         soloist)
                 (equale (e@ ensemble)
                         small-group))))))
``` | ```
(orv
 (ifv (equale (e@ style) jazz)
      (assert!
       (orv (equale (e@ ensemble)
                    orchestra)
            (equale (e@ ensemble)
                    small-group))))
 (ifv (andv (equale (e@ ensemble)
                    orchestra)
            (equale (e@ style)
                    folk))
      (assert! (fail)))
 (ifv (andv (equale (e@ ensemble)
                    soloist)
            (equale (e@ style)
                    symphonic))
      (assert! (fail)))
 (ifv (andv (equale (e@ ensemble)
                    small-group)
            (equale (e@ style)
                    symphonic))
      (assert! (fail))))
``` |

Appendix 3: Declarative Knowledge Available to the Autonomous Agent through soaDM

SemNet Node	Object Properties	Data Properties
home_room	type = origin connected_to = room1 connected_to = room2 connected_to = room3	name = "home_room"
room1	type = destination connected_to = home_room	name = "room1"
room2	type = destination connected_to = home_room	name = "room2"
room3	type = destination connected_to = home_room	name = "room3"
door1	type = door way_in = room1 way_in = home_room way_out = home_room way_out = room1	name = "door1"
door2	type = door way_in = room2 way_in = home_room way_out = home_room way_out = room2	name = "door2"
door3	type = door way_in = room3 way_in = home_room way_out = home_room way_out = room3	name = "door3"
trigger_platec	trigger_platec	name = "trigger_plate" location_x = 3653.0 location_y = 1975.0 location_z = -197.65

References

[1] Anderson, J.: How can the human mind occur in the physical universe?, vol. 3. Oxford University Press, USA (2007)

[2] Anderson, J., Bothell, D., Byrne, M., Douglass, S., Lebiere, C., Qin, Y.: An integrated theory of the mind. Psychological Review 111(4), 1036 (2004)

[3] Anderson, J., Matessa, M.: An overview of the epic architecture for cognition and performance with application to human-computer interaction. Human-Computer Interaction 12(4), 391–438 (1997)

[4] Cesarini, F., Thompson, S.: Erlang programming. O'Reilly Media (2009)

[5] Douglass, S., Mittal, S.: Using domain specific languages to improve scale and integration of cognitive models. In: Proceedings of the Behavior Representation in Modeling and Simulation Conference, Utah, USA (2011)

[6] Douglass, S., Myers, C.: Concurrent knowledge activation calculation in large declarative memories. In: Proceedings of the 10th International Conference on Cognitive Modeling, pp. 55–60 (2010)

[7] Fann, K.: Peirce's theory of abduction. Martinus Nijhoff La Haya (1970)

[8] Gonzalez, C., Lerch, J.F., Lebiere, C.: Instance-based learning in dynamic decision making. Cognitive Science 27(4), 591–635 (2003)

[9] Hwang, M., Zeigler, B.: Reachability graph of Finite and Deterministic DEVS networks. IEEE Transactions on Automation Science and Engineering 6(3), 468–478 (2009)

[10] Keene, S.: Object-oriented programming in Common Lisp: A programmers guide to CLOS. Adison-Wesley (1989)

[11] Kim, T., Lee, C., Christensen, E., Zeigler, B.: System entity structuring and model base management. IEEE Transactions on Systems, Man and Cybernetics 20(5), 1013–1024 (1990)

[12] Klein, G., Phillips, J., Rail, E., Peluso, D.: A data-frame theory of sensemaking. In: Expertise out of context: proceedings of the Sixth International Conference on Naturalistic Decision Making, p. 113. Lawrence Erlbaum (2007)

[13] Ledeczi, A., Maroti, M., Bakay, A., Karsai, G., Garrett, J., Thomason, C., Nordstrom, G., Sprinkle, J., Volgyesi, P.: The generic modeling environment. In: Workshop on Intelligent Signal Processing, Budapest, Hungary, vol. 17 (2001)

[14] Ledeczi, A., Volgyesi, P., Karsai, G.: Metamodel composition in the Generic Modeling Environment. In: Comm. at Workshop on Adaptive Object-Models and Metamodeling Techniques, Ecoop, vol. 1 (2001)

[15] Lee, H., Zeigler, B.: SES-based ontological process for high level information fusion. In: Proceedings of the 2010 Spring Simulation Multiconference, p. 129. ACM (2010)

[16] Lee, H., Zeigler, B.: System entity structure ontological data fusion process integrated with C2 systems. The Journal of Defense Modeling and Simulation: Applications, Methodology, Technology 7(4), 206–225 (2010)

[17] McGuinness, D., Van Harmelen, F., et al.: OWL web ontology language overview. W3C recommendation 10, 2004–03 (2004)

[18] Mittal, S.: DEVS Unified Process for integrated development and testing of Service Oriented Architectures. Ph.D. thesis, Iniversity of Arizona (2007)

[19] Mittal, S.: Net-centric cognitive architecture using DEVS Unified Process. In: Researching and Developing Persistent and Generative Cognitive Models Workshop, Scottsdale, AZ (2010)

[20] Mittal, S., Douglass, S.: From domain specific languages to DEVS components: application to cognitive m&s. In: Proceedings of the 2011 Symposium on Theory of Modeling & Simulation: DEVS Integrative M&S Symposium, pp. 256–265. Society for Computer Simulation International (2011)

[21] Mittal, S., Douglass, S.: Net-centric ACT-R-based cognitive architecture with DEVS Unified Process. In: Proceedings of the 2011 Symposium on Theory of Modeling & Simulation: DEVS Integrative M&S Symposium, pp. 34–44. Society for Computer Simulation International (2011)

[22] Mittal, S., Douglass, S.: DEVSML 2.0: The language and the stack. In: Proceedings of the Spring Simulation 2012 Multiconference, Orlando, FL (2012)

[23] Mittal, S., Risco-Martin, J.: Netcentric System of Systems Engineering with DEVS Unified Process. CRC Press (2012)

[24] Mittal, S., Risco-Martin, J., Zeigler, B.: DEVS-based simulation web services for net-centric T&E. In: Proceedings of the 2007 Summer Computer Simulation Conference. pp. 357–366. Society for Computer Simulation International (2007)

[25] Mittal, S., Risco-Martín, J., Zeigler, B.: DEVSML: automating DEVS execution over SOA towards transparent simulators. In: Proceedings of the 2007 Spring Simulation Multiconference, vol. 2, pp. 287–295. Society for Computer Simulation International (2007)

[26] Mittal, S., Risco-Martín, J., Zeigler, B.: DEVS/SOA: A cross-platform framework for net-centric modeling and simulation in DEVS Unified Process. Simulation 85(7), 419–450 (2009)

[27] Mittal, S., Zeigler, B., Risco-Martin, J.: Implementation of formal standard for interoperability in M&S/systems of systems integration with DEVS/SOA. International Journal of Command and Control 2 (2009)

[28] Molnár, Z., Balasubramanian, D., Lédeczi, A.: An introduction to the Generic Modeling Environment. In: Proceedings of the TOOLS Europe 2007 Workshop on Model-Driven Development Tool Implementers Forum, Zurich, Switzerland (2007)

[29] Newell, A.: Unified theories of cognition, vol. 187. Harvard Univ. Pr. (1994)

[30] Risco-Martín, J., Moreno, A., Cruz, J., Aranda, J.: Interoperability between DEVS and non-DEVS models using DEVS/SOA. In: Proceedings of the 2009 Spring Simulation Multiconference on ZZZ, p. 147. Society for Computer Simulation International (2009)

[31] Rozenblit, J., Hu, J., Kim, T., Zeigler, B.: Knowledge-based design and simulation environment (KBDSE): Foundational concepts and implementation. Journal of the Operational Research Society, 475–489 (1990)

[32] Rozenblit, J., Huang, Y.: Rule-based generation of model structures in multifaceted modeling and system design. ORSA Journal on Computing 3(4), 330–344 (1991)

[33] Rozenblit, J., Zeigler, B.: Representing and constructing system specifications using the system entity structure concepts. In: Proceedings of the 25th Conference on Winter Simulation, pp. 604–611. ACM (1993)

[34] Schvaneveldt, R., Cohen, T.: Abductive reasoning and similarity: Some computational tools. Computer-Based Diagnostics and Systematic Analysis of Knowledge, 189–211 (2010)

[35] Simon, H.: The sciences of the artificial, 2nd edn. The MIT Press (1981)

[36] Siskind, J., McAllester, D.: Screamer: A portable efficient implementation of nondeterministic common lisp. Ircs technical reports series (1993)

[37] Steele, G.: Common LISP: the language, 2nd edn. Digital Press (1990)

[38] Sztipanovits, J., Karsai, G.: Model-integrated computing. Computer 30(4), 110–111 (1997)

[39] Wainer, G., Al-Zoubi, K., Dalle, O., Hill, D., Mittal, S., Risco-Martin, J., Sarjoughian, H., Touraille, L., Traore, M., Zeigler, B.: Discrete Event Modeling and Simulation: Theory and Applications. In: DEVS Standardization: Ideas, Trends and Future (2010)

[40] White, S., Sleeman, D.: Constraint handling in common lisp. Department of Computing Science Technical Report AUCS/TR9805, University of Aberdeen (1998)

[41] Wilson, M.: Six views of embodied cognition. Psychonomic Bulletin & Review 9(4), 625–636 (2002)

[42] Zeigler, B., Chi, S.: Model-based architecture concepts for autonomous systems design and simulation. In: An Introduction to Intelligent and Autonomous Control, pp. 57–78. Kluwer Academic Publishers (1993)

[43] Zeigler, B., Hammonds, P.: Modeling & simulation-based data engineering: introducing pragmatics into ontologies for net-centric information exchange. Academic Press (2007)

[44] Zeigler, B., Luh, C., Kim, T.: Model base management for multifacetted systems. ACM Transactions on Modeling and Computer Simulation (TOMACS) 1(3), 195–218 (1991)

[45] Zeigler, B., Mittal, S., Hu, X.: Towards a formal standard for interoperability in m&s/system of systems integration. In: GMU-AFCEA Symposium on Critical Issues in C4I (2008)

[46] Zeigler, B., Praehofer, H., Kim, T.: Theory of modeling and simulation: Integrating discrete event and continuous complex dynamic systems, 2nd edn. Academic Press (2000)

Semantic Validation of Emergent Properties in Component-Based Simulation Models

Claudia Szabo[1] and Yong Meng Teo[2]

[1] The University of Adelaide,
   Adelaide, South Australia, Australia
[2] National University of Singapore,
   Singapore, Singapore

Abstract. Advances in composable modeling and simulation have facilitated the development and our understanding of more complex models. As a result, the representation, identification and validation of emergence is becoming of increasing importance because emergent properties can have a negative effect on the overall system behavior. Despite a plethora of definitions and methods, a practical approach to identify and validate emergent properties in newly composed simulation models remains a challenge. This chapter reviews current approaches and presents a new approach for identifying emergent properties in component-based systems. Using a simple example of a flock of birds model, we compare and contrast three main approaches: *grammar-based*, *variable-based* and *event-based*. Lastly, building on our previous work on formal semantic validation of model behavior, we present a new *objective-based approach* for semantic validation of emergent properties in composable simulation.

1 Introduction

"The whole is greater than the sum of its parts" - Aristotle

Complex systems often exhibit properties that are not easily predictable by analyzing the behavior of their individual, interacting components [13, 15]. These properties, called *emergent properties*, are increasingly becoming important as software systems grow in complexity, coupling, and geographic distribution [2, 12, 13, 15]. Examples of emergent properties include connection patterns in social network data analysis [7], trends in big data analytics [8], and power supply variation in smart grids due to provider competition [4]. More malign examples of emergent properties in computer systems are Ethernet capture effect [17], router synchronization

A. Tolk (Ed.): Ontology, Epistemology, & Teleology for Model. & Simulation, ISRL 44, pp. 319–333.
springerlink.com   © Springer-Verlag Berlin Heidelberg 2013

problems [9], and load-balancer failures in a multi-tiered distributed system [15] among others.

Because emergent properties may have undesired and unpredictable effects and consequences, and unpredictable systems are less credible and difficult to manage, techniques for the identification and validation of emergent properties pose an interesting challenge. Despite ongoing research interest since the 1970s, most approaches focus mainly on the post-mortem observation of emergence in various biological, social, and AI systems, and less on measuring and advancing our understanding in the cause-and-effect of emergence. A plethora of examples of emergent properties have been identified and classified but few instances have been measured and explained [6, 12, 14, 15].

In this chapter, we present a new approach to identify and validate emergent properties as part of semantic composability validation. In validation, it is important to distinguish between expected behavior that stems from the interactions of the underlying components of a model, and emergence behavior or unexpected behavior arising from seemingly unrelated phenomena. While simulation validation demonstrates that a simulation meets expected behavior, emergent properties validation focuses on showing that the unexpected behavior is valid (or invalid) for a given set of conditions. In section 2, we review three key approaches to identify emergence: *grammar-based*, *variable-based*, and *event-based*. We discuss their advantages and limitations and show how these could be used in a simple example of a bird flocking model. Section 3 presents an objective-based validation approach for the semantic validation of emergence. In this approach, a meta-component describes each sub-component of the system. The meta-component includes among others a specification of the objective that the sub-component achieves. Using our approach, we next compute the entire system state and compare it with an objective-based reconstruction of the system state in the absence of interactions between sub-components. Section 4 summarizes this article.

2 Emergent Properties and Examples

An emergent property can be defined as "a property of an assemblage that could not be predicted by examining the components individually" [2]. Common characteristics of emergence include: radical novelty (features not previously observed in systems); coherence or correlation (meaning integrated wholes that maintain themselves over some period of time); a global or macro "level" (i.e. there is some property of "wholeness"); it is the product of a dynamic process (it evolves); and it is "ostensive" (it can be perceived). The fundamentals behind understanding different types of emergence lie in the assumption that in any component-based system there is a *micro-level*, the abstraction level of each individual component, and a *macro-level*, the abstraction of the composed model as a whole. Micro-level properties are usually measured by observing the component states such as the collection of all variables and their values, of each system component. In contrast, the macro-level properties can be measured either as an aggregation of all the states of the system sub-components, or by observing the overall system behavior and trends such as the change from non-flocking to flocking behavior in birds.

Three main types of emergence have been identified, namely *nominal, strong,* and *weak* [2]. In *nominal emergence*, the macro-level depends on the micro-level in the straightforward sense that the whole is dependent on their constituents. *Strong emergence* is a more powerful definition that assumes nominal emergence, but introduces *downward causation*, which can be informally defined as the influence of the macro-level on the micro-level. In contrast, *weak emergence* states that given the properties of the parts and the interaction rules among them, it is not *trivial* to infer the properties of the whole. In this context, trivial is taken to mean "by-hand" human calculations, and in order to identify weak emergence, one needs a computer model and its simulation.

In the following, we focus on the main techniques and procedures to identify weak emergence in component-based simulation models. In component-based complex systems, emergence validation approaches are classified in three key categories, namely, *grammar-based, variable-based* and *event-based*. Most approaches, such as variable-based and event-based methods, assume that there exists an observation of emergence or irregularity prior to the validation exercise, and aim to identify the cause of emergence. The grammar-based approaches aim to identify emergence on the fly, by computing the difference between a system state obtained by the composition of sub-systems *with* and *without* interactions respectively. This method does not require a-priori observation of the system to identify possible emergent properties or behaviors, which makes it suitable for large systems where such observations are almost impossible. However, the nature of the formalism and the computation of the system states make it difficult to scale, as we will see below.

2.1 Approaches

In this section, we discuss three main types of emergence validation approaches, namely, *grammar-based, variable-based* and *event-based*. We present a theoretical overview of each approach and discuss how it can be applied to a simple model of a flock of birds, also known in the literature as the boid model [18]. Each component abstracts a moving bird, which changes its position based on a set of simple rules that defines its current position and the position of the other birds in the flock. These rules are (i) *separation* - individual bird steer to avoid crowding the other birds in the flock (ii) *alignment*- individual bird stear towards the average herding of local flockmates and (iii) *cohesion* - individual bird moves towards the average position of local flockmates. The boid model has been shown to exhibit emergent behavior of flocking, and flocking after encountering an obstacle, when the flock splits and reunites. Fig. 1 shows a screenshot of flocking in our implementation.

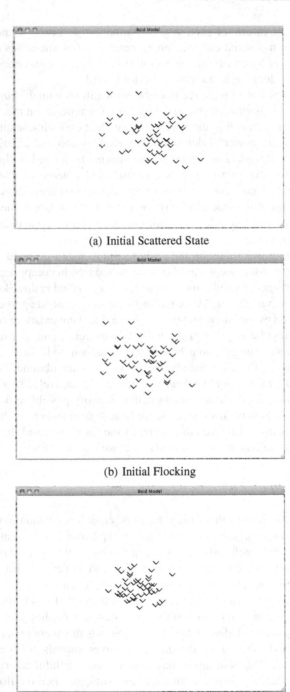

(a) Initial Scattered State

(b) Initial Flocking

(c) Flocking

Fig. 1 Visualization of Flocking in a Boid Model

2.1.1 Grammar-Based Approach

In grammar-based methods, two grammars, L_{WHOLE} describes the properties of the system as a whole, and L_{PARTS} describes the properties obtained from the reunion of the parts [14]. Kubik [14] proposes the use of grammar systems, which are symbolic devices composed from a set of grammars that interact with each other through tapes on which each grammar writes symbols. A formal *grammar* is a set of rules that governs the formation of *words* using a set of *symbols*. This paradigm applies easily to multi-agent systems, where each agent can be represented by a grammar and the behavior of an agent is represented by how it changes the symbols on the common tape.

Emergence is defined as the difference between the properties of the system as a whole L_{WHOLE}, and the reunion of the properties of the system parts, L_{PARTS} [14]. To derive L_{PARTS}, Kubik [14] proposes the superimposition of each agent language defined by the grammar system. Informally, L_{PARTS} is defined by the sum of the changes or conditions the agents bring about the environment if they would act individually in the system. This is obtained by using a superimposition of all the words that the agent grammars produce. In this superimposition, L_{PARTS} is formed using a reunion operator for all possible permutations of words created following rules that give higher priority to the symbols generated by the agent grammar, and less priority to the system symbols. While L_{PARTS} uses the superimposition operator to highlight the behavior of agents without considering the agent interaction with the environment, L_{WHOLE} is obtained by taking all the symbols written by all agents on the tape.

More formally, consider two words $W_1 = a_1 a_2 \ldots a_n$ and $W_2 = b_1 b_2 \ldots b_m$, and their superimposition $W_{supimp} = c_1 c_2 \ldots c_l$. We have the following:

1. if $n \leq m \Rightarrow l = n$ else $l = m$
2. if $a_i \in V_A \Rightarrow c_k = a_i$
3. if $a_i \in V_E$ and $b_j \in V_E \Rightarrow c_k = a_i$
4. if $a_i \in V_E$ and $b_j \in V_A \Rightarrow c_k = b_j$
5. $a_i = \varepsilon \Rightarrow c_k = b_j$
6. $b_j = \varepsilon \Rightarrow c_k = a_i$

Thus, $L_{PARTS} = \{W_1 superimpose(W_2 superimpose(W_3 superimpose \ldots W_n) \cup \ldots W_n superimpose(W_{n-1} superimpose \ldots (W_2 superimpose W_1))\}$. This approach is exemplified using very simple examples of a four-by-four Game of Life glider pattern [10]. However, the calculation of L_{PARTS}, which requires computing the reunion of all permutations of superimposing all of the words produced by agents is computationally expensive and might not scale well.

In the following, we discuss the application of this approach to identify emergent properties in the boid model. The pseudo-code for the grammar-based approach is presented in Fig. 2. As shown, the main difficulty in applying this pseudocode to a boid model lies in identifying the languages generated by each agent in the system, $L(A_i)$. We divide the drawing panel shown in Fig. 1 into a grid that is small enough such that at any point in time, each grid cell is only occupied by a single bird. For

1. Define $L(A_i)$, as each individual agent language
2. Calculate L_{WHOLE}, as the language generated by the agent interaction
3. Calculate L_{PARTS}, as the language generated without agent interaction
4. Calculate emergence, as $Emergence = L_{WHOLE} - L_{PARTS}$

Fig. 2 Pseudo-code for Emergence Validation

simplicity and better visualization, we reduce the number of birds in the flock to five, and the number of cells in the grid to sixteen. The direction of the bird flight is from left to right if the bird is alone in the grid, and the grid is represented as a torus.

The boid model (BM) can then be represented as

$$(V_A = \{B\}), V_E = \{e\}, A_1, \ldots, A_5, \{(v_0,), (v_1,),$$
$$BM = (v_2,), (v_3,), (v_4,), (v_5,), (v_6,), (v_7,), (v_8,), (v_9,), (v_{10},),$$
$$(v_{11},), (v_{12},), (v_{13},), (v_{14},), (v_{15})\})$$

where V_A is a set of agent symbols that represents the position of agents in the cell, B denotes a bird in a cell, V_E is a set of environment variables, (e) denotes an empty cell, and v_i represents the initial state of the system grid. The set, A_i, representing the rule set for each of the birds is shown below:

$$A_i = \begin{cases} \begin{matrix} e\,e\,e\,e \\ e\,e\,B\,e \\ e\,e\,e\,e \\ e\,e\,e\,e \end{matrix} \rightarrow \begin{matrix} e\,e\,e\,e \\ e\,e\,e\,B \\ e\,e\,e\,e \\ e\,e\,e\,e \end{matrix}, \begin{matrix} e\,e\,e\,e \\ e\,e\,B\,e \\ e\,e\,e\,e \\ e\,B\,B\,e \end{matrix} \rightarrow \begin{matrix} e\,e\,e\,e \\ e\,e\,e\,e \\ e\,e\,B\,e \\ e\,B\,B\,e \end{matrix}, \begin{matrix} e\,e\,e\,e \\ e\,e\,B\,e \\ e\,e\,e\,e \\ B\,B\,e\,e \end{matrix} \rightarrow \begin{matrix} e\,e\,e\,e \\ e\,e\,e\,e \\ e\,e\,B\,e \\ B\,B\,e\,e \end{matrix}, \ldots \end{cases}, 1 \le i \le 5$$

Specifically, the rules above show that if an agent (i.e. a bird) is alone in the grid, it will not change its direction of flight (assumed left to right), whereas if there are two or more agents in the vicinity, it will move closer to the center of mass of the "flock", as defined in the second rule above. Similar production rules can be constructed for the rest of the grid and the boid model rules. However, it is important to highlight here that it is difficult to represent a multi-dimensional parameter domain, such as a bird state, that has direction, heading, speed and position among others.

According to the method detailed in [14], a sequence of a rewriting process in the grammar array system for five agents in a 4 x 4 torus is shown next.

$$\begin{matrix} B\,e\,e\,e \\ e\,e\,e\,e \\ B\,B\,e\,e \\ e\,B\,B\,e \end{matrix} \Rightarrow \begin{matrix} e\,e\,e\,e \\ B\,e\,e\,e \\ e\,B\,B\,e \\ e\,e\,B\,B \end{matrix} \Rightarrow \begin{matrix} e\,e\,e\,e \\ e\,e\,e\,e \\ B\,e\,B\,B \\ B\,e\,e\,B \end{matrix} \Rightarrow \begin{matrix} e\,e\,e\,e \\ e\,e\,e\,e \\ B\,B\,e\,B \\ B\,B\,e\,e \end{matrix} \Rightarrow \begin{matrix} e\,e\,e\,e \\ e\,e\,e\,e \\ B\,B\,B\,e \\ e\,B\,B\,e \end{matrix} \Rightarrow \begin{matrix} e\,e\,e\,e \\ e\,e\,e\,e \\ e\,e\,B\,B \\ B\,e\,B\,B \end{matrix} \Rightarrow \begin{matrix} e\,e\,e\,e \\ e\,e\,e\,e \\ e\,e\,B\,B \\ B\,B\,e\,B \end{matrix} \Rightarrow \begin{matrix} e\,e\,e\,e \\ e\,e\,e\,e \\ B\,B\,e\,e \\ B\,B\,B\,e \end{matrix} \Rightarrow \ldots$$

Therefore, from the starting configuration, the group of five agents can generated the language $L(BM)$:

$$L(BM) = \begin{cases} \begin{matrix} B\,e\,e\,e \\ e\,e\,e\,e \\ B\,B\,e\,e \\ e\,B\,B\,e \end{matrix}, \begin{matrix} e\,e\,e\,e \\ B\,e\,e\,e \\ e\,B\,B\,e \\ e\,e\,B\,B \end{matrix}, \begin{matrix} e\,e\,e\,e \\ e\,e\,e\,e \\ B\,e\,B\,B \\ B\,e\,e\,B \end{matrix}, \begin{matrix} e\,e\,e\,e \\ e\,e\,e\,e \\ B\,B\,e\,B \\ B\,B\,e\,e \end{matrix}, \begin{matrix} e\,e\,e\,e \\ e\,e\,e\,e \\ B\,B\,B\,e \\ e\,B\,B\,e \end{matrix}, \begin{matrix} e\,e\,e\,e \\ e\,e\,e\,e \\ e\,e\,B\,B \\ B\,e\,B\,B \end{matrix}, \begin{matrix} e\,e\,e\,e \\ e\,e\,e\,e \\ e\,e\,B\,B \\ B\,B\,e\,B \end{matrix}, \begin{matrix} e\,e\,e\,e \\ e\,e\,e\,e \\ B\,B\,e\,e \\ B\,B\,B\,e \end{matrix}, \ldots \end{cases}$$

As discussed above, $L(BM)$ can be viewed as the language resulting from the interaction of all agents in the system, i.e. L_{WHOLE}.

To determine L_{PARTS}, we employ the superimposition operator in Section 2.1.1. Specifically, consider two agents, A_1 and A_2 that generate the following languages

$$L(A_1) = \left\{ \begin{array}{cccc} e\,e\,e\,e & e\,e\,e\,e & e\,e\,e\,e & e\,e\,e\,e \\ e\,B\,e\,e & e\,e\,B\,e & e\,e\,e\,B & B\,e\,e\,e \\ e\,e\,e\,e\, , & e\,e\,e\,e\, , & e\,e\,e\,e\, , & e\,e\,e\,e \\ e\,e\,e\,e & e\,e\,e\,e & e\,e\,e\,e & e\,e\,e\,e \end{array} \right\} \text{ and}$$

$$L(A_2) = \left\{ \begin{array}{cccc} e\,e\,e\,e & e\,e\,e\,e & e\,e\,e\,e & e\,e\,e\,e \\ e\,e\,e\,e & e\,e\,e\,e & e\,e\,e\,e & e\,e\,e\,e \\ B\,e\,e\,e\, , & e\,B\,e\,e\, , & e\,e\,B\,e\, , & e\,e\,e\,B \\ e\,e\,e\,e & e\,e\,e\,e & e\,e\,e\,e & e\,e\,e\,e \end{array} \right\}$$

with and without interaction respectively. The result of the superimposition of these two languages is

$$L_{sum} = L(A_1) \text{ superimpose } L(A_2) \cup L(A_2) \text{ superimpose } L(A_1) \Rightarrow$$

$$L_{sum} = \left\{ \begin{array}{cccc} e\,e\,e\,e & e\,e\,e\,e & e\,e\,e\,e & e\,e\,e\,e \\ e\,B\,e\,e & e\,e\,B\,e & e\,e\,e\,B & B\,e\,e\,e \\ B\,e\,e\,e\, , & e\,B\,e\,e\, , & e\,e\,B\,e\, , & e\,e\,e\,B \\ e\,e\,e & e\,e\,e\,e & e\,e\,e\,e & e\,e\,e\,e \end{array} \right\}$$

It is important to highlight here that in this case it is evident that the superimposition does not contain interaction. For example, if interaction was considered, the third word in the sequence would contain B symbols on the same line.

The language generated while considering the interaction among agents is richer than individual agents without interactions. L_{WHOLE} is richer both in terms of the number of words and in the density of non-environment symbols, than L_{PARTS}. As such,

$$L(BM) - L_{sum} \neq \emptyset$$

The definition and set difference between $L(BM)$ and L_{sum} provides a straightforward formal method for defining and identifying emergence. However, it is difficult if not impossible to pinpoint which of the elements in the $\{L(BM) - L_{sum}\}$ set represents emergence, and what was its exact cause. Moreover, computing L_{sum} is computationally expensive as the superimposition operator has to consider all the combinations of words generated by all the agents in the system.

2.1.2 Variable-Based Approach

In variable-based methods, a specific variable is chosen to describe emergence. Changes in the values of this variable are said to signify the presence of emergent properties [18]. For example, changes in the centre of mass of a bird flock could be used as an example of emergence in bird flocking behavior, as shown in

[18]. The approach uses Granger causality to establish the relationships between a macro-variable, representing a system property, and micro-variables, representing properties of the system sub-components.

According to the definition of *Granger causality*, a variable Y causes a variable X if the inclusion of past observations of Y reduces the prediction error of X in a linear regression model of X and Y, as compared to a model that only includes X. Another important definition is that of *G-autonomy*, in which the focus is on whether past observations of a variable X influence the current observation of X more than the values of other variables Y in the system. *G-emergence* is defined based on Granger causality and G-autonomy.

A macro-variable M is *G-emergent* from a set of micro-variables m iff (i) M is G-autonomous with respect to variables m and (ii) M is Granger causal with respect to m. In other words, a macro-variable could be G-emergent from a set of micro-variables if there are hidden or latent influences that are not evident in linear regression.

This approach provides a clear and easily measurable process to identify emergence because it looks at measurable quantities found in the system state, which is defined as the reunion of all sub-systems states. However, finding a good variable to describe a system can be a difficult task that requires system expert intervention. Moreover, Granger causality is designed to handle only pairs of variables, and might not apply when the macro-variable depends on more than one micro-variable.

We apply this approach to the flock of birds model shown in Fig. 1. The variable-based approach uses a Matlab software package [1] to calculate G-autonomy ($ga_{M|m}$) and G-causality ($gc_{M|m}$), between a macro variable (M) and a set of micro variables (m). The measure of emergence is then calculated as

$$ge_{M|m} = ga_{M|m}\left(\frac{1}{N}\sum_{i=1}^{N} gc_{m_i \to M}\right)$$

Towards measuring the G-emergence of the center of mass of the flock of birds, we use the pseudo-code presented in Fig. 3, which follows closely the description in [18].

Our aim is to establish whether the coordinates of the center of mass of the flock of birds $CM(CM_x, CM_y)$ are G-emergent from the coordinates of each individual in the flock (x_i, y_i), following a number of observations (obs) of these coordinates. Towards this, we construct *mat_x* and *mat_y*, a matrix with the coordinates of all individuals in the flock, as well as the center of mass, on the x-axis and y-axis respectively. The rows of the matrices represent the variables, and the columns the observations. The data is then pre-processed to reflect the distance from the (x, y) coordinates of the individual to the center of the environment, in *mat_dist*. The pseudo-code returns a matrix in which values closer to one represent high G-emergence, as discussed in [18].

1. Collect mat_x
2. Collect mat_y
3. Calculate mat_dist
4. For each variable i,

   calculate $gc_{m_i \to M}$: `[gci] = cca_granger_regress(mat_dist,obs)`
5. Calculate $ga_{M|m}$: `[ga] = cca_autonomy_regress(mat_dist, obs)`
6. Calculate `ge = 1/N * dot (gc1+gc2+...+ `gc_N`)`

Fig. 3 Pseudo-code for the Calculation of G-emergence

2.1.3 Event-Based Approach

In event-based methods, behavior is defined as a series of events which change a system or sub-system states [5]. The motivating example behind this work is that often when a macro-level property is constructed from the aggregation of sub-system states, there is a loss of information with respect to the *cause* of the emergent behavior. In particular, it is not possible to establish which sub-system interaction is responsible for the current behavior.

Towards this, the authors propose the definition of simple and complex event types. A *simple event* type signifies a change in a sub-system state. It is associated with a transition and has a duration. A *complex event* is defined as being either a simple event or comprises two complex events linked by a relationship. This relationship is a temporal operator (meaning that there is a temporal relationship between the two complex events) that can optionally have descriptions of constraints related to the environment or to the state of the two sub-systems.

Based on the above definitions, emergent behavior is defined beforehand as a sequence of event types, as shown in Fig. 4. A simulation is run and the appearance of the sequence defining emergent behavior is verified. This is formally done by representing the complex event types as a directed multi-graph, where the nodes represent various event instances in the complex event type and the directed arcs denote the relationships between two events. The simulation is also represented as a directed graph S_1. A complex event type is said to appear in the simulation if a sub-graph of the simulation graph can be proved to be isomorphic with the complex type graph. This provides an overview of the sequence of interactions that led to the appearance of an event or property.

For the flock of birds model, a complex event that represents emergence is *bird_flocking*, represented as a temporal sequence of simple events *bird_move_i*, where $i = 1 \dots n$ represents the id of the birds in the flock.

$$bird\_flocking = bird\_move_1 \to bird\_move_2 \to \cdots \to bird\_move_n$$

Running the simulation of the flock of birds clearly identifies this sequence of events. However, a key assumption is that the description of emergence exists

1. Describe components as a collection of simple events s_e
2. Identify complex event $C_e = s_{ei} \otimes s_{ej}$
3. Run the simulation of the complex system and obtain the sequence of transitions from system states as S_1
4. Represent the system state multi-graph from S_1 and identify emergence

Fig. 4 Pseudo-code for Identifying Complex Event Types

beforehand, which is not always the case in real-life scenarios, where emergence is something not seen or predicted before.

2.2 Discussion

The application of emergence validation approaches to the simple flock of birds model highlights a few important issues. The approaches described above can be categorized into *a-priori* and *a-posteriori* methods. In *a-priori* methods such as the variable-based and event-based approaches, there is a need to identify a variable or a complex event that defines the emergent property. The identification of this variable is manual and might not be straightforward for more complex examples. In *a-posteriori* methods such as the grammar-based approaches, a formalism guides the identification of emergence as a difference between the outcome of the interaction among the components in the system, and the outcome calculated if no interaction among components occurs. Ideally, the latter approach would be suitable for large complex systems where a single variable to define emergence is difficult to find. However, the limiting factor in these approaches is the formalism itself, which needs a high level of abstraction, as shown in the example.

It is crucial for the validation of emergence to have a micro-macro separation between the abstractions employed in the system model. The identification of micro and macro variables to describe a system is in most cases difficult to automate. Thus, a variable-based approach is suitable when there is a single variable that can characterize an emergent property, namely, the center of mass in a flock of birds. In this case, it is relatively straightforward to mathematically calculate the causality between this variable and the other parameters of individuals in the system. However, an important assumption is that the variable is identified beforehand to characterize emergence. Moreover, several simplifications need to be established on a case-by-case basis, and as such might make this approach difficult to automate.

Grammar-based approaches seem to solve this issue. However, to the best of our knowledge, current studies only look at applying this approach to the "Game of Life" model, in which a binary state (dead or alive) characterizes each agent. As we have seen in applying this method to the flock of birds example, several limiting abstractions have to be in place, because agents have a more complex state characterized by speed, heading, and position among others. These abstractions might result in a loss of precision in identifying emergence. In addition, the computational complexity of calculating the languages generated by the agents, with and without interaction, seems to be exponential in the number of words generated by each agent. To the best of our knowledge, there is currently no study to analyze the computational complexity of this method.

In the following, we propose an a-posteriori, variable-based method that aims to address these issues. In our objective-based approach, we propose to define each system component in terms of the objective it aims to achieve. These objectives are defined as the outcome of a finite-state machine that models each component and are used in the system simulation. Next, we simulate the complex system and, at each simulation step, analyze the system state. We compare the simulated system state with a calculated system state using reconstructability analysis. In reconstructability analysis, component variables and objectives are used to calculate a system state without considering the interactions among the components, and the components with the environment.

3 Emergence in Component-Based Model Development

In the modeling and simulation community, the possibility of emergent behavior in component-based model development has been highlighted since the 1990's by Page and Opper [16]. They propose a slightly different definition of emergent behavior from systems theory. Let a and b denote two given components and its composition $a \diamond b$, an objective o, and the "satisfies" operator $\vDash$. If $a \vDash o$, then a satisfies objective o. If $a \nvDash o$ and $b \nvDash o$, but $(a \diamond b) \vDash o$, then we can say that the composition is *emergent*. From a more practical perspective, Gore and Reynolds propose to identify the exact cause, with respect to variables and specific position in the source code, of parameter values that have not been encountered before [11]. Programming language techniques such as static analysis are employed to determine the variable(s) in the source code that result in the new value. However, this approach requires an expert to identify and detect a new or un-encountered variable value, as well as direct access to the source code for analysis.

3.1 Objective-Based Approach for Identifying Weak Emergence

We propose to focus on the representation of systems in terms of objectives and properties, to facilitate the automated validation and identification of emergence. The focus of our objective-based approach is two-fold: (a) firstly, a composition is defined as a "sum" of its constituents; and (b) we define objectives to identify emergence. We then propose a semantic validation method to validate the component-based system.

Our objective-based validation approach relies on a definition of a system component that focuses on *what* rather than *how*. Each system sub-component is defined using the objectives it achieves as shown in Fig. 5. For example, a component property for a bird model would be "Fly north-bound with an average speed of 15 km/hour". These objectives, defined as the outcome of a finite-state machine (FSM),

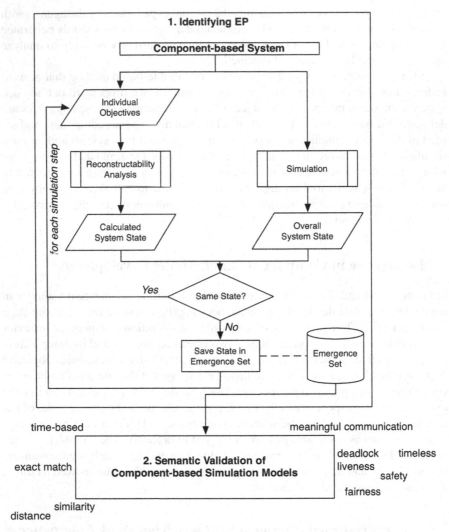

Fig. 5 Objective-based Approach for Identification and Validation of Emergent Properties

defines each component and is used in the system simulation. Each transition in the FSM has a post condition which specifies the objective, in terms of variable values, that should be achieved by the transition. Next, at each time-step during the simulation run, we reconstruct the composed complex system from its sub-systems. This is done using a sub-problem of reconstructability analysis [3], which looks at reconstructing a complex system from variables defining its sub-systems. Next, we compare this theoretical, *calculated state* with the *simulation state*. If there is an unacceptable deviation in the observed parameters, we highlight this state as a possible emergence state and add it to an *emergent set*. We repeat this for the entire simulation. At the end of the process, the emergent set is shown to the user.

The component-based model is then semantically validated. We propose to validate the component-based system model using our deny-validity approach [20, 21]. Towards the semantic validation of composed simulation models, our deny-validity approach subjects the composition to a battery of tests that either discard a composed model as invalid, or increase the credibility of the model that is not eliminated [19]. We first eliminate models that have invalid model properties through a feasible process that uses support from model checking and ontologies. At this stage, the validation process focuses on discarding composed models in a three-step approach. Firstly, the component interoperability with respect to exchanged data is validated, using semantically-sugared attribute values from our proposed component-based ontology. Secondly, we employ model checking to validate all possible interleaved execution schedules. From a practical perspective, we consider timeless transitions. Thirdly, we introduce time and validate a meta-simulation of the composed model, using properties specified by the model composer. However, models that pass the first validation stage may have valid properties but may still be invalid when compared with a reference model. In the next stage, our novel time-based formalism for the representation of the composed model supports the semantic comparison between a composed model and a reference model [21]. A model component is represented as a mathematical function of time and states. We introduce formal definitions of validity that consider closeness with respect to a reference model. We propose a semantic metric relation to evaluate this closeness, considering attribute and state relations in our proposed component-based ontology. The time-based formalism permits the validation of composed models with complex structures, but at increased computational cost. In this process, we incorporate and highlight the states from the emergent set to understand the sequence of events leading to them.

4 Concluding Remarks

This chapter presents state-of-the-art approaches for the identification and validation of emergent properties. The approaches are classified in four categories, namely, *grammar-based*, *variable-based*, *event-based*, and our proposed *objective-based*. Grammar-based approach attempts to formalize emergence as the difference between the language generated by the interaction of the individual components and with its environment, and the language generated by each individual in part. Variable-based methods define a system-wide variable as emergence to determine the causality relation between that variable and individual parameters. In event-based approach, a system-wide complex event is defined to determine the sequence of individual simple events that generate it. Lastly, objective-based methods use simulation and reconstructability analysis to identify differences between the desired system state and the actual system state.

Using a simple example of a flock of birds model, we studied these approaches. We showed that the variable-based approach is suitable when a single variable defines an emergent property such as the center of mass in a flock of birds. In this case, it is relatively straightforward to mathematically calculate the causality between this variable and the other parameters of individuals in the system. However, an important assumption is that the variable is identified *a-priori* to characterize emergence. In addition, several simplifications have to be established on a case-by-case basis, and thus render this approach difficult to automate. In contrast, the grammar-based approach does not need an a-priori identification of an emergent variable. However, to the best of our knowledge, current studies apply this approach only to the "Game of Life" system, using a binary state (dead or alive) to characterize each agent, and thus the system can be easily abstracted an a two-dimensional grid, inline with the nature of a grammar. This is not the case in a more complex model, such as the flock of birds, in which agents are characterized by more than one state such as speed, heading, and position among others. As demonstrated in our example, the complexity of the model results in an additional layer of abstraction, and this abstraction might result in a loss of precision in identifying emergence. Moreover, the computational complexity of calculating the languages generated by the agents, with and without interaction, seems to be exponential in the number of words generated by each agent. To the best of our knowledge, there is currently no study to analyze the computational complexity of this method. Lastly, the event-based approach requires an apriori identification of emergence, defined at the macro-level as a sequence of events from the micro-level.

In conclusion, there has been active research in defining emergence and identifying various real-life examples of emergence, but formally identifying and validating emergent properties, and automating this process remains a key challenge. Current state-of-the-art approaches are limited in most cases to simple systems and have yet to demonstrate their usefulness in more complex systems of practical use.

Acknowledgements. This work is supported by the Singapore Ministry of Education under the grant T1251RES1114.

References

1. Kubik, A.: Causal Connectivity Analysis MATLAB toolbox, http://ccatoolbox.pbworks.com/w/page/5805266/FrontPage (last retrieved January 2012)
2. Bedau, M.: Weak Emergence. Philosophical Perspectives 11, 375–399 (1997)
3. Cavallo, R., Klir, G.: Reconstructability Analysis of Multi-dimensional Relations: A Theoretical Basis for Computer-aided Determination of Acceptable System Models. Int. Journal of General Systems 5, 143–171 (1979)
4. Chan, W., Son, Y.S., Macal, C.M.: Simulation of Emergent Behavior and Differences Between Agent-Based Simulation and Discrete-Event Simulation. In: Proceedings of the Winter Simulation Conference, pp. 135–150 (2010)
5. Chen, C., Nagl, S.B., Clack, C.D.: Specifying, Detecting and Analysing Emergent Behaviours in Multi-Level Agent-Based Simulations. In: Proceedings of the Summer Computer Simulation Conference (2007)

6. Chen, C.C., Nagl, S.B., Clack, C.D.: A Formalism for Multi-level Emergent Behaviours in Designed Component-based Systems and Agent-based Simulations. Understanding Complex Systems (2009)
7. Chi, L.: Transplating Social Capital to the Online World: Insights from Two Experimental Studies. Journal of Organizational Computing and Electronic Commerce 19, 214–236 (2009)
8. Fayyad, U., Uthurusamy, R.: Evolving Data Into Mining Solutions for Insights. Communications of the ACM 45 (2002)
9. Floyd, S., Jacobson, V.: The synchronization of Periodic Routing Messages. In: Proceedings of SIGCOMM, pp. 33–44 (1993)
10. Gardner, M.: Mathematical Games (1970)
11. Gore, R., Reynolds, P.: Applying Causal Inference to Understand Emergent Behavior. In: Proceedings of the Winter Simulation Conference, Miami, USA, pp. 712–721 (2008)
12. Holland, J.: Emergence, From Chaos to Order. Basic Books (1999)
13. Johnson, C.W.: What are Emergent Properties and How Do They Affect the Engineering of Complex Systems? Reliability Engineering and System Safety 12, 1475–1481 (2006)
14. Kubik, A.: Towards a Formalization of Emergence. Journal of Artificial Life 9, 41–65 (2003)
15. Mogul, J.C.: Emergent (mis) behavior vs. Complex Software Systems. In: Proceedings of the 1st ACM SIGOPS/EuroSys European Conference on Computer Systems, New York, USA, pp. 293–304 (2006)
16. Page, E., Opper, J.: Observations on the Complexity of Composable Simulations. In: Proceedings of the Winter Simulation Conference, Phoenix, USA, vol. 1, pp. 553–560 (1999)
17. Ramakrishnan, K.K., Yang, H.: The Ethernet Capture Effect: Analysis and Solution. In: Proceedings of the IEEE Local Computer Networks Conference, Minneapolis, USA (1994)
18. Seth, A.K.: Measuring Emergence via Nonlinear Granger Causality. In: Proceedings of the Eleventh International Conference on the Simulation and Synthesis of Living Systems, pp. 545–553 (2008)
19. Szabo, C., Teo, Y.: An Approach for Validation of Semantic Composability in Simulation Models. In: Proceedings of the 23rd ACM/IEEE/SCS Workshop on Principles of Advanced and Distributed Simulation, New York, USA, pp. 3–10 (2009)
20. Szabo, C., Teo, Y.: On Validation of Semantic Composability in Data-driven Simulation Models. In: Proceedings of the 24th ACM/IEEE/SCS Workshop on Principles of Advanced and Distributed Simulation, Atlanta, USA, pp. 73–80 (2010)
21. Szabo, C., Teo, Y., See, S.: A Time-based Formalism for the Validation of Semantic Composability. In: Proceedings of the Winter Simulation Conference, Austin, USA, pp. 1411–1422 (2009)

Ontological, Epistemological, and Teleological Perspectives on Service-Oriented Simulation Frameworks

Wenguang Wang, Weiping Wang, Qun Li, and Feng Yang

National University of Defense Technology
Changsha, China

Abstract. This chapter investigates service-oriented simulation frameworks from the ontological, epistemological, and teleological perspectives. First, we give an overview of various specific frameworks that imply particular referential ontological, epistemological, and teleological perspectives for real world systems. Then we combine the partial considerations derived from the review into a unifying framework. It inspects the crossover between the disciplines of M&S, service-orientation, and software/systems engineering. From a methodological perspective, we show its ontological, epistemological, and teleological implications for abstract approaches. The unifying framework can, in turn, facilitate the classification, evaluation, selection, description, and prescription of the known or proposed frameworks. Thus, the referential and methodological perspectives build a systematical philosophical foundation of the service-oriented simulation paradigm.

Keywords: Ontology, Epistemology, Teleology, Service-oriented simulation, Service-oriented architecture (SOA), Software engineering, Systems engineering, Referentiality, Methodology, Composability, Interoperability.

1 Introduction

With the prevalence of net-centric environments, the modeling and simulation (M&S) community is highly demanded to offer agile capabilities (e.g. for intelligent applications) by providing, reusing, and composing heterogeneous resources. Service-Oriented Architectures (SOA) [1] as well as its implementation techniques (Web Services etc.) provides such an opportunity. Therefore, the use of SOA to extend the capabilities of M&S frameworks has attracted increasing attention [2]. Various service-oriented simulation frameworks have been proposed or implemented by different institutes using different formalisms and techniques. These include formalism-based [3], model-driven [4], interoperability protocol based [5], Open Grid Services Architecture (OGSA) based [6], and ontology driven

A. Tolk (Ed.): Ontology, Epistemology, & Teleology for Model. & Simulation, ISRL 44, pp. 335–358.
springerlink.com

[7] frameworks, as well as the eXtensible Modeling and Simulation Framework (XMSF) [8]. All of these imply particular ontological, epistemological, and teleological considerations that direct and shape the diversities of implementations by developers.

Ontology, epistemology, and teleology build the philosophical foundation of a discipline [9,10]. Ontology is the study of what exists, often captured as a finite set of concepts and their relations. Epistemology focuses on the way we define knowledge, especially how we come to know new knowledge. Teleology emphasizes on the study of purpose and action, i.e. purposeful behavior while seeking of a goal. To gain maturity and evolve as a new simulation paradigm/ discipline, service-oriented simulations must build a solid philosophical foundation from the ontological, epistemological, and teleological perspectives. Therefore, we must study (1) the purpose of the paradigm/discipline (teleology); (2) the philosophical foundations and the basic concepts/elements thereof (ontology); (3) its formation and evolution (epistemology); (4) a systematic methodology that include the formation, way of thinking, and assessment of particular methods (epistemology and teleology); (5) ways to find and solve new problems/gaps (epistemology and teleological activities).

Two key challenges still need to be addressed in the research of service-oriented simulation. First, developers of each framework use particular ontological, epistemological, and teleological assumptions that lead to varieties of formalisms, designs, techniques, and implementations. It is thus necessary to undertake a review to facilitate the classification, evaluation, and selection of the reviewed or future frameworks. Second, to the best of our knowledge, there is no efficient way to connect and compare one framework to another. Therefore a general (or high level) systematic philosophical foundation of the paradigm derived from the partial perspectives is needed.

This chapter has two interrelated goals. The first is to undertake an overview of various service-oriented simulation approaches that imply particular ontological, epistemological, and teleological considerations. The second goal is to combine the partial considerations derived from the state-of-the-art into one unifying framework that reflect a systematical philosophical foundation of the service-oriented simulation discipline. Ontology, epistemology, and teleology have both referential (for real world) and methodological (for abstract methods) categories [11]. In this chapter they exhibit referential properties in the particular approaches, while reveal methodological characteristics in the unifying framework.

The rest of the chapter is organized as follows. In Section 2, we give a comprehensive survey of various service-oriented frameworks. Deriving from the review, we propose a novel unifying methodology in Section 3 and show its ontological and epistemological implications. Driven by teleology, we use the unifying frame to describe, compare, and prescribe the reviewed approaches in Section 4. In the last section we describe the contributions and recommend future work.

2 Particular Ontological, Epistemological, and Teleological Perspectives on Specific Frameworks

Different people have different perceptions, understandings, and assumptions of the service-oriented simulation paradigm. Thus, the diversity of their ontological, epistemological, and teleological perspectives leads to various design and implementations of the reviewed frameworks.

2.1 Formalism-Based Framework

The category has some common ontological and epistemological assumptions. They all depend on certain formalisms in a theoretical or mathematical way e.g. the Discrete Event System Specification (DEVS). More or less, they cover the domain ontologies of M&S (referent, model, simulator, and experimental frame), service-orientation, and software/system engineering.

   (1)   DEVS Unified Process framework (DUNIP)

From the teleological perspective, DUNIP was proposed by Mittal and his colleagues [3] for the integrated development and testing of service-oriented architectures. Additionally, the applied projects [3] such as the Joint Close Air Support (JCAS) Model, DoDAF-based Activity Scenario have more real world teleological characteristics. From the ontological and epistemological viewpoints, the authors use DEVS as a unified model specification, take simulator as services while models as resources, and propose a bifurcated model-continuity methodology for system engineering.

   (2)   DEVS framework for service-oriented computing systems (SOAD)

From the teleological perspective, SOAD [12] was proposed to extend the DEVS with basic SOA concepts for modeling and simulation of service-oriented computing systems. A DUNIP Web enables the DEVS framework as a service-oriented framework, but the M&S objectives are not necessary service-oriented systems. While a SOAD may not necessarily be service-oriented itself, the M&S objectives are service-oriented systems. From the ontological and epistemological viewpoints, the conceptual framework of an SOAD is reported in [12]. The research on SOAD concerns the three roles of a SOA, messaging patterns, primitive and composite service composition, and hardware models for router links.

   (3)   Web services based Cell-DEVS framework (D-CD++)

From the teleological perspective, The D-CD++ [13] is proposed to Web-enable the DEVS formalism that defines spatial models as cell spaces. From the ontological and epistemological viewpoints, the set of service interfaces in D-CD++ includes session management, configuration, simulation modeling and control, and retrieving data interfaces. The execution of D-CD++ conforms to parallel DEVS simulation protocols and adopts a global conservative time management strategy.

2.2 Model-Driven Framework

From the ontological and teleological perspective, a framework of this type utilizes high level abstract models as the start and basis for the analysis, design, implementation, deployment, and maintenance in the entire lifecycle of service-oriented software development. The Dynamic Distributed Service-Oriented Simulation Framework (DDSOS) [4, 14] is a typical example. From the ontological and epistemological viewpoints, they define a Process Specification and Modeling Language for Services (PSML-S) to model SOA systems. They hold that dynamic rebinding, re-composition, and re-architecture are the merits of the framework. Therefore some agent services and mechanisms are proposed to support these characteristics. Service-oriented systems engineering (SOSE) and MDA provide whole lifecycle support.

2.3 Interoperability Protocol Based Framework

From the ontological and teleological perspective, this approach utilizes the some interoperability protocols (e.g., HLA) as the standard simulation bus for service integration and information exchange. A typical example is service-oriented HLA (SOHLA) [5]. From the epistemological perspective, researchers of this category suggest to web enable the HLA at four layers, i.e. communication layer (such as Web-Enabled RTI), interface specification layer (e.g., HLA Evolved Web Service API and Unified Architecture [15]), federate interface layer (such as the HLA Connector [15]) and the application layer (e.g., HLA Island). The BOM and modular FOM [16] can facilitate model interoperations, and the FEDEP can provide a system engineering basis. A recent PhD dissertation is reported by Wang [17] that aims to improve the service ability and composability of the HLA framework based on some unifying theories.

2.4 EXtensible Modeling and Simulation Framework (XMSF)

From the ontological and teleological perspective, XMSF [18] is defined as a composable set of standards, profiles, and recommended practices for Web-based M&S. The practice of XMSF includes the Web-Enabled RTI and the project using XMSF to connect Navy Simulation Systems, Simkit, and CombatXXI, for joint modeling and analysis sponsored by SAIC. From the ontological and epistemological perspective, Web/XML, Internet/Networking and M&S are regarded as the major focus areas of XMSF.

2.5 Open Grid Services Architecture Based Framework (OGSA)

From the ontological and teleological perspective, the Grid is used to integrate various distributed resources as a 'Grid' in support of the sharing of collaborative resources and problem solutions for virtual organizations. Resource sharing is the essence of the Grid. Grids can be classified into computing, storage, data, knowledge, and service Grids according to the properties of the resources at the nodes. From the epistemological perspective, researchers and practitioners regard the HLA/RTI services/components as Grid services. These frameworks include the Cosim-Grid [6], SOAr-DSGrid [19], G-HLAM [20], and SOHR [21].

2.6 Other Service-Oriented Simulation Frameworks

Besides the above classical frameworks, Northrop Grumman's Service Integration/Interoperation Infrastructure (Si3) [22] was proposed to support simulation-based transformation. Ontology-driven framework [7,23,24] uses ontologies or semantic Web to improve the communication between users and Web services that use different terminologies.

2.7 A Summary and Overall Comparison

The reviewed approaches exhibit common M&S, Service-orientation, and software/system engineering ontologies more or less. We give an overall comparison in Table 1. We compare the six reviewed categories of approaches, listed as rows, with respect to the metrics for typical examples and three important dimensions. The advantages and limitations of each framework are listed in Table 2. In particular, we specifically check the model specification, M&S standards, and simulation protocols in the M&S dimension; resources that are published as services and interfaces, dynamic composition, fault-tolerance, QoS management, and semantic UDDI of services in the SOA dimension; and lifecycle support in the engineering dimension.

3 Ontological and Epistemological Perspectives on a Unifying Framework

Based on the partial ontological, epistemological, and teleological assumptions that lead to the ad-hoc frameworks, this section proposes a unifying framework or methodology (i.e., a three-dimensional reference model) derived from the review. It has upper ontological, epistemological, and teleological characteristics compared with all the specific methods. It also reveals the common functionalities and totality of research issues in the service-oriented simulation paradigm.

3.1 Principle of the Unifying Methodology

Based on our previous detailed review [25] and the explanations from the ontological, epistemological, and teleological viewpoints above, we identify (at least) three distinct, yet related fundamental dimensions (domains or viewpoints): M&S, service-orientation, and software/systems engineering. We regard the three dimensions as independent or orthogonal conceptual domains (sub-ontologies), since each has its own relatively complete and mature set of theory, approaches, standards, techniques, practices, and applications.

Table 1 Overall comparison of classical service-oriented simulation methods

Methods	Examples	M&S	Service-orientation	System Engineering
F1	DUNIP, DEVS/SOA, SOAD, D-CD++	Unified DEVS model specification. DEVSML for platform independent models. SOAD can model & simulate service-based software & hardware systems. DEVS simulation protocol.	Simulators as services, models as resources in DUNIP. No coordinator services. Session management, configuration, simulation modeling & control, and retrieving data service interfaces in D-CD++.	DUNIP has bifurcated model-continuity systems engineering methodology.
F2	DDSOS	PSML-S can model SOA systems. RTI as runtime infrastructure. Optimistic time synchronization.	Systems/environment simulation agent services & RTI services. Support dynamic rebinding, re-composition, and re-architecture.	MDA and service-oriented systems engineering (SOSE) support.
F3	Service oriented HLA, HLA Evolved Web Service API etc.	BOM & modular FOM facilitate interoperability levels of models. Low bandwidth, uncertainty & dynamic properties need considering.	Web-Enabling HLA for communication, HLA interface specification, federate interface & application layers. HLA Evolved XML Schema, smart update rate and fault tolerant mechanisms.	FEDEP needs to be modified to reflect the idea of Web centric and support of reuse, composition, and collaboration of services.
F4	XMSF and profiles	The M&S focus area of XMSF.	The Web/XML, Internet/Networking focus area of XMSF.	N/A
F5	Cosim-Grid, SOAr-DSGrid, G-HLAM, SOHR	Simulation components, HLA/RTI services, computing & storage resources can be Grid services.	Focus on management of distributed computing resources. Based on Grid middleware.	Not clear
F6	Si3, ontology/semantic driven framework	HLA/RTI simulation engine in Si3. Service description, semantic service matchmaking, not focusing on simulation execution in ontology approach.	Si3 packaging models, simulation, applications, tools, utilities & databases as services. Ontology method focuses on service UDDI, composition & fault-tolerant.	Have some development and usage procedures.

F1=Formalism based, F2=Model driven, F3=Interoperability protocol based, F4=XMSF, F5=OGSA based, F6=Other approaches.

Table 2 Advantages and Limitations of classical service-oriented simulation methods

Methods	Advantages	Limitations
F1	Mature formalism with long history. Strong presentation capability to various systems. Rigorous theoretical basis and mathematical semantics.	Too abstract & hard to follow by users. Have not been widely recognized by industrial & academic standards. Primarily for educational use. Simplicity, convenience & performance need to be improved.
F2	Model-driven, excellent dynamic composability and SOSE support.	Focus on service oriented software development. Limited simulation capabilities. Theory, efficiency & applications to be improved.
F3	Worldwide recognized IEEE standards. Solid research and practice foundations. HLA Evolved new standards	Revision while not the revolution of HLA may constrain further development. Lower levels of interoperability. Conflicts between SOA & HLA in service granularity.
F4	Pioneer technical frame; separate focus areas, issues & techniques.	Lack of concrete standards, products & systems engineering support. Has been terminated.
F5	Resource dynamic allocation, load balancing & fault tolerance. Transparency.	Needs grid middleware. M&S theoretical basis, systems engineering, performance & full use of SOA to be improved.
F6	Integration & interoperation of heterogeneous applications in Si3. UDDI & semantic composition in ontology methods.	Few publications & not mature. Need further research in the M&S dimension especially VV&A, states & time management of simulation service.

F1=Formalism based, F2=Model driven, F3=Interoperability protocol based, F4=XMSF, F5=OGSA based, F6=Other approaches.

The three dimensions comprise a reference model for a service-oriented simulation (Figure 1). The M&S dimension is our focused basic domain. The SOA is a new paradigm/technology that impacts highly on the M&S, while the software/system engineering dimension can benefit the other two from a management view. Besides the dimensions, the elements in each dimension can also be derived from the review [25] and are inspired by the ontology of each discipline. To reveal the upper ontology and epistemology, we inspect the crossover of the three dimensions aggressively from a 1D, 2D, and 3D perspectives. Our 3D reference model is also inspired by, but differs from, the methodology of Morphological Analysis [26] and 3D morphology of systems engineering [27]. We pay more attention to the coverage of "functionality morphology" in the 2D or 3D space, while not being constrained by the single cell focus of the Morphological Analysis method.

3.2 One-Dimensional Ontological Implication

Ontologies are often captured by a set of concepts and their relations. A 1D view enables us to look at each fundamental sub-ontology/dimension individually. The source system is located at the origin. It stands for an existing or proposed system that we intend to observe or develop.

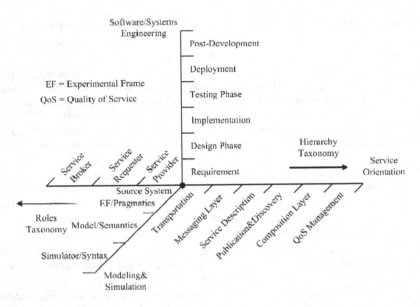

Fig. 1 A reference model for service-oriented simulation

3.2.1 M&S Dimension/Ontology

Besides the source system, the basic concepts in M&S [28] include an experimental frame (EF), model, and simulator. Modeling and simulation are the fundamental relationships. The EF–Model–Simulator comprises a general conceptual frame that explains nearly all the issues in the M&S domain well. The concepts implicated in the review also identify the three basic concepts.

Additionally, other views in M&S, in particular composability and interoperability are necessary complementary. The challenges and contributions lead to a hierarchical structure, in which we define three levels, i.e., Pragmatics-Semantics-Syntax. Pragmatics focuses on the use of information or artifacts within or across M&S solutions. Note that the EF is associated with pragmatics because it is the operational formulation of the M&S objectives. Semantics concentrates on the meaning of information or artifacts. It is the way in which we conceptualize our world as models. Syntax stresses formats and structures. It represents the way we implement and execute IT based simulation. The syntactic and semantic composability [29,30], the LCIM [31-33], the layers of M&S [23], and the interoperability challenges of model-based information systems (e.g., complex military simulation systems) [34] can also be mapped to these three levels with some reformulation or different interpretations.

Consequently, the EF/Pragmatics - Model/Semantics - Simulator/Syntax comprise the M&S dimension from both an object-oriented view and the perspective of linguistic/conceptual information exchange. This gradually moves from conceptualization focused modeling views to implementation focused simulation views. With the complement of the Pragmatics-Semantics-Syntax, the

M&S dimension is powerful enough to explain net- or Web-based alignment needs for distributed M&S services at different levels of composability and interoperability.

3.2.2   Service-Orientation Dimension/Ontology

Service-orientation is an increasingly state-of-the-art and promising approach for designing simulation systems. With the appealing characteristics of agility, reusability, and interoperability, services have been successfully incorporated in systems analysis, design, development, and integration [35]. An implementation-independent service description can be published by a service provider via a service broker. Based on the published information, a service requestor can discover and compose requested services with other services. Service-oriented approaches can benefit business systems and others in addressing the requirements of agility and flexibility, while allowing for changes in the requirements themselves. The SOA [35] is a conceptual framework for the design of business enterprise systems, while Web services is the prevailing technology for implementing a SOA. Previous work [35] provides a detailed review of approaches, technologies, and research issues in service-oriented approaches.

Service-orientation dimension has two taxonomies that come from the conceptual structure of SOA and the implementation hierarchies of Web services, respectively. The two taxonomies are complementary and the combination of them can better facilitate the analysis and implementation of service-oriented applications.

One of the taxonomies, from the viewpoint of roles, is structured as a triangle that consists of a service provider, requester, and broker. We use this particular order for this scale because the service provider and requester are more fundamental roles than the service broker. The service provider must provide its service earlier than the requestor's demand so as to compose a successful application.

The other taxonomy, from the perspective of Web service stack, is where the hierarchies of transportation, messaging, service description, service publication and discovery, composition and collaboration, and quality of service (QoS) management appear. Transportation, messaging and service description are the core layers that constitute the basis for static SOA. Service publication and discovery, composition and collaboration levels enhance the dynamic capabilities for dynamic SOA. QoS management makes services more dependable and robust by focusing on QoS requirements such as performance, reliability, scalability, interoperability, and security. We sequence the elements by their decreasing importance on the scale in Figure 1.

3.2.3   Software/Systems Engineering Dimension/Ontology

Simulation systems usually include software, at least in part [36]. The "Simulation as software engineering" mode of simulation practice [37] is applicable for teams of modelers and researchers, projects with lengthy lifecycles, and complex projects. For example, this model dominates military simulation due to the large scale models, long period of development, and expectation to be reused over a long period. The research and techniques for software engineering, especially software architecture and lifecycles, are of great use in simulation systems. The investigation of McKenzie et al. [36] show that there are no fundamental differences at the architectural level between simulation systems and general software systems. Formal and informal software architecture design methods can also be widely used in the M&S community.

Additionally, systems engineering can also benefit service-oriented simulation as a valuable complement in the hardware, optimization, trade-off, decision making, and other aspects that fall beyond the scope of software engineering. Systems engineering is a multidisciplinary methodology that comprises several logical phases that are independent of ad-hoc techniques. In general, the phases define that each system goes through a lifecycle, and certain steps need to be followed to ensure that the objective is supported. The better our system is managed in the phases, the smoother it runs.

The lifecycle of software/systems engineering may be assigned to different ontologies from multiple viewpoints [38]. In this work, we use the taxonomy of requirement, design (e.g., description, design, and analysis), implementation, testing, deployment, and post-development (e.g., maintenance, evolution, reuse, and retirement). In fact, the activities included in the engineering dimension are often cyclic or concurrent.

The research and practice of software/systems engineering are reported in [39,40]. Note that design and implementation often receive preferential treatment in general research and practice.

3.3   Two-Dimensional Epistemological Implication

Epistemologies study how we come to know, define, represent, and convey knowledge. In contrast with the ontological properties of the 1D view, a 2D view has an epistemological nature. It inspects the domains consisting of the Cartesian product of two sub-ontologies/dimensions to reveal the known or unknown knowledge in the cross-discipline landscape.

For a given specific framework that is compatible with the reference model, the issues/knowledge resulting from the reference model can be categorized as the following three categories:

(1) **core** issues/knowledge (C), the fundamental nature of service-oriented simulation; if they are not present, the framework cannot be called a service-oriented simulation framework;

(2) **supporting** issues/knowledge (S), the important characteristics of service-oriented simulation; if they are missing, the framework will be heavily affected; and

(3) **nice-to-have** issues/knowledge (N), the complementary functions of service-oriented simulation; if they are not present, the framework may be slightly affected.

This classification can be applied to 1D, 2D and 3D views. The crossover between research disciplines is identified and analyzed in Tables 3, 4, and 5. The 2D tables can be used for a cross-consistency assessment [26] process. They identify the logical and empirical meaning of each cell that consists of a pair of elements from the compared dimensions.

Table 3 Narrow service-oriented simulation (M&S vs. Services)

	S1	S2	S3	S4	S5	S6	S7	S8	S9
MS1	N	N	N	N	N	N	N	N	N
MS2	S	C	C	C	C	C	S	S	N
MS3	S	C	C	C	C	C	S	S	N

The increasing gray intensity of the cells identifies nice-to-have (N), supporting (S), and core issues (C), respectively. S1=Broker, S2=Requester, S3=Provider, S4=Transport, S5=Messaging, S6=Description, S7=Publish&Discovery, S8=Composition, S9=QoS, MS1=EF/Pragmatics, MS2=Model/Semantics, MS3=Simulator/Syntax.

Table 4 M&S engineering (M&S vs. Engineering)

	SE1	SE2	SE3	SE4	SE5	SE6
MS1	N	N	N	N	N	N
MS2	S	C	C	S	S	N
MS3	S	C	C	S	S	N

The increasing gray intensity of the cells identifies nice-to-have (N), supporting (S), and core issues (C), respectively. SE1=Requirements, SE2=Design, SE3=Implementation, SE4=Testing, SE5=Deployment, SE6=Post-development, MS1=EF/Pragmatics, MS2=Model/Semantics, MS3=Simulator/Syntax.

Table 5 Service-oriented engineering (Services vs. Engineering)

	S1	S2	S3	S4	S5	S6	S7	S8	S9
SE1	S	S	S	S	S	S	S	S	N
SE2	S	C	C	C	C	C	S	S	N
SE3	S	C	C	C	C	C	S	S	N
SE4	S	S	S	S	S	S	S	S	N
SE5	S	S	S	S	S	S	S	S	N
SE6	S	S	S	S	S	S	S	S	N

The increasing gray intensity of the cells identifies nice-to-have (N), supporting (S), and core issues (C), respectively. S1=Broker, S2=Requester, S3=Provider, S4=Transport, S5=Messaging, S6=Description, S7=Publish&Discovery, SE1=Requirements, SE2=Design, SE3= Implementation, SE4=Testing, SE5=Deployment, SE6=Post-development.

3.3.1 Narrow Service-Oriented Simulation

The Cartesian product of the M&S and service-orientation onlotogies/dimensions let us come to know service-oriented simulation in a narrow sense (Table 3). This is the fundamental domain for service-oriented simulation, which we refer to as the "narrow approach" since it may lack rigorous engineering principles or processes. Some ad-hoc research or practices [22] belong to this category. This 2D space has two epistemological implications that reveal the two directions of SOAs for M&S and vice versa: an approach that enables the extension of traditional M&S artifacts by service-oriented principles, and an approach that models or simulates service-oriented systems by means of M&S. For example, on the one hand, we can use SOA artifacts to publish a model as a service; on the other hand, we can also model SOA artifacts for analytical purpose. As mentioned previously, the Cartesian product of differently sequenced dimensions provides different directions. This principle is an extension of the non-directional cross-consistency assessment process [26].

Capturing M&S and SOAs as discrete ontological elements and crossing them, produces some interesting epistemological observations. (i) From a 1D view, the headings of the first column in Table 3 identify the discrete ontological elements together with their relationships in the M&S dimension. This principle also works in the SOA dimension. From a 2D view, a cell in the 2D table reflects a sequential pair of elements from the crossover of the two dimensions. For example, the simulator is where the simulation relation is captured. The simulator can certainly be a service with all core SOA capabilities. (ii) Furthermore, from the composability view of Pragmatics-Semantics-Syntax, if the assumptions and constraints regarding service description differ, we will not be able to discover the services. If we use different semantics to describe the services, we cannot compose them to work correctly. (iii) Moreover, the M&S dimension can be further discretized as a conceptual model, simulation model, and context [41]. A SOA also has other detailed taxonomies. The crossover of further discrete elements with their new relations can facilitate further research on the reusability and composability of M&S services.

3.3.2   M&S Engineering

The Cartesian product of the M&S and software/systems ontologies/dimensions provides an M&S engineering epistemology (Table 4) that applies engineering principles to traditional M&S as in, for example, the classical IEEE HLA Federation Development and Execution Process and VV&A standards. This is the traditional M&S engineering domain that does not necessarily refer to service-oriented simulation.

On the one hand, M&S engineering demands all elements of EF/Pragmatics - Model/Semantics - Simulator/Syntax to be addressed in each phase of the software/system engineering. For instance, testing in net-centric environments needs to be conducted simultaneously at the pragmatic, semantic, and syntactic levels [23].

On the other hand, M&S engineering also demands each element of EF/ Pragmatics - Model/Semantics - Simulator/Syntax to be supported and aligned in and between all phases of the engineering process. For example, conceptual views in the requirements phase will influence the reuse of the system in the post-development phase. This allows requirements (e.g., composability and interoperability) that come up in later phases to be formulated and supported in earlier phases. Otherwise, we are disconnected if our metrics for success when we define the system are different from the metrics for success when we test the prototype and later the real system. Note that the EF/pragmatics changes over the phases of the systems engineering process. It first specifies the objectives, assumptions, and constraints for requirements, becomes a development context later, then turns into a reference for testing cases, and finally becomes the context for VV&A and post-development.

3.3.3 Service-Oriented Engineering

The Cartesian product of service-orientation and software/systems ontologies/ dimensions creates a service-oriented engineering epistemology (Table 5). Here, engineering principles are applied to a service-orientation community. Although the basic engineering principles seem still unchanged (along the classical engineering dimension), new requirements and challenges are introduced by the SOA paradigm. For example, services are key elements, service interfaces, reuse and composition are paid more attention to, and the development style is mainly model driven. Service-oriented engineering is a new emerging domain. Typical examples include service-oriented systems engineering [42] and service-oriented software engineering [43]. In particular, these authors discussed the impact of the SOA paradigm on classical software/systems engineering principles and practices.

3.4 Three-Dimensional Epistemological Implication

Despite the partial ontological and epistemological perspectives on the 1D and 2D interpretation, the 3D view illustrated in Figure 2 provides a complete multi-perspective epistemology of a service-oriented simulation. The whole 3D space consists of the Cartesian product of all three ontologies/dimensions. The 3D space can be illustrated as a cube, with each cell representing part of our knowledge. The importance of each cell is identified according to the core, supporting, and nice-to-have classification. The coverage of cells indicates our active areas of the totality of research issues/knowledge in service-oriented simulation. This cube represents 'service-oriented M&S engineering', also called 'general service-oriented simulation' because it applies engineering principles to the whole development lifecycle of service-oriented simulation systems. The cube identifies several axes for necessary alignment, and is able to show and explain nearly all the challenges in service-oriented simulation, within and across phases (software/system engineering), solutions (services), and concepts (M&S). The 3D model can facilitate communications in and across organizational or disciplinary boundaries, in particular among managers in engineering, implementers of solutions, and specialists in M&S. In summary, to evolve as a new and mature M&S paradigm, the philosophical foundations of service-oriented simulation must cover the whole 3D space demanded by the 3D model.

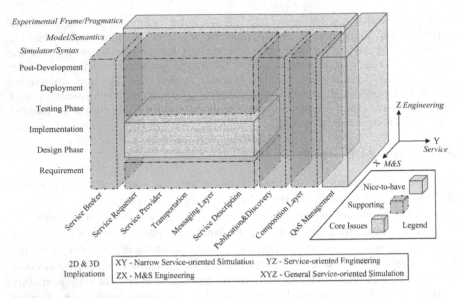

Fig. 2 Three-dimensional reference model for service-oriented simulation

The 3D reference model can be applied to separate concerns and used as a taxonomy to find the similarities and differences of the existing service-oriented simulation frameworks. Moreover, it can aid domain experts to define clearer and more specific knowledge or activities. Some sub-phases or steps can be added by multiple discipline experts using Cartesian products so that potential new knowledge or research issues can be discovered. Examples of possible new knowledge/research problems generated by the crossover of the service-orientation and M&S dimensions include how to encapsulate the capability of models, simulators, and experimental frames as services, and how to manage, use, and implement them in their respective layers. From an engineering point of view, the properties, design, and implementation problems should be considered as complements to the above issues.

3.5 Descriptive and Prescriptive Roles

Engineering methods and their ontology, epistemology, and teleology distinguish characterization (description) and mandatory (prescription) [11, 44]. The 3D reference model for service-oriented simulation can serve both functions. In its descriptive role, the 3D model describes the properties and functional morphology of service-oriented simulation within an existing ad-hoc framework. In its prescriptive role, the 3D model prescribes a set of net-centric M&S requirements that must be satisfied during the engineering of a proposed specific framework. The two roles can be combined to show the potential and possible future directions of classical frameworks. The first role shows what cells have been covered in the 3D space, while the other shows what cells need to be filled in. The 3D model emphasizes applying rigorous engineering principles and methods to embrace the full potential of service-oriented simulation.

4 Teleological Perspectives on the Unifying Framework

Teleology is the study of purpose and purpose-driven actions that result in methods. Teleology has both the referential and methodological characteristics. In context of this chapter, the former focuses on real world systems to be simulated, while the latter emphasizes on modeling and simulation techniques. Based on the review and the 3D reference model, this section shows the teleology-driven activities/practice of the unifying methodology to describe, compare, and prescribe various frameworks.

4.1 Teleology Driven Recommended Practice

The unifying framework can be built from existing approaches by merging the similar, equivalent, or complementary capabilities they provided in all three dimensions. The teleology-driven recommended practice of the unifying methodology is listed as follows.

(1) Define objectives of a service-oriented simulation

The purpose of this step is to identify user needs and develop objectives. The direction of the service-oriented simulation should be determined based on the problem to be simulated and the proposed simulation mechanism. If the problem is service oriented but the mechanism is not, then it belongs to M&S for SOAs. Vice versa, if the mechanism is service oriented but the problem is not, then it belongs to SOAs for M&S. If both are service oriented, then service-oriented approaches are used to simulate service-oriented applications. If neither is service oriented, it belongs to classical M&S that beyond the scope of this chapter.

(2) Develop capability requirements

Based on a conceptual analysis of the problem, this step is intended to identify all the required capabilities in the 3D space.

(3) Select candidate frameworks for reuse

The purpose of this step is to determine if an existing reusable framework meets or partially satisfies the requirements. The descriptive and prescriptive roles of the 3D model can be used to identify the capabilities provided and gaps left by the current frameworks. Section 4.2, 4.3, and 4.4.1 describe this activity in detail.

(4) Expand or compose known capabilities of candidate frameworks

This step is intended to close the gaps by merging the existing part solutions of all necessary dimensions. We give a detailed explanation in Section 4.4.2.

(5) Align system activities between all dimensions

The purpose of this step is to dissolve the conflicts of system activities between all dimensions when a best framework is reused or some candidate capabilities are composed. Section 4.4.2 presents this activity in detail.

(6) Establish new research topics or develop new capabilities

This step is intended to identify and suggest research and practice topics for the missing gaps that have not been covered by any contribution. We give a detailed explanation in Section 4.4.2.

The teleology and teleology-driven activities are revealed in the lifecycle of the recommended practice above. It facilitates to identify what do we need (purpose and required capabilities in step 1 and 2), What do we have (capabilities provided by the existing frameworks in step 3), What do we miss (gaps in the frameworks, plus missing alignments if the frameworks address different dimensions in step 3), and how do we close the gaps (expanding or merging the capabilities of frameworks in step 4, alignments of system activities between all dimensions in step 5, and identifying remaining gaps for future research in step 6).

The recommended practice can be tailored to meet specific user needs/teleology. We apply the steps to the reviewed frameworks in the following subsections. Note that we focus only on the methodological teleology that covers the both directions and the whole 3D space demanded by service-oriented simulations. Therefore we would not mention step 1 and 2 in the following subsections that can be found in specific referential teleology of the ad-hoc frameworks [3,4].

4.2   Contributions of Existing Frameworks

The 3D reference model can be utilized as metrics to find and compare the capabilities provided by the existing frameworks. We fill the 3D space with existing blocks of capabilities and make a detailed analysis from all the 1D, 2D, and 3D views in the Appendix Tables of our previous work [25]. In this chapter, we only show the primary 3D views in the Appendix Tables A1-A3. Using the descriptive role of the 3D model, we can identify the contributions of existing frameworks from each ad-hoc framework in particular and the union of all the frameworks in general.

From the viewpoint of each ad-hoc framework, the formalism-based approach has a rigorous theoretical basis and a number of important properties such as modular and hierarchy composition. This approach has an extensive coverage (especially in the M&S dimension) in the 3D space. With similar wide coverage, the model-driven method pays more attention to the direction of "M&S for SOAs" (e.g., service-oriented software engineering and dynamic properties). Although the coverage seems inadequate, the interoperability protocol based approach has a mature basis of international standards, wide applications and promising potential. In spite of weak coverage, the XMSF is the earliest approach amongst others that outlines the techniques framework for Web-based simulation. The OGSA-based method supports dynamic management, reuse, and transparent access for various M&S resources. The coverage of this method is moderate and needs further investigation. The Si3 and ontology-driven frameworks have the advantages of service integration and semantic interoperability respectively. Their coverage indicates the emphasis on the publication, discovery, and composition of services.

From the perspective of the union of all the frameworks, the existing capability blocks (Appendix Tables A1-A3) are intensively distributed and overlapped in the core area. The supporting space has a moderate coverage, and the nice-to-have field is distributed by some sparse capability units. The united capabilities of all the frameworks lead to a better coverage of the whole 3D space. This indicates the existing frameworks are developing aggressively from core, to supporting, and nice-to-have regions. In the future, the frameworks or the union of them are expected to provide full capabilities that can fill in the 3D space completely. Meanwhile, different frameworks have different concerns. The unique or scarce capabilities they provided indicate their competitive edge. Note that there is a sharp distinction between the directions of "SOAs for M&S" and vice versa. Different directions or objectives may make the semantics of the cells different, and bring difficulties to the merging and alignment activities. Although our 3D model can cover both directions, we pay more attention to the use of SOAs for M&S.

4.3   Gaps of Existing Frameworks

Using the prescriptive role of the 3D model as well as the analysis in the review, we can identify the gaps of existing frameworks. The formalism-based approach is limited in terms of standardization and ease of use. The gaps in the 3D space show room for improvements such as the publication and discovery, the composition, the broker, the QoS, testing, and post-development. The model-driven method has limited capability in terms of M&S. Its gaps represent that the requirements, SOAs for M&S, conceptual interoperability aspects amongst others can be further improved. The wide gaps of interoperability protocol based approach demand the enhancement of model services, higher levels of composability, and full potential of SOAs. The XMSF needs concrete standards and implementations, and the XMSF study group has been dismissed. The OGSA-based method requires a Grid middleware infrastructure. The gaps indicate the M&S aspects and full potential of SOAs require further research. Regarding the Si3 and ontology-driven frameworks, the gaps show that the M&S domain, the VV&A of services, and full lifecycle support need to be improved.

4.4   Teleology Driven Selection of Frameworks and Gaps Filling

4.4.1   Frameworks Selection

Based on the teleological capability requirements, and the contributions and gaps of existing frameworks, some recommendations can be made for the selection of frameworks. Technical constraints (e.g., reusability, VV&A, standardization) and managerial constraints (e.g., security, availability, preference, and mandate) should be considered before the selection process. In general, the framework which meets the requirements with maximum capabilities under all the constraints is the best choice. Otherwise, a set of frameworks that partially satisfies the requirements can be considered as candidates for composition and alignment.

In particular if the theory or education purposes are in a dominate position, the formalism-based approach should be considered the first. If the problem and objects are SOA systems, and the dynamic properties of service-oriented software engineering are emphasized, then we can choose the model-driven framework. If the governments or managers mandate mature standards for interoperation and compatibility with legacy systems, the interoperability protocol based approach like HLA is an appropriate choice. If we highlight the sharing of resources and problem solutions for virtual organizations, we can give the first priority to the OGSA-based approach. We would not recommend the XMSF approach because it has been ceased. The ontology driven framework can be considered if conceptualization of domains, semantics of services, brokers, publication and discovery are emphasized. Note that the ontology and Si3 frameworks are not as mature as the others thus far. In sum, the formalism-based approach is the most mature one from an M&S theory perspective, while the interoperability protocol based method has the most potential in the practice.

4.4.2 Gaps Filling

After the selection process of candidate frameworks, this subsection discusses the ways to fill the remaining gaps by the possible expansion/composition, alignments, and recommendations for future research.

(1) Expanding or merging of candidate part solutions

There are two possible ways to fill the missing gaps. One is the extension of the best candidate framework itself. For example, the SOHLA framework could be extended to fill the missing gaps by providing object models as services. The other is the merging of part solutions provided by a set of candidate frameworks. Some application-independent capabilities of candidate frameworks can be reused as common services, such as the runtime infrastructure services from the SOHLA, and the broker service from the model driven framework. The merging of the two frameworks can benefit the SOHLA from the publication and discovery of its service description. The reference model can facilitate to identify the expansion or merging path in the 3D space. In this step, the merging of model services and disposing of duplicated/similar services are difficult problems that need further research.

(2) Alignment of system activities between all dimensions

In course of expansion or merging candidate frameworks, the alignments between system activities take place. This step is intended to check and align the compatibility and consistency between capability units or system activities in terms of objectives, assumption, and constrains. It is more important when heterogeneous capability blocks are composed. The 3D reference model acts as a checklist for alignment along each column and row in 1D, 2D, and 3D views. From the 1D view, the composition and artifacts of capability units are aligned along the M&S, service-orientation, and engineering dimensions for compatibility and consistency. From the 2D view, system activities between all dimensions are checked. For example, the EF/Pragmatics is aligned from the requirements to post-development phases, and the pragmatics, semantics, and syntactic are adjusted simultaneously in the testing phase. From the 3D view, all the capability units are adjusted and harmonized across phases, solutions, and concepts. The alignments at the syntactic, semantics, and pragmatics levels by using the data, process, and assumption-constraint engineering are reported by Tolk et al. [45].

(3) Remaining gaps for future research

There are still some gaps (unknown knowledge or practice) that have not been covered by any contribution. This precludes the union of existing frameworks from a full coverage of the 3D space. In the M&S dimension, the gaps indicate that the capability of "models as services" falls some short. In the service-orientation dimension, the gaps of brokers, publication and discovery, dynamic properties, composition, and QoS still have room for improvements. In the engineering dimension, the full lifecycle support can also be further enhanced, in particular the phases of requirement (e.g., the semi-automatic generation of models, EFs, or generation of testing frames from requirements), testing, deployment, and post-development. The gaps also reveal that the higher levels of conceptual interoperability are inadequate. Therefore the M&S services are not so well annotated to facilitate intelligent agents to find, understand, orchestrate, and compose services meaningfully and automatically.

5   Conclusions and Future Work

Service-oriented M&S is the interdisciplinary field of M&S, the service-oriented paradigm, and software/systems engineering. It addresses the interoperability and composability challenges of distributed M&S services and represents the current focus and future direction of M&S in the prevailing net-centric environments.

In this chapter, we investigate service-oriented simulation frameworks from the ontological, epistemological, and teleological perspectives. We propose a unifying framework derived from the review of specific frameworks. With a referential (for real world) nature, the reviewed particular ontology, epistemology, and teleology of specific frameworks lead to various formalisms, techniques, and implementations. With a methodological (for abstract methods) nature, the ontology, epistemology, and teleology of the unifying framework build a systematical philosophical foundation for the service-oriented simulation paradigm. The unifying framework is first applied to two specific approaches [46]. Afterwards it is extended and further applied to some other prevailing methods [25]. It shows the unifying framework can, in turn, facilitate the classification, evaluation, selection, description, and prescription of the known or proposed frameworks.

Besides our former recommendations for the service-oriented simulation paradigm [25], there is also much future work from the context of the chapter. The referential and methodological nature, as well as the description and prescription roles of the ontology, epistemology, and teleology perspectives of the frameworks can be further investigated on more real world systems and abstract methods. Despite some specific frameworks, such as the DUNIP and DDSOS, support net-centric intelligent M&S applications, a better understanding of ontology, epistemology, and teleology of the frameworks are also necessary.

Acknowledgments. The authors thank Prof. Dr. Andreas Tolk for the invitation of the book chapter contribution. This work was partly supported by the National Natural Science Foundation of China [grant numbers 61074107, 60974073, 60974074, 91024015, 71031007]. The chapter is based on the author's paper "Wang WG, Wang WP, Zhu YF, Li Q (2011) Service-oriented simulation framework: An overview and unifying methodology. SIMULATION 87 (3):221-253." copyright © 2011 by the authors, the Society for Modelling and Simulation International, and Sage Publications Ltd. Reprinted by Permission of SAGE. We rewrite the chapter from the ontological, epistemological, and teleological perspectives.

References

1. Erl, T.: Service-Oriented Architecture: Concepts, Technology, and Design. Prentice Hall PTR, United States (2005)
2. Chen, Y.: Modeling and Simulation for and in Service-Orientated Computing Paradigm. Simulation 83(1), 3–6 (2007), doi:10.1177/0037549707079218
3. Mittal, S.: DEVS Unified Process for Integrated Development and Testing of Service Oriented Architectures. Ph.D Dissertation, The University of Arizona, United States – Arizona (2007)
4. Fan, C.: DDSOS: A Dynamic Distributed Service-Oriented Modeling and Simulation Framework. Ph.D Dissertation, Arizona State University, United States – Arizona (2006)
5. Wang, W.G., Yu, W.G., Li, Q., Wang, W.P., Liu, X.C.: Service-Oriented High Level Architecture. In: European Simulation Interoperability Workshop, Edinburgh, Scotland. Simulation Interoperability Standards Organization (2008)
6. Li, B.H., Chai, X.D., Di, Y.Q., Yu, H.Y., Du, Z.H., Peng, X.Y.: Research on Service Oriented Simulation Grid. In: Proceedings of Autonomous Decentralized Systems, ISADS 2005, Chengdu, China, pp. 7–14 (2005), doi:10.1109/ISADS.2005.1452008
7. Zhang, T.: Research on Key Technologies of Service-Oriented Semantically Composable Simulation. Ph.D Dissertation, Changsha: National University of Defense Technology, China (2008)
8. Naval Postgraduate School MOVES Institute: eXtensible Modeling and Simulation Framework (XMSF) (2004), http://www.movesinstitute.org/xmsf
9. Tolk, A.: M&S Body of Knowledge: Progress Report and Look Ahead. SCS Magazine 2(4) (2010)
10. Turnitsa, C., Padilla, J.J., Tolk, A.: Ontology for Modeling and Simulation. In: Winter Simulation Conference, pp. 643–651 (2010)
11. Hofmann, M., Palii, J., Mihelcic, G.: Epistemic and Normative Aspects of Ontologies in Modelling and Simulation. Journal of Simulation 5(3), 135–146 (2011)
12. Sarjoughian, H., Kim, S., Ramaswamy, M., Yau, S.A.: Simulation Framework for Service-Oriented Computing Systems. In: Winter Simulation Conference, pp. 845–853 (2008)
13. Wainer, G.A., Madhoun, R., Al-Zoubi, K.: Distributed Simulation of DEVS and Cell-DEVS Models in CD++ Using Web-Services. Simul. Model. Pract. Theory 16(9), 1266–1292 (2008), doi:10.1016/j.simpat.2008.06.012
14. Tsai, W.T., Fan, C., Chen, Y., Paul, R.: A Service-Oriented Modeling and Simulation Framework for Rapid Development of Distributed Applications. Simul. Model. Pract. Theory 14(6), 725–739 (2006), doi:10.1016/j.simpat.2005.10.005

15. Möller, B., Löf, S.: Mixing Service Oriented and High Level Architectures in Support of the GIG. In: Spring Simulation Interoperability Workshop, San Diego, California. Simulation Interoperability Standards Organization (2005)
16. Wang, W.G., Xu, Y.P., Chen, X., Li, Q., Wang, W.P.: High Level Architecture Evolved Modular Federation Object Model. Journal of Systems Engineering and Electronics 20(3), 625–635 (2009)
17. Wang, W.G.: Service-Oriented Composable Simulation: Theory and Application for HLA Evolved. Ph.D Dissertation, Changsha: National University of Defense Technology, China (2011)
18. Brutzman, D., Zyda, M., Pullen, M., Morse, K.L.: Extensible Modeling and Simulation Framework (XMSF) Challenges for Web-Based Modeling and Simulation. In: XMSF 2002 Findings and Recommendations Report: Technical Challenges Workshop and Strategic Opportunities Symposium (2002)
19. Chen, X., Cai, W., Turner, S.J., Wang, Y.: SOAr-DSGrid: Service-Oriented Architecture for Distributed Simulation on the Grid. In: Proceedings of the 20th Workshop on Principles of Advanced and Distributed Simulation, PADS 2006, pp. 65–73 (2006)
20. Rycerz, K., Bubak, M., Malawski, M., Sloot, P.: A Framework for HLA-Based Interactive Simulations on the Grid. Simulation 81(1), 67–76 (2005)
21. Pan, K., Turner, S.J., Cai, W., Li, Z.A.: Service Oriented HLA RTI on the Grid. In: 2007 IEEE International Conference on Web Services, ICWS 2007, pp. 984–992 (2007)
22. Strelich, T.P., Adams, D.P., Sloan, W.W.: Simulation-Based Transformation with the Service Integration/Interoperation Infrastructure. Technology Review Journal 13(2), 99–115 (2005)
23. Zeigler, B.P., Hammonds, P.E.: Modeling & Simulation-Based Data Engineering: Introducing Pragmatics into Ontologies for Net-Centric Information Exchange. Academic Press, New York (2007)
24. Yilmaz, L.: A Strategy for Improving Dynamic Composability: Ontology-Driven Introspective Agent Architectures. Journal of Systemics, Cybernetics, and Informatics 5(5), 1–9 (2007)
25. Wang, W.G., Wang, W.P., Zhu, Y.F., Li, Q.: Service-Oriented Simulation Framework: An Overview and Unifying Methodology. Simulation 87(3), 221–253 (2011), doi:10.1177/0037549710391838
26. Ritchey, T.: Problem Structuring Using Computer-Aided Morphological Analysis. The Journal of the Operational Research Society 57(7), 792–801 (2006)
27. Hall, A.D.: Three-Dimensional Morphology of Systems Engineering. IEEE Transactions on Systems Science and Cybernetics 5(2), 156–160 (1969), doi:10.1109/TSSC.1969.300208
28. Zeigler, B.P., Praehofer, H., Kim, T.G.: Theory of Modeling and Simulation: Integrating Discrete Event and Continuous Complex Dynamic Systems, 2nd edn. Academic Press, USA (2000)
29. Petty, M.D., Weisel, E.W.: Composability Lexicon. In: Spring Simulation Interoperability Workshop. Simulation Interoperability Standards Organization (2003)
30. Szabo, C., Teo, Y.M., See, S.A.: Time-Based Formalism for the Validation of Semantic Composability. In: Winter Simulation Conference, pp. 1411–1422 (2009)
31. Tolk, A., Muguira, J.A.: The Levels of Conceptual Interoperability Model. In: Fall Simulation Interoperability Workshop, Orlando, Florida. Simulation Interoperability Standards Organization (2003)

32. Tolk, A., Diallo, S.Y., Turnitsa, C.D.: Applying the Levels of Conceptual Interoperability Model in Support of Integratability, Interoperability, and Composability for System-of-Systems Engineering. Systemics, Cybernetics, and Informatics 5(5), 65–74 (2008)

33. Tolk, A., Bair, L.J., Diallo, S.Y.: Supporting Network Enabled Capability by Extending the Levels of Conceptual Interoperability Model to an Interoperability Maturity Model. JDMS (2012), doi:10.1177/1548512911428457

34. Hofmann, M.A.: Challenges of Model Interoperation in Military Simulations. Simulation 80(12), 659–667 (2004)

35. Papazoglou, M.P., van den Heuvel, W.J.: Service Oriented Architectures: Approaches, Technologies and Research Issues. The International Journal on Very Large Data Bases 16(3), 389–415 (2007), doi:10.1007/s00778-007-0044-3

36. McKenzie, F.D., Petty, M.D., Xu, Q.: Usefulness of Software Architecture Description Languages for Modeling and Analysis of Federates and Federation Architectures. Simulation 80(11), 559–576 (2004), doi:10.1177/0037549704050185

37. Robinson, S.: Modes of Simulation Practice: Approaches to Business and Military Simulation. Simul. Model. Pract. Theory 10(8), 513–523 (2002)

38. IEEE (2008) Systems and Software Engineering - System Life Cycle Processes. 15288-2008

39. Mei, H., Shen, J.R.: Progress of Research on Software Architecture. Journal of Software 17(6), 1257–1275 (2006)

40. Jamshidi, M.: System of Systems Engineering—Innovations for the 21st Century. John Wiley & Sons, New York (2009)

41. Yilmaz, L.: On the Need for Contextualized Introspective Models to Improve Reuse and Composability of Defense Simulations. JDMS 1(3), 141–151 (2004)

42. Tsai, W.T.: Service-Oriented System Engineering: A New Paradigm. In: Proceedings of the 2005 IEEE International Workshop on Service-Oriented System Engineering, SOSE 2005, Beijing, China, pp. 3–6 (2005), doi:10.1109/SOSE.2005.34

43. Tsai, W.T., Bai, X.Y., Chen, Y.N.: On Service-Oriented Software Engineering. Tsinghua University Press, Beijing (2008)

44. Wang, W.G., Tolk, A., Wang, W.P.: The Levels of Conceptual Interoperability Model: Applying Systems Engineering Principles to M&S. In: SCS Spring Simulation Multiconference, SpringSim 2009, San Diego, CA, USA, pp. 375–384. ACM (2009)

45. Tolk, A., Diallo, S.Y., King, R.D., Turnitsa, C.D.: A Layered Approach to Composition and Interoperation in Complex Systems. SCI, vol. 168, pp. 41–74. Springer (2009)

46. Wang, W.G., Wang, W.P., Zander, J., Zhu, Y.F.: Three-Dimensional Conceptual Model for Service-Oriented Simulation. Journal of Zhejiang University Science A 10(8), 1075–1081 (2009)

Appendix Tables

Table A1 Comparison of frameworks from model/semantics's perspective (3D view)

	S1	S2	S3	S4	S5	S6	S7	S8	S9
SE1			F1	$F2^T$	F1	F1, M1			
SE2	$F2^T$, O2	F1 (User), $F2^T$, M1(PSML^T)	F1, $F2^T$, M1(PSML^T), G1, O1	F1, $F2^T$, M1(PSML^T)	F1, $F2^T$, M1(PSML^T)	F1(DEVSML), $F2^T$, M1(PSML^T), O1, O2	$F2^T$, O 2	F1(SES, static), $F2^T$ (static), M1(PSML^T), O2	$F2^T$, O 2
SE3	$F2^T$, O2	F1 (User), $F2^T$, M1(PSML^T)	F1, $F2^T$, M1(PSML^T), G1, O1	F1, $F2^T$, M1(PSML^T)	F1, $F2^T$, M1(PSML^T)	F1(DEVSML), $F2^T$, M1(PSML^T), O1, O2	$F2^T$, O 2	F1(SES, static), $F2^T$ (static), M1(PSML^T), O2	$F2^T$, O 2
SE4			F1, M1			F1			M1
SE5			F1, M1	F1, M1	F1, M1	F1, M1		M1	M1
SE6			M1	M1	M1	M1		M1	M1

The increasing gray intensity of the cells identifies nice-to-have, supporting, and core issues, respectively. Elements marked with a superscript 'T' (transposition) identify M&S for SOAs; normal elements identify SOAs for M&S. S1=Broker, S2=Requester, S3=Provider, S4= Transport, S5=Messaging, S6=Description, S7=Publish&Discovery, S8=Composition, S9=QoS, SE1= Requirements, SE2=Design, SE3=Implementation, SE4=Testing, SE5=Deployment, SE6=Post-development, F1= DUNIP, F2=SOAD, M1=DDSOS, G1=OGSA based frameworks, O1=Si3, O2=Ontology driven frameworks, SES=System Entity Structure, DEVSML=DEVS Modeling Language.

Table A2 Comparison of frameworks from simulator/syntax's perspective (3D view)

	S1	S2	S3	S4	S5	S6	S7	S8	S9
SE1					F1				
SE2	M1, O2	F1 (User), M1	F1, F3, M1, X1, I1, G1, O1	F1, M1, X1, I1, G1	F1, F3, M1, X1, I1, G1	F1, F3, M1, X1, I1, G1, O1, O2	M1, O2	M1, O2	M1, O2
SE3	M1, O2	F1 (User), M1	F1, F3, M1, X1, I1, G1, O1	F1, M1, X1, I1, G1	F1, F3, M1, X1, I1, G1	F1, F3, M1, X1, I1, G1, O1, O2	M1, O2	M1, O2	F3, M1, O2
SE4			I1						
SE5			F1, M1	F1, M1	F1, M1	F1, M1		M1	M1
SE6			M1	M1	M1	M1		M1	M1

The increasing gray intensity of the cells identifies nice-to-have, supporting, and core issues, respectively. S1=Broker, S2=Requester, S3=Provider, S4=Transport, S5=Messaging, S6= Description, S7=Publish&Discovery, S8=Composition, S9=QoS, SE1=Requirements, SE2=Design, SE3=Implementation, SE4=Testing, SE5=Deployment, SE6=Post-development, F1= DUNIP, F3=D-CD++, M1=DDSOS, I1=SOHLA, X1=XMSF, G1=OGSA based frameworks, O1=Si3, O2=Ontology driven frameworks, SES=System Entity Structure, DEVSML=DEVS Modeling Language.

Table A3 Comparison of frameworks from EF/pragmatics's perspective (3D view)

	S1	S2	S3	S4	S5	S6	S7	S8	S9
SE1						F1			
SE2	O2		M1	F1, M1	F1, M1	F1, M1	O2	M1	$F2^T$, F3, M1, G1, O2
SE3	O2		M1	F1, M1	F1, M1	F1, M1	O2	M1	$F2^T$, F3, M1, G1, O2
SE4									
SE5				F1	F1	F1			
SE6									

The gray intensity of the cells identifies nice-to-have issues. Elements marked with a superscript 'T' (transposition) identify M&S for SOAs; normal elements identify SOAs for M&S. S1=Broker, S2=Requester, S3=Provider, S4=Transport, S5=Messaging, S6=Description, S7=Publish&Discovery, S8=Composition, S9=QoS, SE1=Requirements, SE2=Design, SE3=Implementation, SE4=Testing, SE5=Deployment, SE6=Post-development, F1= DUNIP, F2=SOAD, F3=D-CD++, M1=DDSOS, I1=SOHLA, X1=XMSF, G1=OGSA based frameworks, O2=Ontology driven frameworks.

Epilogue – Modeling and Simulation as a Humble Approach

Andreas Tolk

Old Dominion University
Norfolk, VA, United States

1   Introduction

In August 2006, a group of internationally recognized experts in various domains came together in the Consistorial Hall of Copenhagen University to talk about the fundamental concepts of matter and information in their domains. Physicists, biologists, philosophers, and theologians presented their view on matter and information with respect to the question of ultimate reality. The extended position papers are captured by Davies and Gregersen in a book [1] that brings as diverse facets and viewpoints together as we tried with this book.

The idea behind the driving concept for such an approach is called the *Humble Approach*. This initiative is supported by the John Templeton Foundation that serves to this purpose as a philanthropic catalyst for discoveries relating to the big questions of human purpose and ultimate reality. The approach is driven by interdisciplinary research that remains sensitive to disciplinary nuances while looking for theoretical linkages and connections. In other words: it allows for the various canons of research and differences in the foundational philosophies of science to bring the results together as complementary insights on a higher dimension.

This is exactly what we are trying with intelligent modeling and simulation (M&S). No single viewpoint is sufficient to capture all details and interpretations. When looking at the current efforts to define the Body of Knowledge for M&S (BoKMS), this becomes immediately obvious[1]. The BoKMS is understood as the comprehensive and concise representation of concepts, terms, and activities needed to explain a professional domain by representing the common understanding of relevant professionals and professional associations, but due to the ubiquity of such applications, just identifying the relevant professionals and professional associations is a nearly impossible task. The necessary approach must

---

[1]   One of the leading experts of this effort in service of the community is my friend and mentor Professor Tuncer Ören, who also contributed two chapters to this book. He hosts as website with current information and contributions to the BoKMS as a living compilation:
http://www.site.uottawa.ca/~oren/MSBOK/MSBOK-index.pdf

be inclusive, but focused on BoKMS relevant issues, which requires allowing for the various canons of research and differences in the foundational philosophies of science as stated before.

In their introduction to [1], Davies and Gregersen describe a long tradition of using the pinnacle of current technology as a metaphor for the universe on the search for universal truth. They observe:

> *"In ancient Greece, surveying equipment and musical instruments were the technological wonders of the age, and the Greeks regarded the cosmos as a manifestation of geometric relationships and musical harmony. In the 17th century, clockwork was the most impressive technology, and Newton described a deterministic clockwork universe, with time as an infinitely precise parameter that gauged all cosmis change. In the 19th century the steam engine replaced clockwork as the technological icon of the age and, sure enough, Clausius, von Helmholtz, Boltzman, and Maxwell described the universe as a gigantic, entropy-generating heat engine, sliding inexorably to a cosmic heat death. Today, the quantum computer serves the corresponding role. Each metaphor has brought its own valuable insights; those deriving from the quantum computation model of the universe are only just being explored."* [1, p. 3-4]

I was driven by this idea when I started to recruit authors and convince Springer that a book on ontology, epistemology, and teleology of M&S is needed. I think that the idea that we can define a concept – whether it has a real world referent or not – by its axioms and rules as an executable simulation and 'bring it to life' using animation and visualization, and potentially using emergent environment to make the user being part of this creation, is a powerful approach to understand things that are, that could be, or that could not be. For me, quantum computation will be a powerful technology that will support the next generation of intelligent M&S applications, but it will be the application that will drive our imagination and the understanding of the universe. Intelligent M&S applications will become the pinnacle of technology of our epoch.

The reason why I see intelligent M&S applications in such a favorable position is that all our understanding is connected with models. If we really understand what the attributes, characteristics, and behaviors of something are, and how these things are interrelated with each other and trigger responses, we can build a model. If we cannot build a model, we did not yet really understand it. If we try to gain a common understanding in a group, we are building models together. We may call them business plans, a common operational picture, or mission and vision; at the end, these are models we agree upon. Models are the essence of understanding, learning, and teaching. As stated by van Dam [2] during his lecture at Stanford:

> *"If a picture is worth a 1000 words, a moving picture is worth a 1000 static ones, and a truly interactive, user-controlled dynamic picture is worth 1000 ones that you watch passively."* [2]

We can not only recreate what we understand, we can use such virtual creations to understand better. Although it is the nature of models to be simplified and not complete, and even if it is the nature of computability to add significant constraints, intelligent M&S applications is pivotal to what we can understand. We can even create our on approaches of "what-if" worlds and use visualization and animation to emerge into them. If we can imagine something, we can build a model of it and bring it into virtual being.

As such, I see our approach with this book as a small but significant contribution to better understand M&S, so that we can use M&S in the search for universal truth. In the introduction, I started the discussion about whether M&S is a tool or a discipline, and if this discipline is engineering or science. The answer seems to be 'all of the above.' The important thing is that all is necessary to significantly contribute to searching for answers to the big questions of human purpose and ultimate reality. This is the ultimate goal of intelligent M&S applicant, and that is why ontology, epistemology, and teleology are so important to understand the philosophical, computational, and conceptual foundations.

References

[1]   Davies P, and NH Gregersen (eds.) (2010). *Information and the Nature of Reality: From Physics to Metaphysics.* Cambridge University Press, Cambridge, United Kingdom
[2]   van Dam A (1999). "Education: the unfinished Revolution." *ACM Computing Surveys (CSUR)* 31(4) Article No. 36

Biographical Sketches of Contributors

Ipek Bozkurt is Assistant Professor in the Engineering Management program and the Systems Engineering program at the University of Houston, Texas. She holds a PhD in Engineering Management and Systems Engineering and a Master in Engineering Management from Old Dominion University, and a B.Sc. in Chemical Engineering from Hacettepe University in Ankara, Turkey. Her research interests include research methodologies, organizational behavior, and complex problem solving including agent based model approaches in support of such efforts. She is an Associate Editor for *Engineering Management Journal,* has served as reviewer in numerous journals and conferences, chaired technical sessions, and judged local and international competitions.

Sergio de Cesare has a PhD in Management Information Systems from LUISS Guido Carli in Rome (Italy) and a Laurea (cum laude) in Economics from the University of Bari (Italy). He is currently a lecturer at Brunel University (U.K.) where he teaches Information Systems Development (ISD) and Semantic Web technologies. His current research focuses on Ontology-Driven Information Systems Engineering (ODISE) and investigates the practical and formal application of ontologies to all phases of the software development lifecycle. Sergio has authored over 50 peer-reviewed publications in several areas related to the modeling of information systems and their development. Sergio has an active presence within the international research community. He is chair of a successful series of ODISE workshops co-located with prestigious conferences such as OOPSLA/SPLASH and CAiSE. He is also engaged in the wider Information Systems community as Managing Editor of the European Journal of Information Systems (EJIS), Editorial Board Member of the International Journal on Organizational Design and Engineering (IJODE) and program committee member of numerous conferences.

Andrew Collins is a Research Assistant Professor at the Virginia Modeling Analysis and Simulation Center (VMASC) where he applies his expertise of game theory and agent-based modeling and simulation to a variety of projects including foreclosure and entrepreneur modeling. He earned his Ph.D. and M.Sc. in Operational Research from the University of Southampton in the United Kingdom (UK). He has spent the last 10 years, while conducting his Ph.D. and as an analyst for the UK's Ministry of Defense, applying game theory to variety of practical problems. His is principle investigator on a federal M&S standards governance project and also the principle analyst on an award winning investigation which applies agent-based modeling to the foreclosure crisis. Other recent research areas include entrepreneurship modeling, transportation, visual rhetoric and bio-terrorism.

Simon Deichsel studied philosophy and economics in Bayreuth and Bologna since 2001. In 2006, he obtained the degree of a Master of Arts (with distinction) with a thesis about model-platonism in economics. His specialization in philosophy of science and institutional economics set the stage for his PhD project in philosophy of economics at the University of Bremen where he got a position as research assistant of Prof. Dr. Dagmar Borchers in 2006. Simon was co-supervised by Prof. Dr. Andreas Pyka, and finished his PhD (with summa cum laude) in October 2009. He currently holds a post-doctoral position at the Institute of Philosophy in Bremen.

Saikou Y. Diallo is a Research Assistant Professor at the Virginia Modeling Analysis and Simulation Center (VMASC) of Old Dominion University. He received his M.S. and Ph.D. in Modeling & Simulation from ODU. His research focuses on the theory and practice of interoperability and the advancement of M&S science. Dr. Diallo has authored or co-authored over eighty publications including a number of awarded papers and articles in conferences, journals and book chapters. He participates in a number of Modeling and Simulation related organizations and conferences and is currently the co-chair of the Coalition Battle Management Language drafting group, an M&S IEEE standard development group.

Alan Dorin is an academic at Monash University, Melbourne, Australia where he is an artificial life researcher and generative artist working in electronic media. His interests include ecosystem simulation and agent-based modeling, artificial chemistry, self-assembling systems, the evolution of complexity, the history and philosophy of science (especially Artificial Life) and art, and the links that bind all fields together. Alan's qualifications include: PhD Computer Science (Monash University), Postgrad. Dip. Animation and Interactive Media (R.M.I.T.), B. Computer Science (Hons) (Monash University) and a B. Science (Applied Mathematics).

Scott A. Douglass Scott A. Douglass is Research Psychologist with the 711/HPW Cognitive Models and Agents Branch (RHAC), US Air Force Research Lab, Wright-Patterson Air Force Base, Ohio. He holds a Ph.D. (2007) in cognitive psychology from Carnegie Mellon University. Working with John R. Anderson at CMU, he acquired expertise in cognitive architectures and the modeling and simulation of complex situated cognitive processes. He is currently leading a AFRL Large-Scale Cognitive Modeling research initiative investigating ways to increase the scale and interoperability of cognitive models and agents. His research interests include large-scale cognitive modeling, artificial intelligence, knowledge engineering, multi-formalism modeling, and intelligent tutoring systems. He is a member of the Society for Modeling and Simulation International (SCS).

John Z. Elias completed his BS in Molecular Biology and BA in English at the University of California, Los Angeles (UCLA) and his MS in Modeling and Simulation at the Institute for Simulation and Training (IST) at the University of Central Florida (UCF). He is currently pursuing his PhD in Philosophy as a Marie Curie Fellow at the University of Hertfordshire in the UK. His research interests include embodiment, social interaction, normativity, and language.

Nicholas Geard is a Research Fellow in the School of Population Health, University of Melbourne and an Adjunct Research Fellow in the Faculty of Information Technology, Monash University. He obtained his PhD in computational simulation from the University of Queensland and has extensive experience in the application of agent-based models and other simulation techniques to problems in complex biological and social systems.

Brian L. Heath is a Consultant in Operations Research and Analytics for Cardinal Health, a large distribution company in the USA. He obtained his PhD in Engineering with focus in Industrial and Human Systems and his M.S. in Industrial and Human Factors Engineering from Wright State University. His interests include simulation philosophy, validation, agent-based modeling, and work measurement. He is a member of the Institute of Industrial Engineers and the Institute of Operations Research and Management Science.

Marko Hofmann is project manager and scientific senior researcher at the Institute for Technology of Intelligent Systems, an affiliated institute of the University of the Federal Armed Forces in Munich, Germany. He holds a PhD and M.S. in Computer Science, both from the University of the Federal Armed Forces in Munich and has written his habilitation on simulation-based decision making in complex systems. He gives lectures in operations research, computer science and mathematics at the University of the Federal Armed Forces in Munich and at the University of Applied Science in Kufstein, Austria.

Paul H. Humphreys is Professor of Philosophy at the University of Virginia and a member of the university's Cognitive Science Committee. He has degrees in philosophy and statistics from Stanford University and in logic and physics from the University of Sussex. He is the author of numerous books and articles, including Extending Ourselves (Oxford) and the Oxford Bibliography Online article on computational science, and co-edited (with Mark Bedau) the collection Emergence (MIT Press). His current research interests include computational science, emergence, probability theory, complexity theory, and philosophical methods.

Ross A. Jackson teaches economics and management in the School of Business at the University of Phoenix, and is a Visiting Assistant Professor at the Naval Postgraduate School. He received his PhD from Walden University and the M.A. from Ohio University. He graduated with distinction from the Air War College distance learning program, and holds professional cost estimating certifications from the Society of Cost Estimating and Analysis and from the International Society of Parametric Analysts. His current research applies Critical Management Studies approaches to strategy development and cost estimating. Deconstruction and postmodern analysis approaches are applied to better understand how strategic plans and cost estimates, as texts, are potentially misunderstood, co-opted, and distorted as these documents progress through organizational hierarchies.

D'An Knowles Ball is a faculty member of Old Dominion University's Art Department, instructing students in art criticism, art theory and digital design. She is also the Marketing and Communications Manager at the Virginia Modeling, Analysis and Simulation Center (Suffolk, VA). With a B.F.A. in Graphic Design from Appalachian State University (Boone, NC) as well as an M.A. in Humanities from Old Dominion University (Norfolk, VA), she has worked professionally in the design and communications industry for over a decade. Ms. Ball is a Humanities scholar with a focus on visual rhetoric and new media communications. After completing a thesis entitled *Duchampian Authenticity and The Readymade Consumer,* her current research interests include visual rhetoric and visualization, mass communication recomposition and deconstruction, as well as the analysis of visual representations of wit.

Kevin B. Korb is a Reader in the Clayton School of Information Technology, Monash University. He received his PhD in philosophy of science from Indiana University in 1992. His research interests include Bayesian philosophy of science, causal discovery algorithms, Bayesian networks, agent-based models, and evolutionary artificial life simulation. He is author of "Bayesian Artificial Intelligence" (CRC, 2010) and "Evolving Ethics" (Imprint Academic, 2010) and co-founder of the journal Psyche, the Association for the Scientific Study of Consciousness and the Australasian Bayesian Network Modeling Society. He was an invited speaker at the Singularity Summit (Melbourne, 2010 and 2011), Causality and Probability in the Sciences (Kent, 2008), the 13th International Congress of Logic, Methodology and Philosophy of Science (Beijing, 2007), and Causality, Uncertainty and Ignorance (Konstanz, 2004).

Qun Li is a Professor at College of Information Systems and Management, National University of Defense Technology (NUDT), China. He holds a Ph.D in Systems Engineering from NUDT in 1999. He is currently the Director of Systems Simulation Lab, NUDT. He has nearly twenty years of experience in systems modeling and simulation community. His research interests include systems simulation, simulation based acquisition, service-oriented simulation, and simulation composability and interoperability etc. He is the principal investigator of the National Natural Science Foundation of China project on "Service-oriented Simulation" under grant number 60674069 and 60974074.

Andrew Mitchell is a Principal Ontology Consultant for BORO™ Solutions. He has a First-class honors degree in Engineering Studies from Portsmouth University. Recent work has involved visualizing, validating and verifying semantic models; these models are used in projects such as software modernization and semantic interoperability. Other work has focused on analysis for the military domains of underwater warfare; littoral maneuver and command and control within air defense, including system integration and decision support projects. He has five years' experience of ontology and 15 in the defense science industry.

Saurabh Mittal is Research Scientist at Cognitive Models and Agents Branch, US Air Force Research Lab (711 Human Performance Wing) for L-3 Communications, Link Simulation and Training at Wright-Patterson Air Force Base, Ohio. He holds a Ph.D. (2007) and an M.S. (2003), both in Electrical and Computer Engineering from the University of Arizona, Tucson. He is a recipient of US Joint Interoperability Test Command's highest civilian contractor recognition 'Golden Eagle' award and is an author of several journal articles and chapter contributions. His research focuses on Discrete Event modeling using DEVS Formalism, SES theory and applications, net-centric system of systems engineering with DEVS Unified Process, executable architectures, simulation-based computing and large scale M&S infrastructure using Service Oriented Architectures. He is a member of Institute of Electrical and Electronics Engineers (IEEE), Association of Computer Machinery (ACM), and Society for Modeling and Simulation International (SCS).

Tuncer Ören is a Professor Emeritus at the University of Ottawa in Canada. He has been involved with simulation since 1965. His research interests include: advanced methodologies; agent-directed simulation, in particular agents for cognitive and emotive simulations (including representations of human personality, emotions, emotional intelligence, understanding, and conflict management training); quality assurance, failure avoidance, and ethics; the M&S body of knowledge; and terminology of simulation. He has over 450 publications, has contributed to over 400 conferences and seminars held in over 30 countries and has delivered almost 200 invited talks. Dr. Ören has been recognized, by IBM Canada, as a pioneer of computing in Canada. He received plaques and certificates of appreciation from organizations including ACM, AECL, AFCEA, NATO, and Turkish General Staff. He is a distinguished lecturer of SCS (Society for Modeling and Simulation International), received Information Age Award from the Turkish Ministry of Culture, and was inducted to SCS Modeling and Simulation Hall of Fame (Lifetime Achievement Award).

José J. Padilla is a Research Scientist with the Virginia Modeling, Analysis and Simulation Center (VMASC) at Old Dominion University, Suffolk, VA. He received his Ph.D. in Engineering Management from Old Dominion University. He also holds a B.Sc. in Industrial Engineering from la Universidad Nacional de Colombia, Medellín, Colombia and a Master of Business Administration from Lynn University, Boca Raton, Florida. His research interest is on the nature of the processes of understanding and interoperability and their implications in Human Social Culture Behavior (HSCB) modeling.

Chris Partridge is an expert in the pragmatic implementation of semantic data (particularly ontologies) for information systems and enterprise architectures. He has been involved in projects implementing these techniques since 1987. He also regularly publishes and speaks on the subject. He is the prime developer of the BORO™ products – the BORO Methodology™ and the BORO Foundation, which have been used in a number of standards (ISO 15926, IDEAS, DoDAF 2.0 and MODEM). He is the author of the book *Business Objects: Re-engineering for re-use* (published by Butterworth-Heinemann in 1996), one of the first books to describe business information re-engineering using business objects and ontologies. He regularly publishes papers, speaks at workshops and conferences and runs training workshops. Since 2000, he has been working as Chief Ontologist for the BORO Solutions. He has a MA (Oxon) in Mathematics and Philosophy.

Andreas Pyka graduated in Economics at the University of Augsburg in 1998 and spent afterwards two years as a Post Doc in Grenoble, France, participating in a European research project on innovation networks. Following the Post Doc he worked as an assistant professor at the chair of Prof. Dr. Horst Hanusch at the University of Augsburg. His fields of research are Neo-Schumpeterian Economics and Evolutionary Economics with a special emphasis on numerical techniques of analyzing dynamic processes of qualitative change and structural development. From October 2006 to March 2009 he worked at the University of Bremen as Professor in Economic Theory. Since April 2009, Andreas Pyka holds the chair for innovation economics at the University of Hohenheim, Stuttgart.

Roger Smith is the Chief Technology Officer for the Nicholson Center for Surgical Advancement at Florida Hospital. He previously served as the Chief Technology Officer for U.S. Army Simulation, Training and Instrumentation (PEO STRI) and Research Scientist for Texas A&M University. He is applying simulation and related technologies to surgical education and military training. He has published three books on simulation, created multiple commercial simulation courses, published over 150 papers and presentations on technical and management topics, and has served on the faculties of four universities. Dr. Smith holds a B.S. in Applied Mathematics, M.S. in Statistics, Master's and Doctorate in Business Administration, and Ph.D. in Computer Science.

Claudia Szabo is an Associate Lecturer at the School of Computing at the University of Adelaide. She holds a PhD in Computer Science from the National University of Singapore and a B.Sc. in Computer Science from the Politehnica University of Bucharest (PUB) in Romania. Her research interests are component-based software engineering, distributed and cloud computing, and verification and validation of distributed systems. Her research focuses on the understanding of the behavior and performance of complex software systems through modeling and simulation. Her work on a time-based formalism for the validation of semantic composability has been awarded by the ACM SIGSIM as best student paper in 2009.

Yong Meng Teo is an Associate Professor of Computer Science at the National University of Singapore. He leads the Computer Systems Group and the Information Technology Unit. His is also a Visiting Professor at Shanghai Advanced Research Institute, Chinese Academy of Sciences. He currently serves as external grant evaluator for the European Research Council's Ideas Specific Program. He was a Fellow of the Singapore-Massachusetts Institute of Technology Alliance. He holds a PhD and MSc in Computer Science, both from the University of Manchester, UK.

Andreas Tolk is Professor for Engineering Management and Systems Engineering at Old Dominion University in Norfolk, VA. He holds a joint appointment with the Modeling, Simulation, and Visualization Engineering department. He is affiliated with the National Centers for System of Systems Engineering and the Virginia Modeling Analysis and Simulation Center. He received the Frank Batten Excellence in Research award, the Technical Merit Award of the Simulation Interoperability Standards Organization, and the Outstanding Professional Contribution award of the Society for Modeling and Simulation. He holds a PhD and M.S. in Computer Science, both from the University of the Federal Armed Forces in Munich, Germany.

Klaus G. Troitzsch took his first degree as a political scientist in 1972. From 1974 to 1978, he was a member of Hamburg's Parliament. In 1979, after receiving a PhD in Political Science from the University of Hamburg, he returned to academia as a postdoc in an election research project. Since 1986, he is full professor of computational social science at the University of Koblenz-Landau, Germany. His main interests in teaching and research are social science methodology and modeling and simulation in the social sciences. He was among the founders of the SimSoc Consortium, which publishes the Journal of Artificial Societies and Social Simulation (JASSS), and of the European Social Simulation Association (ESSA). He is author, co-author, and co-editor of a number of books on simulation, author of a number of articles in social simulation, and he organized a number of national and international conferences in social simulation.

Weiping Wang is a Professor at College of Information Systems and Management, National University of Defense Technology (NUDT), China. He holds a Ph.D in Systems Engineering from NUDT in 1997. He is the Vice Dean of the Graduate School, NUDT. He is the founder and director of Systems Simulation Lab, NUDT. He has over twenty years of experience in systems modeling and simulation community. He is a Ph.D supervisor. His research interests include systems simulation, system of systems engineering, systems of systems simulation, simulation based acquisition, and simulation composability and interoperability.

Wenguang Wang is a lecturer at College of Information Systems and Management, National University of Defense Technology (NUDT), China. He holds a Ph.D in Control Science and Engineering from NUDT in 2011. He is a member of the SCS, SISO and CASS (Chinese Association for System Simulation). He is also an invited reviewer for several international journals including Simulation Modeling Practice and Theory. His research interests include Interoperability for system of systems, service-oriented composable simulation, High Level Architecture, and simulation composability and interoperability.

Paul Weirich is a Curators' Professor in the Philosophy Department at the University of Missouri. UCLA awarded him a Ph.D. in philosophy. He specializes in decision theory and game theory and is the author of *Equilibrium and Rationality: Game Theory Revised by Decision Rules* (Cambridge, 1998); *Decision Space: Multidimensional Utility Analysis* (Cambridge, 2001); *Realistic Decision Theory: Rules for Nonideal Agents in Nonideal Circumstances* (Oxford, 2004); and *Collective Rationality: Equilibrium in Cooperative Games* (Oxford, 2010). His current research includes construction of normative models that explain the rationality of acts, in particular, acts serving the public's interests, such as a government agency's imposition of a regulation to reduce a risk.

Feng Yang is an associate professor at College of Information Systems and Management, National University of Defense Technology (NUDT), China. He holds a Ph.D in Management Science and Engineering from NUDT in 2003. He has nearly twenty years of experience in systems modeling and simulation community. His research interests include systems simulation, simulation based acquisition, and cognition evolutionary computing. He is the principal investigator of the National Natural Science Foundation of China project under grant number 61074107.

Levent Yilmaz is Associate Professor of Computer Science and Software Engineering and holds a joint appointment with the Industrial and Systems Engineering at Auburn University. He received his MS and PhD degrees from Virginia Tech. His research interests are in Modeling and Computer Simulation, Agent-Directed Simulation, Complex Adaptive Systems. He serves as the Editor-in-Chief of Simulation: Transactions of the Society for Modeling and Simulation International and is the founding organizer and General Chair of the annual Agent-Directed Simulation conference series.

Author Index